Aktivhaus
The Reference Work

Aktivhaus
The Reference Work

From Passivhaus to Energy-Plus House

Manfred Hegger
Caroline Fafflok
Johannes Hegger
Isabell Passig

Birkhäuser
Basel

Authors:
Manfred Hegger
Caroline Fafflok
Johannes Hegger
Isabell Passig

Translation: Raymond Peat, David Koralek
Copy editing and proofreading: Monica Buckland
Layout (based on the layout of the German edition by Martin Fräulin): Kathleen Bernsdorf
Cover (based on the layout of the German edition by Anzinger | Wünschner | Rasp):
Kathleen Bernsdorf
Printing: Holzhausen Druck GmbH, Wolkersdorf, Austria

Library of Congress Cataloging-in-Publication data
A CIP catalog record for this book has been applied for at the Library of Congress.

Bibliographic information published by the German National Library
The German National Library lists this publication in the Deutsche Nationalbibliografie;
detailed bibliographic data are available on the Internet at http://dnb.dnb.de.

This publication is also available as an e-book
(ISBN PDF 978-3-03821-486-1; ISBN EPUB 978-3-03821-689-6).

P. O. Box 44, 4009 Basel, Switzerland
Part of Walter de Gruyter GmbH, Berlin/Boston

Printed on acid-free paper produced from chlorine-free pulp. TCF ∞

Printed in Austria

ISBN 978-3-03821-643-8

9 8 7 6 5 4 3 2 1 www.birkhauser.com

Contents

Foreword

It is now generally accepted that buildings are responsible for a major part of the energy consumption in many countries; the proportion often is 40% or more. The environmental impacts of heating, cooling, ventilation and lighting these buildings, and supplying them with electricity, are also correspondingly large. Whereas in other areas of life, the polluter pays the costs of these detrimental effects on the environment, this has not as yet applied to buildings and their users. But this is changing. Laws and regulations in more and more countries demand that every house creates a proportion, even though it may be a small proportion, of its energy demand regeneratively. The requirements of the new EU Energy Performance of Buildings Directive 2020 and targets set in other countries go much further. Energy autonomy is what governments will demand of future buildings. Building should largely fulfil this criterion in the foreseeable future, with the public sector leading the way.

The precise legal requirements for its implementation are yet to be formulated. Not all of today's building projects will be able to fulfil this ambitious goal of energy autonomy. However, the methods, technologies and tools for designing and constructing buildings able to achieve extreme energy savings, or in many cases generate surplus energy, are at our disposal today. This book gives information about them and about some pioneering projects.

The change to climate-neutral buildings can succeed because buildings, in contrast to most other goods, have characteristics that favour self-sufficiency. They protect people from the inclemencies of nature and the weather. Efficient protection is the first and foremost reason for making buildings independent of external energy supplies. A skilfully designed shape, a balanced ratio of openness and closure, of transparency and mass, and insulation and thermal storage capacity make their contributions to its achievement. This is the first necessary step: to make use of all the passive characteristics offered by a building and its envelope. The Passivhaus standard and the associated technologies have set important prerequisites. However, the rigid benchmarks underlying them fail to give adequate consideration to the diversity of building projects. As a result, they can give rise to constraints that detrimentally affect living and working atmospheres or lead to high additional costs without any tangible economic advantages: for example, over-thick walls, shaft-like windows, heating systems that detract from comfort or have other negative characteristics.

This is where the active measures associated with an Aktivhaus building can achieve their potential. Buildings are objects standing in the open air. They can therefore make use of natural energy sources: from the ground on which they stand, from the wind flowing around them, and from the daylight enveloping them. Buildings allow direct active use of regenerative energy sources where they are present on the building itself or in its grounds. Solar radiation, environmental heat, wind or geothermal heat can be transformed into heat energy and electricity. These energy sources are free and secure for the future, which cannot be said of our traditional energy carriers. The technologies for the use of these energy sources are becoming ever cheaper. When considered overall, creating regenerative energy at the building is increasingly economic and competitive with passive measures.

Aktivhaus is the contemporary further development of previous building energy standards. It is based on the principles of minimising the building's energy losses and internal energy consumption, and exploiting the direct passive use of solar radiation by the building itself.

The Aktivhaus not only saves energy: it is also designed to generate energy from the building envelope, the parts in contact with the ground and the immediate environment. The Aktivhaus uses the potential for self-sufficiency offered by the immediate environment.

For architects, this is a new challenge. Creative design is given new impetus by this energy dimension. The *genius loci*, the building project's link to the particular place and to a specific programme, widens by a further dimension: the adroit consideration of the environment, weather and utilities supply situation of the site.

While the architect used to deal primarily with providing protection from these influences, he or she now exploits them to the advantage of the building users, for their wellbeing and safety, and to reduce the economic burden of their building operating costs.

The complexity of the design criteria increases at the same time. There are no standard solutions here. Far more in demand are solutions that combine economy and appropriateness to the location, an aim that requires intensive cooperation between architects and engineers. This means breaking long-established habits. It requires the integrated development of a solution from the start of the design. The engineer should not play the part of an extra, but should be the creative partner of the architect. The new requirements demand giving up patterns of behaviour, codes of practice and certainties accepted for decades. However, the change is unavoidable, when faced with the challenges of the transition to renewable energy, climate protection and security of supply.

This book provides a guide for clients, architects and engineers on the route from a Passivhaus to an Energy-Plus House. The book covers the general rules of energy-efficient building, future-oriented codes of practice, the latest discussions, the views of various experts and detailed support in the design process.

The value of sustainable and resource-conserving building and the contribution of concepts such as the Aktivhaus are discussed and explained along with examples of reculatory frameworks.

How Aktivhaus buildings are designed and what interdependencies exist between passive and active measures are discussed in the chapter on "Aktivhaus design". Based on clear examples of an actual new build and a refurbishment project, an integral design is reconstructed step by step: from the beginning of the concept to the design of the building shell and the envelope, and the energy supply from active systems.

The book goes on to describe the tools, the individual technologies and the ways they can be used. In addition to measures for energy conservation and passive energy gain, it deals in particular with active systems that collect, transform and store renewable energy, and finally release it to the building in some useful way.

The examples from practice shown are in the Central European region and are therefore located in the temperate climate zone. They demonstrate worthwhile ways in which the theoretical approaches may be linked to one another. The projects extend from detached and semi-detached houses to apartment blocks and non-residential buildings. They are new builds and refurbishment projects; smaller and larger buildings showing that implementing an Aktivhaus concept is possible and can be done successfully, whatever the scale.

The outlook shows further possibilities for development in the future. In addition to the development of building standards, consideration of the urban design context is becoming increasingly important. The energy linking of urban modules, energy autonomy of neighbourhoods and eventually cities, opens the way to undreamt-of new images of a city of the future.

Supplemented by an extensive glossary, the publication is a comprehensive reference work. It is intended to stimulate imitation and provides the necessary detailed information.

Why have we published this book? We wished to show that a considerable contribution to the transition to renewable energies can be made by sustainable and energy-efficient building, and by integrating energy-creating building elements technically and aesthetically into the architecture. Buildings and their users are freed from their roles as energy consumers and become creators of energy, and do so right there, in the place where it is needed. This requires new solutions that enrich building and free the building industry from its conservative reputation. They show that architecture and the building sector are once more in a position not only to set social responsibilities, but also assume a leading role in the future: in defining the way society should develop with a focus on sustainability, and the transition to renewable energy.

Prof. Manfred Hegger
Darmstadt University of Technology, Germany

The Aktivhaus principle

Passivhaus technology is today a common standard for energy-saving construction. Nonetheless, it holds systemic disadvantages that consequentially led to development of the Aktivhaus, which enables the shortcomings of Passivhaus technology to be overcome.

Especially the first oil crisis in the early 1970s, and the first Report to the Club of Rome [001], published shortly beforehand, changed the relationship between humanity and the environment at a broader level than ever before. The natural environment has been increasingly seldom seen as a resource given over to exploitation by humankind. On the contrary, the recognition has increasingly taken hold that humanity is part of a complicated total system that sometimes can be hardly be understood, even by science. Vital resources and raw materials such as crude oil were recognised as being finite and the problem that developed countries like Germany had with their dependence on imports moved into public consciousness. This insight and knowledge about the climate-changing impact of burning coal, oil and gas prompted lawmakers to take action in the following years to reduce energy consumption and resulting emissions. In the building sector, this led among other things to introduction of various iterations of the German Energy Saving Ordinance (EnEV), which, viewed in retrospect, were important and correct measures.

Passivhaus versus Aktivhaus

These developments more or less took the construction industry by surprise. This was also evident in that, at the time the EnEV was introduced, no comprehensive methodological approach existed for implementing the requirements in a building. Due in part to resulting pressures for implementation, the initially broad search for problem-solving approaches quickly led to development of so-called Passivhaus technology. Passivhaus technology can in essence be described by the building's absolute airtightness and concomitant forced ventilation, by abundant exterior insulation, and by reducing surfaces that induce losses (typically the window areas). Today it can be identified as the standard technology for energy-saving construction. The success of Passivhaus technology is also based on a series of legislative and funding measures that initially seemed wise, but which, upon closer inspection, can be recognised as acting to stifle innovation. In particular, neither the construction industry nor lawmakers nor the planners and designers recognised that a Passivhaus, due to its systemic rudiments, must always be regarded as a sub-optimal outcome. The main reason for this is that the Passivhaus has constant, which is to say invariant, physical properties. Thus it cannot react to changes from outside (such as temperature changes, rain and wind conditions, or solar radiation intensities – which are dependent on the time of day or season), nor to changes from inside (such as the presence or absence of inhabitants). In the late 1990s, criticism springing from this knowledge led to the development of Aktivhaus technology, which found its first rigorous application in the R128 House in Stuttgart [002].

Aktivhaus technology is associated with the implementation of control or regulation systems. These are of course capable of being deactivated, so that every Aktivhaus can always be used as a building that is controlled in a traditional manner. Normally, however, daily residential routines are subject to a control and regulation mechanism that should be influenced by the user only to a limited extent. The implementation of control and regulation systems also implies the integration of sensor capabilities together with actuator capabilities, both of which are meanwhile typically displayed and integrated through a building automation system, even at the residential level. The introduction of building automation systems for residential use was opposed by (increasingly dwindling) resistance on the part of planners, contractors, and users. This was because, for one thing, there was a reluctance to employ technologies which had not been used previously and are often not understood in detail, and which are often also not yet robust enough nor designed to be sufficiently economical, and, for another, there was a fear that the human everyday world would be dominated by, in colloquial terms, a "computer" in the room. The latter argument is more than understandable in an age of rapidly increasing penetration of the private data sphere. For instance, official government action for the purposes of the war against terrorism, for example, or even the systematic, unquestioned collection, processing, and resale of personal data by so-called social networks and similar companies increasingly impinges on personal spheres of life. A concern about the loss of familiar elements in the living environment itself was added to the argument of the loss of the privacy of personal data space. This was strikingly brought to light in a discussion comparing the architectural approaches of Sobek and Wittgenstein [003].

Wise use of energy

The criticism of Passivhaus technology nevertheless goes beyond the criticism of the elementary fallacy of responding to a continually changing exterior and interior with a building envelope of invariant physical properties. There is, on the one hand, the relationship between energy consumption in the begin-of-life phase, the usage phase, and the end-of-life phase as well as, on the other, the resource consumption necessary for the production of a Passivhaus and the recyclability of such a building. Neither aspect has thus far been considered or discussed with sufficient intensity, by neither lawmakers nor construction research. The magnitude of the energy input expended in the begin-of-life phase is remarkable, especially in relation to the energy consumption during the usage phase. While a residential building constructed in Germany in the 1980s still possesses embodied energy amounting to approximately 20 or 30 times the annual heating energy demand, this ratio increasingly aspires to become virtually limitless as modern buildings continue to consume less energy in the usage phase. But this raises the question of whether, on the basis of producing energy principally from fossil fuels, it makes any sense today to install additional insulating materials, that is, to consume the bulk of the employed energy before even occupying the building. Or whether it would be more sensible to minimise the sum of the embodied energy, the energy consumption in the usage phase, and the energy consumption in the end-of-life phase. Aside from the fact that this minimisation of the total energy consumption across all phases in the life of a building is the only scientifically acceptable approach, upon closer inspection it also turns out that it is the only economically sensible approach: in a period of transition from a fossil energy economy to a solar energy economy, it does indeed make sense to postpone energy consumption. This is especially true because humankind will have no more energy problems in an age of solar energy production.

A comparison of the energy needed to produce thermal insulation with the amount of energy saved by this thermal insulation over a prolonged period reveals that with increasing frequency, more energy is put into the production of thermal insulation than can be saved with it in the short term. Therefore it is important to use thermal insulation systems with low embodied energy. Otherwise today's insulation requirements already prove

D10 south of Ulm (DE): This Aktivhaus demonstrates how sustainable construction and aesthetics can go hand in hand.

R128 in Stuttgart (DE): The world's first Triple Zero Building had already demonstrated the potential of control systems for comfort and energy efficiency as early as the year 2000.

to be too high. Plus there is a second aspect: the thermal insulation composite systems used today in rapidly increasing magnitude consist of more than just a considerable portion of petroleum-based materials. Such systems are typically made of an inseparable composite of various layers of different materials, so from today's perspective they are nothing but future special waste. This second problem is not, however, necessarily a consequence of using Passivhaus technology. Due to the massive deficit of suitable alternative technologies and the increasing numbers of passive houses and correspondingly retrofitted existing buildings, however, it occurs with increasing frequency.

With Aktivhaus technology, the shortcomings of Passivhaus technology can be overcome. But today's expanded approach also calls for more than solely minimising the total energy consumption across all phases in the life of a building. In addition, a means of construction is needed that is accompanied by a reduction in the amount of building materials employed and which guarantees the ability to completely reuse all the installed materials in either technical or natural cycles [004]. An example of such a building is Haus F87 in Berlin, an Efficiency House Plus combined with electromobility that was developed by this author and his employees in 2011 on behalf of the German Federal Government [005], [006].

Prof. Werner Sobek
University of Stuttgart, Germany

F87 in Berlin (DE): The building uses renewable sources to produce enough electricity for its entire needs, including electromobility.

Floor assembly in F87: Excellent thermal and sound insulation properties with full recyclability of all the building materials used.

Design

The following four chapters describe the basis for the development of the Aktivhaus idea and provide guidelines for the designer. From the fundamentals of sustainable and energy-efficient building, standard regulations in German-speaking Europe, to design tools and technical details, the text explains what makes an Aktivhaus, how a design is developed, and what components should be considered in bringing it to realisation.

The first part deals with the role of energy in our society and in sustainable development. The focus, of course, is primarily on the use of energy in buildings, but the many possibilities for energy gains using the building and its immediate surroundings are discussed. From this emerge strategies for buildings that take into account not only energy consumption but also energy creation and storage. After describing energy balance parameters, the book goes on to discuss specific building energy standards. This section concludes by setting out the basic requirements for an Aktivhaus and how they relate to external conditions (such as site, climate), and internal conditions (such as users, equipment). The final chapter in this part of the book provides an overview of energy supplies and building technology, including a detailed consideration of constructional and technical measures, and their scope of use.

Principles

The idea of the Aktivhaus projects the development of the principles of building and building standards logically into the future. It takes the need for sustainability in building fully into account. In addition to an increase in efficiency, this also involves switching to the use of environmentally compatible technologies (eco-effectiveness), in particular the provision of energy, and a rethink in the direction of moderate behaviour (sufficiency). These three sustainability strategies are yardsticks in the development of Aktivhaus concepts.

Basic needs of building and living

The house builder has always had a key function in the history of humankind. The "being" of human is, like the etymological relationship of the words "building" and "being", inseparably linked with the process of building.

In order to be, people always require the protection offered by a formed building, and not only in our latitudes. A building offers safety from external influences, in particular from the adversities of the climate, from changing and sometimes unpredictable weather, from dangers of all kinds. Since humans left their original homeland, east Africa, where the climate ideally suited them, the function of protection by a building has been at the focus of human activities. Without this protective third skin, it would not be possible to survive in our latitudes.

Building and being housed are therefore fundamental requirements of humankind, on the same level as other basic needs such as food and clothing. These are defined as human rights in the United Nations Charter.

The quality of buildings, and thus protection from the rigours of the weather, have been considerably further developed since the beginning of building in the form of the original house. It is a long way from a simple leaf roof to the timber and stone house, many at first without windows, to today's technically complex buildings offering high levels of comfort.

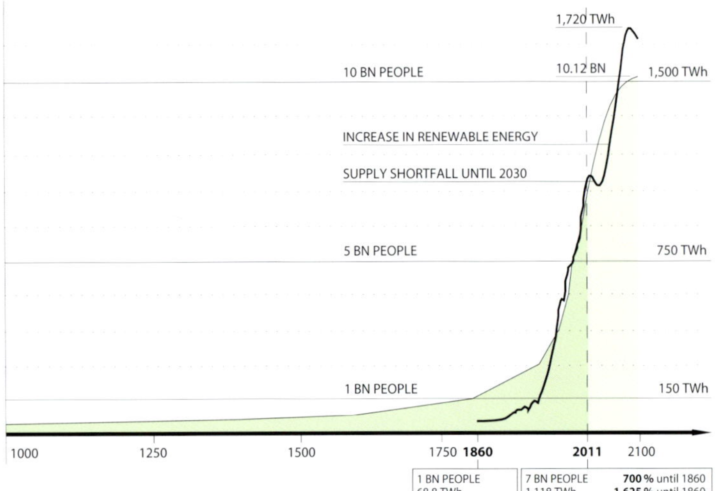

Comparison of world population development and worldwide primary energy demand

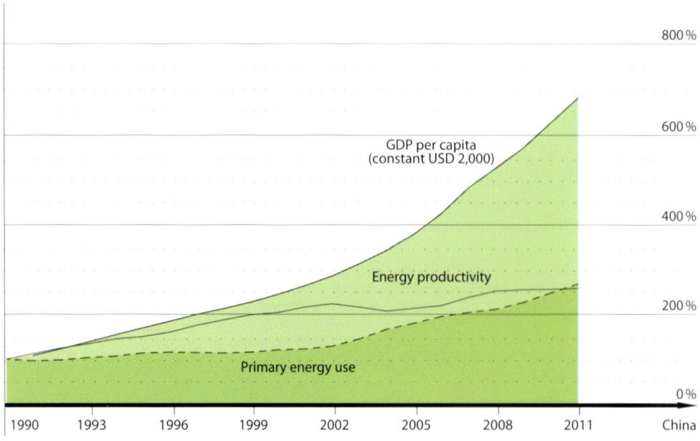

Comparison of growth gross domestic product/person and primary energy consumption per person 1990–2011 in Germany and USA (1990 = 100 %)

The role of energy

The development process has accelerated, particularly since the Industrial Revolution. House building has developed at a rapid pace since those times, when energy was cheap, available in large quantities and seemingly inexhaustible, since the days when raw materials for building – also driven by cheaply available energy – were available in all sorts of forms and likewise apparently inexhaustible. The demand for personal comfort and comfortable conditions within buildings has increased greatly over the same period.

Against the background of easily available resources, the world's population has grown by a factor of 7 over almost 150 years. People in the developed countries have largely been able to obtain comfortable living space and a diverse range of buildings in which to live. The newly industrialised countries are following suit. In Central Europe today, we have much more residential floor area than we had at the start of the Industrial Revolution. In many countries, the available residential floor area per person has greatly multiplied in the last 50 years alone. In parallel with this, the range of buildings and facilities for work, consumerism and leisure has expanded considerably.

So it is no wonder that energy use has increased much more than the population since the start of industrialisation. In most countries, energy consumption is still a key indicator of the standard of living. Only in recent years has it seemed possible to decouple the achievement of living standards from energy consumption. This became clear from a comparison of the development of gross domestic product and energy consumption during recent decades. Perhaps this is also an indicator for the predicted development of a post-material society that favours the use of services over the possession of goods. This can contribute to a reduction of resource consumption.

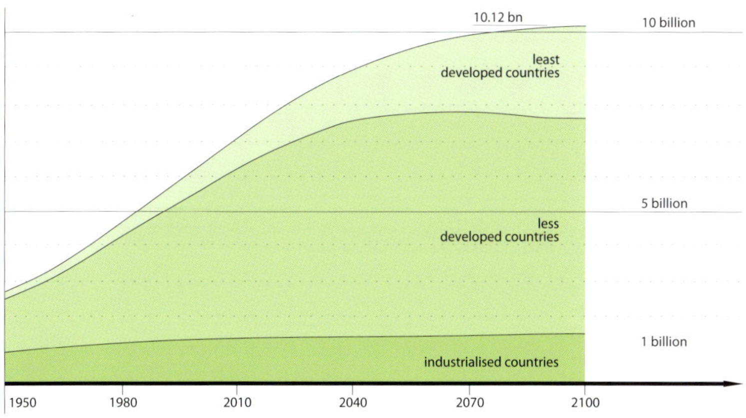

Global population development up to 2100

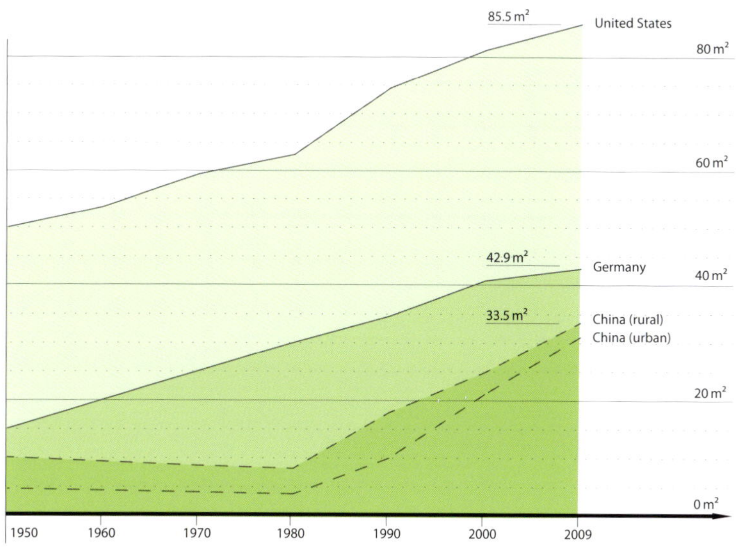

Comparison of growth of residential floor area per person 1950–2009 in China, Germany and USA

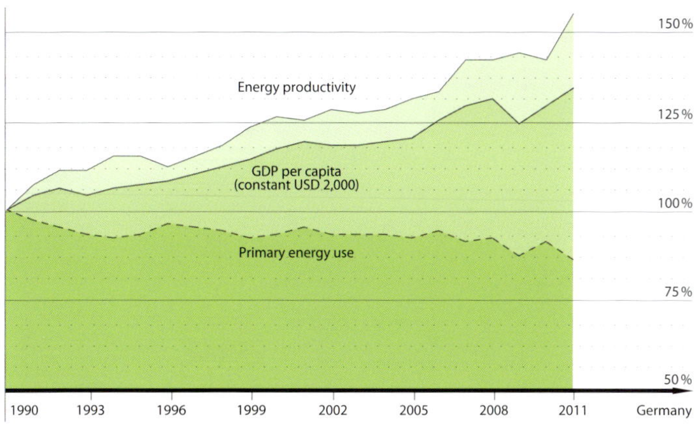

Comparison of growth gross domestic product/person and primary energy consumption per person 1990–2011 in China and USA (1990 = 100 %)

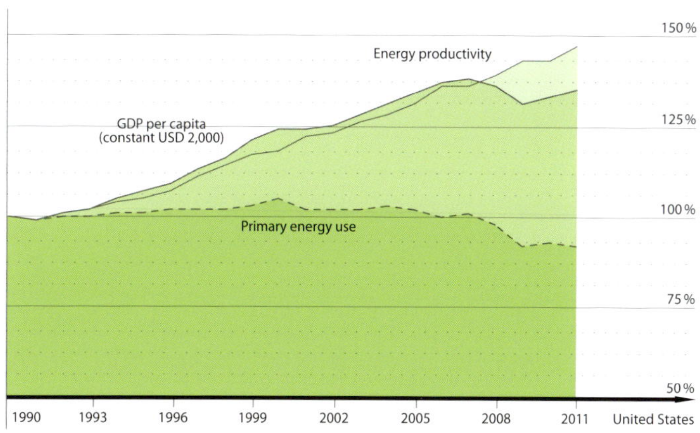

Comparison of growth gross domestic product/person and primary energy consumption per person 1990–2011 in China and Germany (1990 = 100 %)

Consequences of energy use

This is another way for a sustainable global society to develop, and it requires a completely new approach to dealing with energy and resources. The production and use of raw materials and especially the energy required for our current lifestyle are causing more and more problems. We have certainly managed to contribute to improved air quality through measures such as replacing coal-fired heating with oil and gas systems. Phenomena such as smog, which was still responsible for many respiratory illnesses and the absolutely unbearable environmental conditions occurring in our metropolises in the middle of the 20th century, have been almost eliminated in the developed countries. Comparable phenomena are reoccurring in the rapidly developing new metropolises of the newly industrialised countries; but even here, there is an expectation of some relief as a result of progressive development.

We have not yet been able to control the growing worldwide CO_2 emissions and many other environment impacts linked to the rapidly rising consumption of resources and energy. CO_2 emissions have gone sky-high in parallel with population development. Worldwide CO_2 emissions rose by about 300 % from 1900 to 2011; by 50 % in the brief 18-year period between 1993 and 2011 alone. These emissions are probably the main cause of climate change. Since the beginning of industrialisation, the mean temperature on the earth has risen by an average of 1 °C. This process cannot be reversed in the short term. The temperature may rise – if nothing is done to counter it – by a further 6 °C by the end of this century. This would mean many regions of the earth being uninhabitable, new and uncontrollable weather events occurring, and harvests endangered.

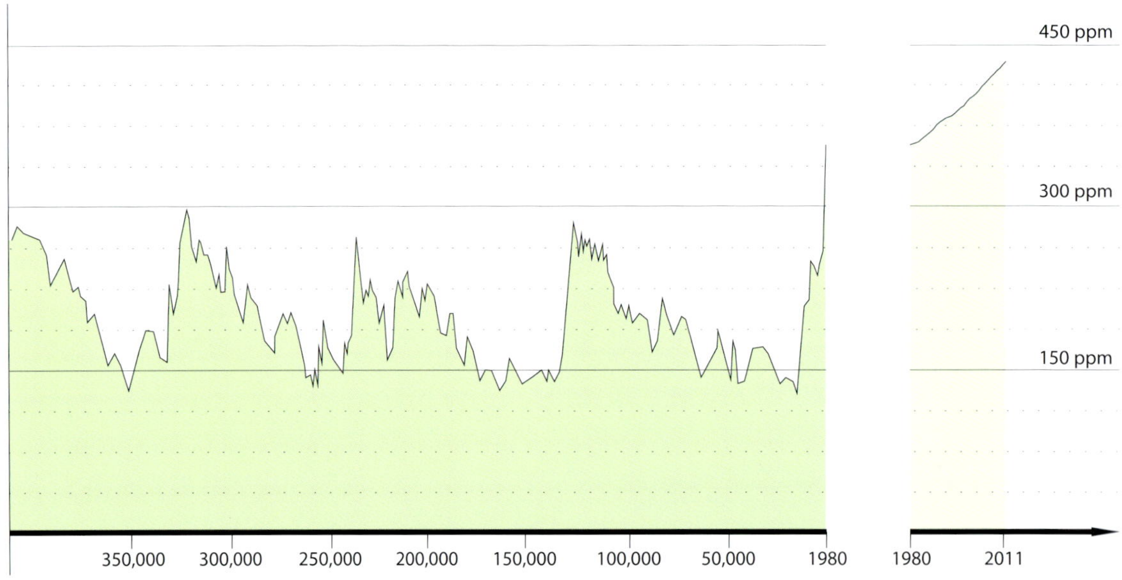

Global CO_2 concentration in the last 420,000 years

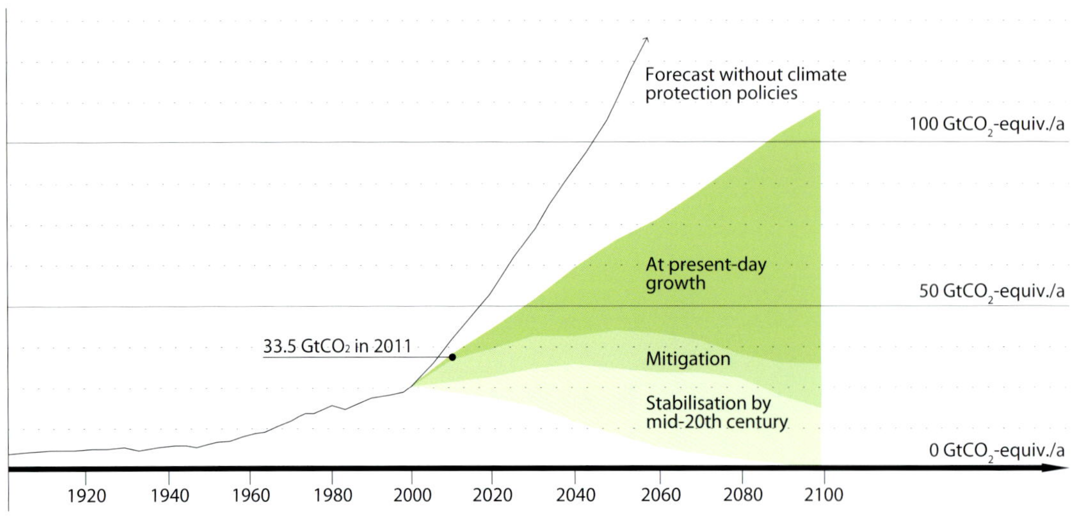

Global CO_2 concentration in the recent years and forecast of CO_2 emissions up to the year 2100

Population growth and resource conservation

After the large rise since the beginning of industrialisation, the world's population continues to grow; the forecasts indicate a rise from 7 to 9 billion by 2050 and to 10–11 billion by 2100. An ever larger proportion of this population feels entitled to achieve good living conditions similar to those prevailing in the developed parts of the world. The urbanisation of the globe accelerates this development process because, for many people, the city is more than the bearer of hope. It secures their survival, offers work and promises prosperity. Every second person in the world today already lives in a city. By 2050, with the growth in population, this is expected to be 70 %. A consequence of this development will be a considerable increase in the consumption of resources.

Concern about the security of supply of conventional energy in the world will therefore be greater. This concern has several sources. First, the lion's share of these energy carriers is imported. Many of the main supplier countries are politically unreliable or less stable partners. Second, forecasts about the extent of the globally available resources is hardly encouraging. In particular, crude oil, natural gas and uranium are likely to remain available for not much longer than one lifetime. Even if this period is extended by further discoveries, these reserves can be expected to become considerably more expensive as a result of their scarcity.

The burning of coal, crude oil and natural gas also consumes a finite raw material that is valuable in the manufacture of useful everyday products. Over time these will include consumables, consumer and capital goods such as body-care products, fertilisers, synthetic resins, plastics and fibres. Fossil resources are therefore far too valuable to burn.

Linked with this is the concern for the survival of humankind in a world that is in danger of making itself extinct as a result of burning fossil energy carriers. The dwindling of these energy reserves and the associated move towards regenerative energy sources will not bring about rapid relief because some of the environmental effects of generating energy from fossil fuels take a long time to appear.

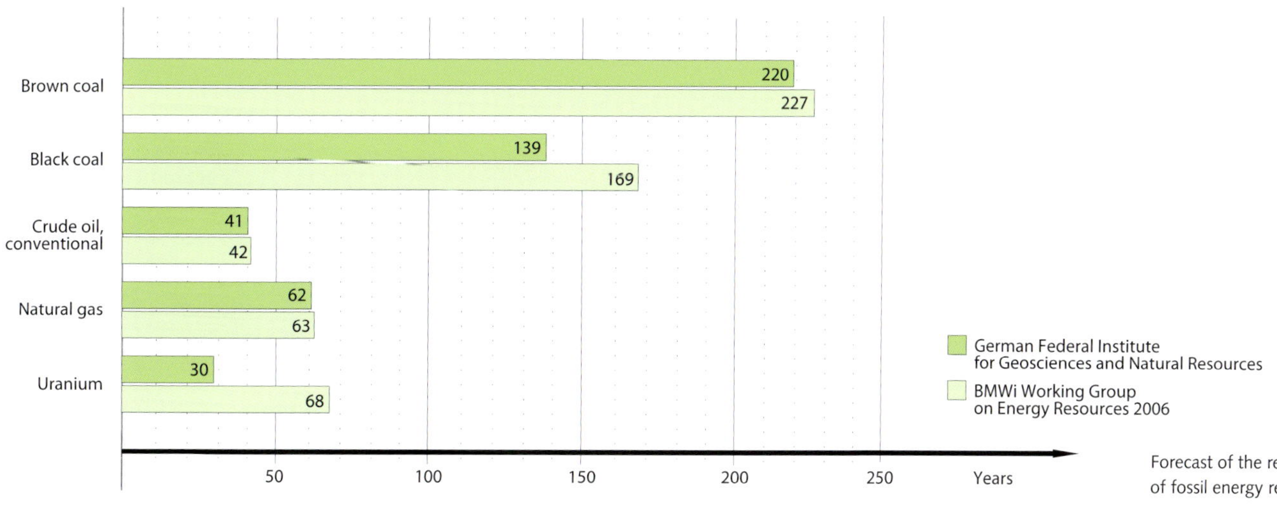

Forecast of the reserves of fossil energy resources

The price of energy

The world finds itself caught in a complex development trap: humankind's existence is endangered by the uncontrolled use of fossil energy sources. At the same time, these non-renewable energy sources are declining. The prices of energy are rising much more steeply than those of other goods. Thus, oil prices rose between 500 % and 1,700 % in various countries from 1970 to 2011, while the development of gas prices was much more variable, with a rise of 85% to a 10-fold increase of over 900%. The general consumer price index rose over this period by about 300%. The worsening shortage of oil and gas, the gradually widening gap between supply and demand, will further accelerate this development. The most valuable and versatile energy form, electrical energy, recorded a rise of "only" between 125 and 425%, depending on the particular country. It can be generated from different energy forms – and increasingly these are regenerative.

If nothing is done, this continuing trend will result in more and more population groups being unable to maintain their standard of living. The planned change to renewable energy carriers as part of the transition to regenerative energy use has initially steepened price rises and will drive prices for some years into the future. In the medium to long term, it will lead to a noticeable price stabilisation.

One indicator of this is that regenerative energies and their creation are becoming increasingly less expensive. Technological development, above all the more efficient conditions of production, have led to a situation in which more and more regenerative energy supplies are approaching or achieving "grid parity". This means the cost of creation from renewable energy sources results in prices below the market price for electrical energy. Further evidence for this is provided by the prices of photovoltaic modules. In 1970, they were almost USD 90/Wpeak; in 2012 they can be bought for much less than USD 1/Wpeak.

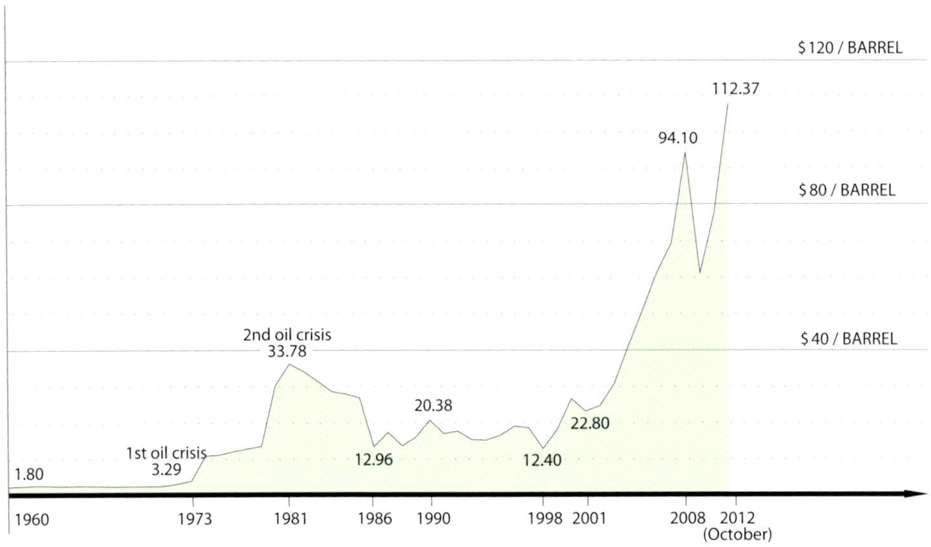

Crude oil world price from 1960 to 2011

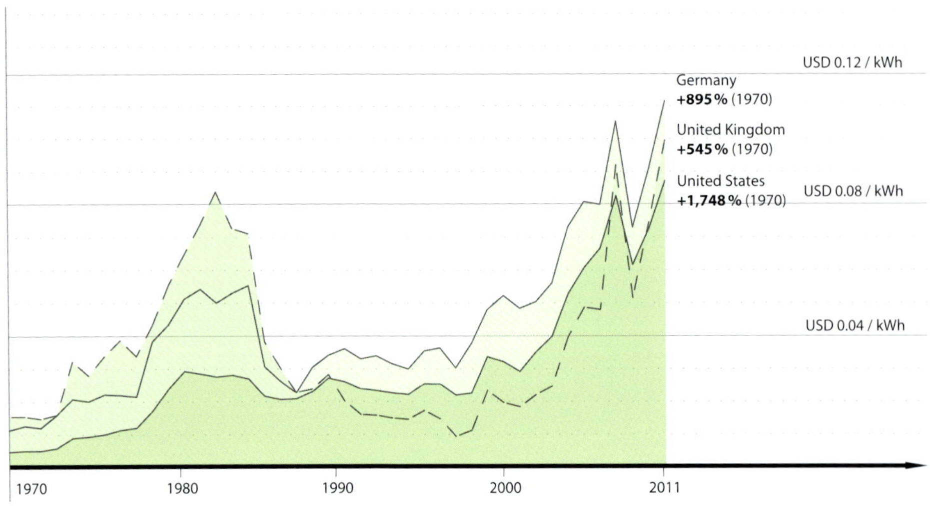

Comparison of oil prices from 1970 to 2011 in Germany, USA and UK

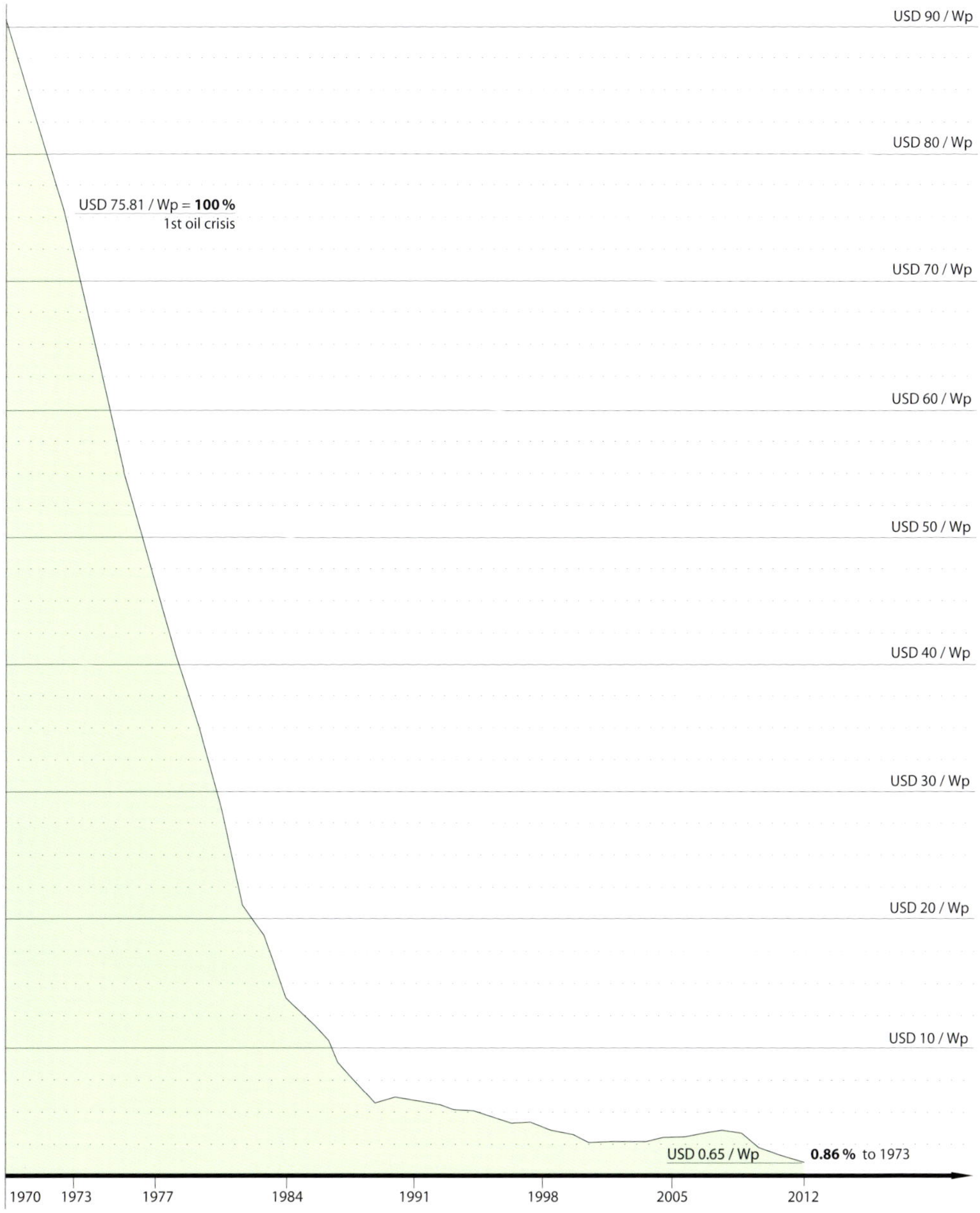

decline in prices: 1995–2012: approx 87 %
2005–2012: approx 80 %

PV module prices from 1970 to 2012

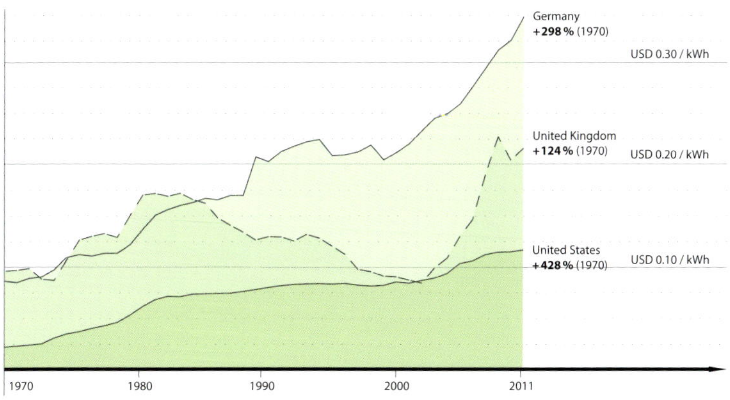

Comparison of domestic electricity prices from 1970 to 2011 in Germany, USA and UK

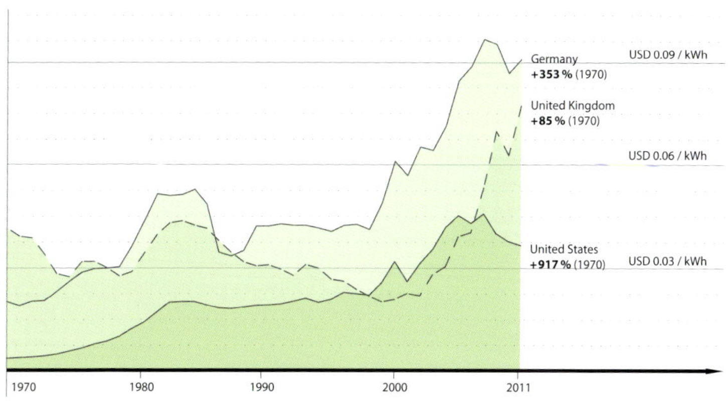

Comparison of gas prices from 1970 to 2011 in Germany, USA and UK

Energy in the building sector

In the temperate climate zones, the operation alone of all buildings, that is to say their heating and cooling, the production of hot water and the electricity for lighting, equipment and technical services plant, devours about 40 % of the total final energy demand. Not included here is the energy used in constructing the building itself: for the extraction of raw materials, the manufacture of building materials and components, for the ongoing maintenance and repairs, and for demolition at the end of the building's life. The manufacture of cement alone represents about 5 % of worldwide energy use and causes a correspondingly high proportion of global CO_2 emissions. Construction is therefore the sector with the highest energy demand, followed by industry and transport.

Especially in Central Europe, concern about rising prices, poor security of supply and the environmental problems described above has led to stricter requirements for the energy standards of construction. The potential for saving is huge. The heating of buildings consumes a great deal of energy and this use of energy alone makes up about one third of Central Europe's total energy demand. It was therefore self-evident for efforts to save energy to be concentrated first on this area.

The successes of the past decades in this field have been considerable. A new residential building can have a much reduced energy demand: as low as a twentieth of an unrefurbished older building. Even a refurbished older building, in favourable circumstances, can achieve a similarly remarkable result. The occupants of these buildings feel more comfortable at the same time.

On the other hand, many other savings are rendered less worthwhile by rising living standards. In recent decades, higher building standards and intelligent solutions have contributed substantially to reducing space heating demand per square metre of usable floor area. The benefit of this development has been cancelled out almost completely by higher floor areas per occupant.

The advantages of new, much more energy-efficient appliances and lighting are similarly counteracted by their increasing numbers installed in our buildings. This rebound effect shows that the transition to greater renewable energy use can be accomplished only alongside a change in our approach to energy. The issue is much more one of how we implement a sustainable economy in all its facets. And this means changes in lifestyle, for consumers and for producers. Success can be achieved only by adopting a holistic viewpoint and a comprehensive consideration of the situation. Building and buildings have a part – and a very special one at that – to play in this.

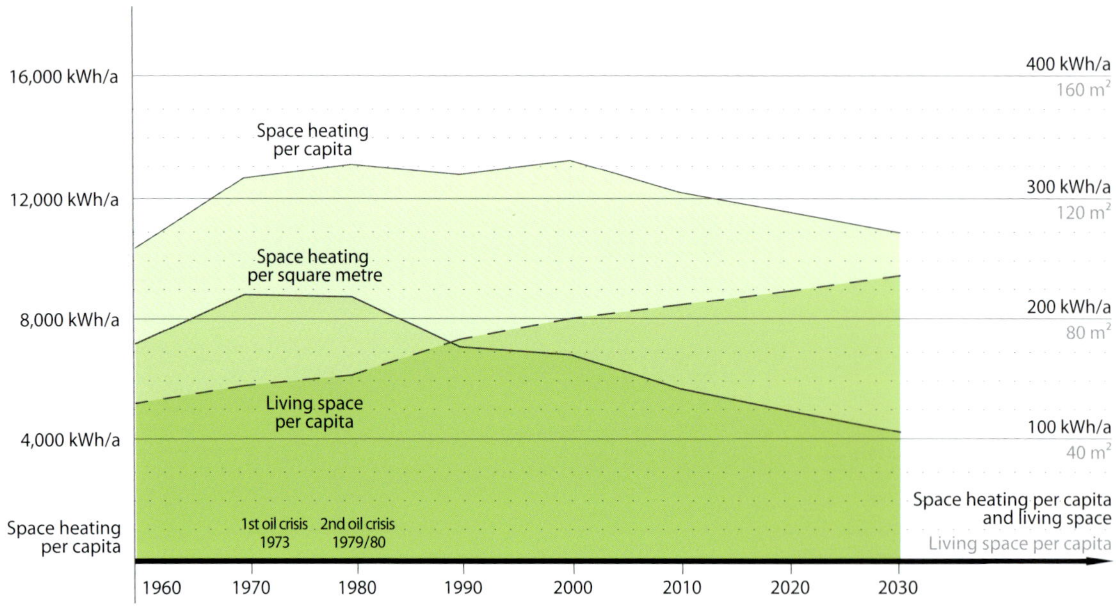

Historical and forecast residential floor area and heat demand per person in the USA

Building energy standards in selected countries

In recent years, the pressure to take action in response to climate change and shortage of resources has also increased internationally and has led to the development of new building energy standards. All these standards have the same objective: to design the provision of energy to our built environment to be as efficient as possible over the long term and to promote the use of renewable energy. The standards are comparable only to a limited extent, because they differ not only in their system boundaries but also often in their calculation principles and parameters. In addition, the limits of observation and, in some instances, the focus of each standard are very different. For this reason it is generally impossible to compare standards across national borders. The section below introduces the various standards, which are discussed in detail in the chapter "Energy balance".

Germany

Since the energy crisis in the 1970s, endeavours have strengthened to build energy-efficiently on the individual level as well as through regulations and financial subsidies. The Energy Saving Act (EnEG 1976) was replaced by the Thermal Insulation Ordinance (WSchV 1977) and the comprehensive Energy Saving Ordinance (EnEV); both have been published in three editions, each with more stringent requirements. The currently applicable EnEV is accompanied by additional incentives, such as the KfW Efficiency House programme offered by the federal state promotional bank, the Kreditanstalt für Wiederaufbau, which seeks continuously to raise standards through subsidies and favourable rates of interest on credit. The most widespread standard is Efficiency House Plus, which combines energy saving and local energy creation, and must create more energy than it needs when considered over the whole year. In the private building sphere, it is the Passivhaus standard that drives forward the previously described passive building design measures to the current maximum feasible level and achieves a very small heating heat requirement of 15 kWh/m²a. The maximum permissible primary energy consumption for heating, cooling, auxiliary power and household power is 120 kWh/m²a.

Switzerland

The Swiss MINERGIE® standards for highly efficient building systems were developed parallel to the German standards. In addition to the basic MINERGIE standard and the much more ambitious MINERGIE-P, a further standard was developed, MINERGIE-A, which as well as a highly insulated building and the associated low energy demand, also considered and evaluated energy created from renewable resources. The object of this standard is at least to cover the demand with the energy created. Each of the standards can also have an ECO quality seal variant. This extends the system into the operating phase of a building and also considers the energy used in producing the building, often referred to as embodied energy.

Italy

Influenced by the developments in Switzerland, the building label CasaClima (in English: ClimateHouse) was established in 2002 in the province of South Tyrol as part of a programme of work on energy-efficient buildings. The CasaClima Agency was founded in the autonomous province of Bolzano in 2006. This body is responsible for the further development, publications and certification of the standard. In principle, there are three different classes of CasaClima, which are based on their energy demand (heating and hot water): CasaClima B: heating energy demand less than 50 kWh/m²a (5-litre house); CasaClima A: heating energy demand less than 30 kWh/m²a (3-litre house); CasaClima Gold: heating energy demand less than 10 kWh/m²a (1-litre house). The CasaClima initiative has established an additional label, CasaClima Nature, which goes beyond the consideration of energy alone to include the careful use of resources and the energy used in the manufacture of the building materials. CasaClima Nature sets out ground rules, for example, for avoiding the use of fossil fuels, synthetic insulation, pollutants and tropical woods.

Austria

In Austria, energy standards similar to the German Niedrigenergiehaus (low-energy house), Niedrigstenergiehaus (lowest-energy house) and Passivhaus standards were introduced in line with international developments and EU directives. The assessments are made primarily using the heating energy demand in accordance with the Austrian standard ÖNORM H5055. In addition to this assessment based purely on energy, the Austrians also developed the label "klima:aktiv haus". It expands on the energy qualities and the calculation methods of the Passivhaus by adding considerations other than just the operating energy. Over and above purely building measures to save energy on new builds and refurbishments, separate aspects of the initiative consider and control the use of renewable energy, transport and barrier-free design, and integration into the surrounding built environment.

United Kingdom

Developments in this field do not stop at the northern limits of continental Europe. At the moment, the United Kingdom is working on a standard for a house that not only covers its own energy requirements but also compensates for its effects on the environment caused by carbon dioxide emissions from energy creation. This "zero carbon home" is CO_2-neutral in operation. The ambitious objective is for all new homes built in the United Kingdom to meet this standard by 2016. The scope of the standard is even widened beyond the system boundaries of an individual building to include settlements and cities. All these new construction works are also expected to be CO_2-neutral.

Energy in a broader sense

Most energy standards have concentrated up to now on the reduction of heating energy demand, because it currently makes up the greater part of the energy used in a building. It can be reduced without too much constructional or technical effort. In many cases, they also take into account the auxiliary energy demand used in the creation and storage of the heat, as well as the energy required to prepare hot water. This applies in particular to housing. The overall assessment, however, requires other energy supplies to be included.

Electrical energy

With the creation and retention of heat in buildings increasingly under control, other energy consumers come into focus. With a multifamily housing block built to contemporary standards (KfW 40 or with a Passivhaus-standard envelope), the proportion of the total end-energy demand represented by heating energy is only about 15 %. Added to this is about 15 % to 20 % for hot water preparation. The largest energy demand by far is in the area of electrical energy, particularly for domestic electricity, which is a significant factor at about 60 % of the total end-energy demand. This figure already assumes the use of efficient domestic appliances (A+++) and lighting (LED). The remainder (approx. 5 %) is made up by auxiliary electricity for pumps, fans etc. It should be borne in mind that for an energy mix with a high proportion of electricity generated from fossil fuels, the proportion of primary energy used is higher still. Consequently, the consumption of electricity will be the subject of much more attention in the future. This applies all the more to other building uses (e.g. offices, shops, production and research facilities) where electricity consumption is proportionally and in absolute terms much higher than in housing.

Embodied energy

With increasing efficiency in the operation of buildings, another energy criterion comes into the frame: embodied energy. This is the energy required to extract and process the building materials and elements, to produce a building, to carry out maintenance and modernisation during its lifetime, and finally the energy required for demolition. With increasing success in reducing operating energy, attention shifts further towards embodied energy. Even when long service lives are assumed for buildings, the energy consumption due to embodied energy calculated back to a reference year can be higher than the total energy consumed in operating the building. A number of strategies are available for minimising embodied energy. They range from consciously using renewable raw materials or building materials with a low degree of processing or energy content, to using recycled or completely recyclable materials, to completely lightweight construction to minimise material quantities.

Ensuring a long service life is another effective strategy. In this way, existing buildings can achieve their best, because most of their embodied energy remains within them over the extended period of use. With new buildings, as much value is placed on designing for ease of reconfiguration and multiple changes of use as a building's location and low maintenance requirements – from an aesthetic as well as a technical point of view.

Average annual demand for supplied primary energy (e.g. from the public electricity grid) for residential buildings built to different energy standards (period of observation 50 years). The reduction of the heat requirement will reach its conclusion by 2020 with the EU's Net Zero-Energy Building (NZEB). This building will cover its own demand of energy for heating, hot water, auxiliary and user electricity on average over the year. Buildings will then only have a primary energy demand for production, maintenance and disposal of the building construction.

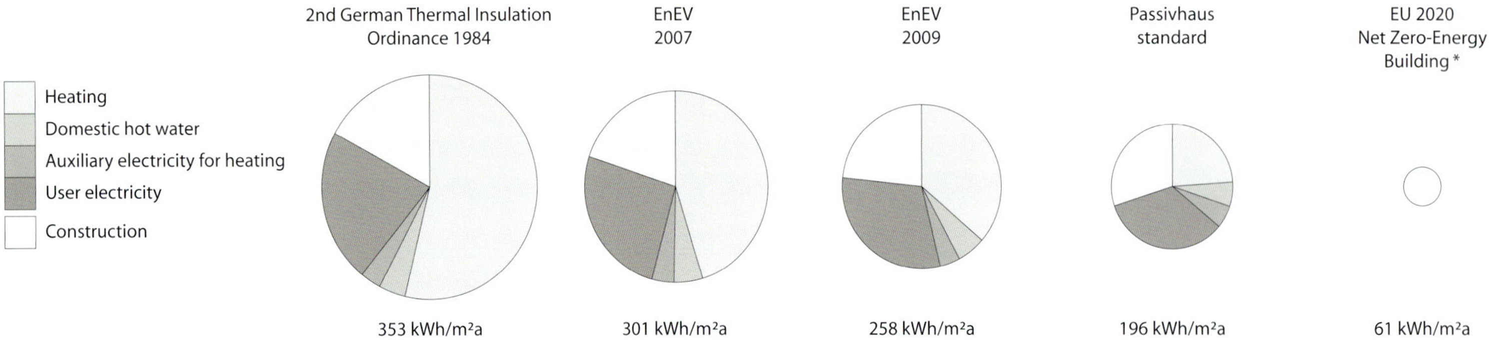

* BMVBS definition dated August 2011

New apartment block in Riedberg in Frankfurt (DE) with 20 residential units designed as an Energy-Plus House in accordance with the guidelines of the Efficiency House Plus

Building shape, compactness and orientation are directed towards maximising the use of daylight, natural ventilation and solar gain. The building's single-pitch roof faces south to exploit energy from the environment. This creates particularly attractive rooms with inclined ceilings and galleries on the two upper floors. At the same time, the arrangement achieves high yields from the photovoltaic panels integrated into the roof. Similar panels are also integrated into the south facade. In conjunction with the use of geothermal energy, this results in a surplus over the year. The linking of buildings and the operation of electrically powered vehicles as well as new methods of storage of thermal and electrical energy will considerably increase the degree of self-produced energy in a building's energy supply.

Architect: HHS Planer + Architekten AG, Kassel (DE)

Contributing to sustainable development

The Aktivhaus concept takes the development towards sustainable building in a more consistent way. It starts from the increasingly acknowledged basic conflict between economic growth and the environment, and recognises that there is no substitute for nature. The concept follows a comprehensive strategy that brings together the elements of a better use of the available resources and the transition to environmentally compatible technologies. At the same time, it also accepts the premise that nothing will go any further without a change in thinking and a desire to move in the direction of sufficiency and adequacy. That will be possible only if we elevate sustainability as defined in these terms to a philosophy for life and a lifestyle. Aktivhaus connects all three strategies of sustainability.

Efficiency

A sustainable economy depends on economic and ecological efficiency. It follows the principle of achieving as much as possible with the least possible use of resources. The path of efficiency reacts to the recognition that global resources and natural sinks for pollutants are finite. It suggests that rising efficiency allows society to push the limits of the growth economy in relation to the use of material resources as far as it would like into the future. The Aktivhaus is conceivable only as a highly efficient building, highly productive with respect to floor area, building form, use of materials and building services, and is at the cutting edge of technology. But the strategy of increasing efficiency alone is not enough for society to be able to master the tasks facing it.

Eco-effectiveness

Eco-effectiveness is the second pillar upon which the transformation to the use of environmentally compatible resources rests. This applies to material use in construction as much as it does to energy use in operation and this opens up a possible conflict with efficiency. If renewable energy is almost infinite and its use does not damage the environment, if wood largely regenerates as it is used, then there can be nothing against using it extravagantly. The technologies required to extract these resources certainly consume resources themselves, which are largely non-renewable. Entropy increases. This sets limits and suggests, for the time being, that these resources should be used sparingly. For the Aktivhaus, this means that the building obtains the energy required for its operation as far as possible from renewable energy sources, and uses renewable raw materials and/or is completely renewable. Efficiency and eco-effectiveness lead to an optimised mastery of nature. They embody the principle of hope, even in the face of a growing world population and rising living standards, of manufacturing more for everyone using fewer resources and in this way circumventing social questions of restriction or even redistribution.

Sufficiency

The third pillar, sufficiency, opposes the social and economic viewpoint of "more and more". Sufficiency questions the degree. It seeks to set limits to overconsumption of resources while implanting sufficiency and adequacy in the social consensus. Sufficiency is accused, because of its origins in the "small is beautiful" movement and its approach based on renunciation, of being backward-looking, pessimistic and unrealistic. The sufficiency route demands, first, the answer to the basic question of whether a new build, a building to cover a critically examined need for space, is required at all. If the answer is yes then the next question follows – as referred to above – the question of degree.

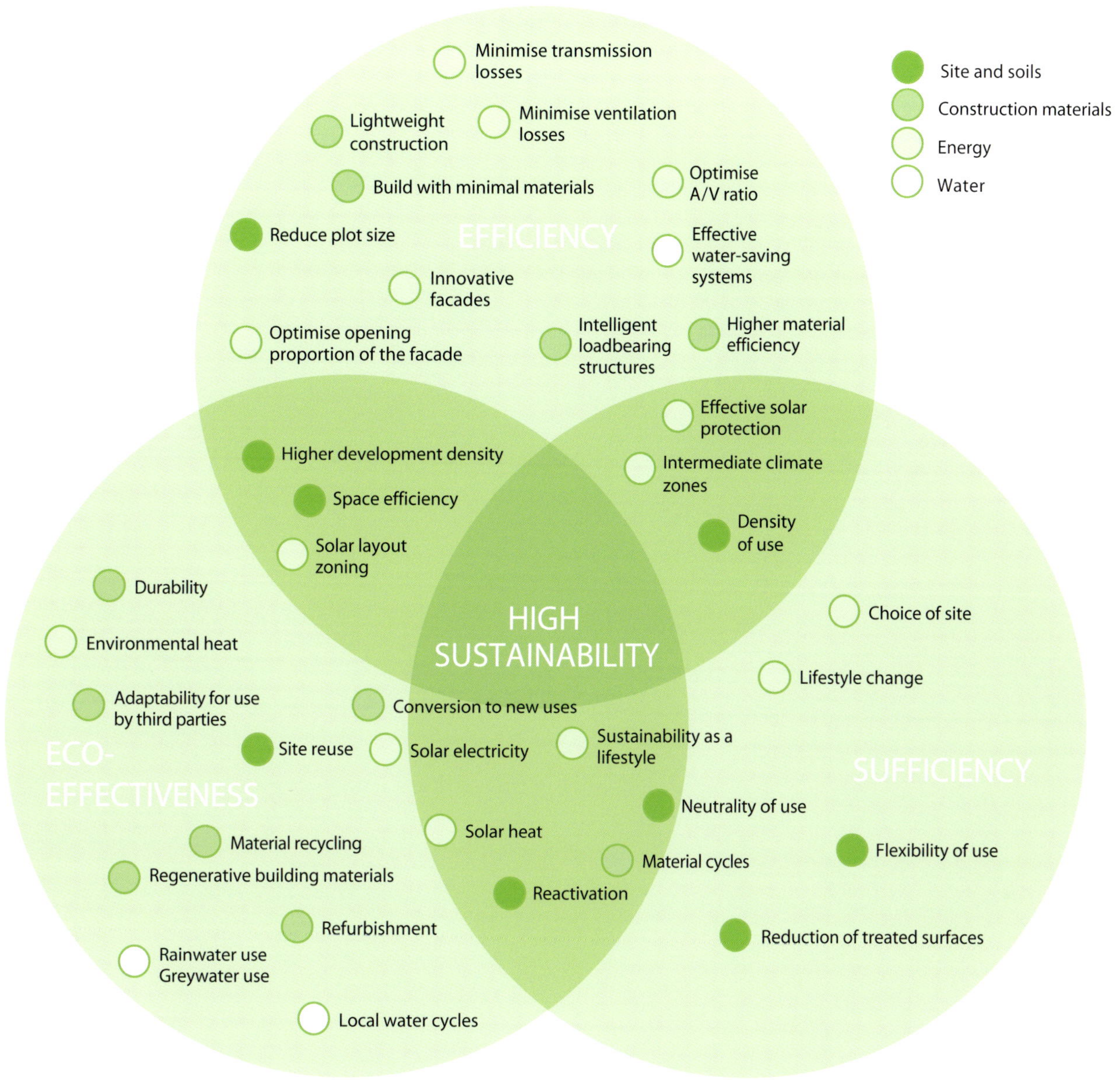

Strategies of building to conserve resources, sustainability map

Aktivhaus

Aktivhaus is the contemporary further development of previous building energy standards. It is based on the principles of minimising the building's energy losses, its internal energy consumption and exploiting the direct passive use of solar radiation by the building itself. Of course, these principles are normally not enough in an average year to provide the building with pleasant living conditions all year round and to supply heating, cooling, ventilation, lighting and electricity.

These measures are therefore supplemented by the active use of energy from regenerative sources, where these are accessible from the house or its grounds, for example, from the conversion of solar radiation, ambient heat, wind flows or geothermal energy into heat and electricity for the building. Therefore the Aktivhaus not only saves energy, but is also designed to generate energy from its building envelope, its parts in contact with the soil and its immediate environment.

The Aktivhaus therefore no longer relies on the customary major external energy supply systems. It does not matter whether they are operated using fossil fuels, such as coal, oil or gas, or whether the energy is produced regeneratively in large plants, such as the North African solar arrays or offshore wind farms. The Aktivhaus uses the high self-sufficiency potential of the immediate environment. It strives for the objectives of the *small is beautiful* and the simple – without being backward or pre-industrial. The Aktivhaus integrates into the neighbourhood and city with other buildings and institutions. In this cooperative arrangement, supply and demand increasingly balance one another out and therefore reduces the present high need for storage of renewable energies. At the same time, the self-sufficiency potential of the city improves and contributes to the security of supply and self-confidence of the city. At first somewhat sporadically, Aktivhaus buildings can be networked and play an important role in the self-sufficiency of urban quarters and eventually cities. They can offer, in parallel to the developing urban landscape, a new role model for the construction and development of the city.

Strategies

What kind of buildings should be built to satisfy the above criteria? Four levels of consideration point the way to Aktivhaus.

Programme

In view of the stagnating population in Central Europe, any plans for the development of new sites should be considered judiciously. This approach reduces traffic and avoids the need for new infrastructure. Building in existing fabric, renewing the city continuously within its existing footprint as far as possible, secures the future of the city and maintains urban life. It makes retaining the existing infrastructure easier – whether roads, cabling or waste water systems – but also the social and cultural institutions, and public utilities.

This extends to living areas, working areas and land for many other uses. They will have to be built and operated to a far higher standard than was the case only a few decades ago. However, more space does not always result in better living quality. Instead, the aim should be to create better space that combines high-quality locations with attractive spatial qualities, the best energy performance and therefore high comfort.

Foresight in the selection of location and a carefully designed room schedule reduce the demand for space and can best answer the call for sufficiency. Reasonable solutions contribute considerably to stabilising our cities and communities in the long term as well. Through the principle "Quality not area", adequate solutions create efficient spaces for high energy standards and sustainability.

Construction measures

The second strategic level of consideration involves the building construction. Its potential plays a decisive role on the way to energy-efficient buildings. A compact building form proves to be particularly effective, the use of solar radiation through suitably oriented and correctly sized windows, a sealed and well-insulated building envelope and adequate thermal mass to compensate for sharply fluctuating temperatures, heat absorption or reflection. This is the focus of architecture – in essence, form and articulation, mass and transparency, texture and colour. Key here is the creativity of the architect, who, in the best cases, produces an energy-efficient building form without unnecessary additional expenditure and with a low technology content. The measures adopted are passive, because they can be achieved by the architecture without the involvement of technology, the geometry of the house and the character of its envelope. The logic of action at this level is obvious. With new builds in particular, it has taken us a long way into a sustainable energy future.

Building services

A third level is building services. In temperate climate zones and in its simplest form, it consists of a combined solution for heating and hot water and a ventilation plant with heat recovery. Within more complex buildings, it could be an air-conditioning plant with refrigeration units, thermal storage, emergency power system, an uninterruptible power supply, and much more. The building services design most suitable for the location, use and building arrangement should always be adopted. It should offer the users pleasant conditions and thermal comfort. The energy required for this should be expended in the most efficient way. The design should ensure intelligent interaction of building and technology. The building services must react to the types and times of use, to the division of the space, to the type of construction (light- or heavyweight), the passive qualities of the building, and many other characteristics. Simplicity and robustness are preferred for several reasons: building services technology changes at a more rapid pace than many other building components, and each piece of equipment requires maintenance, while the operation of interlinked systems can very quickly exceed the competence of the client and even the experts. This can result in easily understandable but sub-optimal operational settings being applied.

The energy demand can be substantially reduced only through intelligent architecture combined with a constructional concept and building services, fine-tuned to the characteristics of the building.

Energy generation

The fourth level of consideration – after the development of an optimised design in conjunction with a building services system attuned to its characteristics – is energy generation by the building envelope and in the grounds of the building. The use of environmental energy right there where the energy is consumed appears sensible in many respects. It makes the consumer into a generator (prosumer), and reduces the dependence on external systems beyond his or her control, while avoiding some of the considerable cable and conversion losses. It reduces the investment and transport costs of fossil and biomass fuels, gas and electricity. The use of renewable energy is fostered, in particular from the sun, wind, flowing and standing water, and geothermal sources. Environmental energies can make a considerable contribution to the planned transition to energy from renewable sources. Finally, the generator-consumer symbiosis creates a considerably raised awareness of the availability of energy and a more conscious use of energy.

One planning strategy – no energy standard

The term "Aktivhaus" (when not referring to the Energy-Plus building) does not primarily describe a quantitatively defined standard, but rather a design strategy. It pays attention to the principles of passive solar construction in the design. It develops the building starting with the climate, creates a stable system even without technology, and integrates the environment as directly and extensively as possible into the provision of pleasant living conditions. It falls back on traditional construction principles, refined and improved locally over the centuries, which were also derived from the climate.

This strategy turns away from the idea that technical systems alone can perform the task of space conditioning, which has predominated building since the beginning of industrialisation, and in particular since the age of classical modern architecture and its fascination with technology. This should also raise the profile of the prudent principles of climate-compatible construction in the public mind, because architecture is able to create comfort to the greatest extent by design and construction. This demands a new way of developing solutions. Their discovery and form-finding can result from well-known but often overlooked natural laws as much as from completely new ideas.

An Aktivhaus building is not forced to stay below the Passivhaus' rigid limit of a mere 15 kWh/m^2a for heating energy demand, which can only be achieved at high economic cost. Sometimes it is more economic to compensate for a slight difference here with the active use of renewable energy. This applies in particular to buildings with unfavourably large building envelope surface area to volume (A/V) ratios. The way is open to activating the amply available surfaces instead of installing a disproportionate amount of extra insulation. However, this slightly reduced standard of thermal insulation in the building skin must not lead to a loss of comfort.

The direct and active use of environmental energy, in particular solar radiation, frees the Aktivhaus from the constraints of purely passive strategies. This avoids the overly thick walls that result from the mandatory large thicknesses of insulation. In the Passivhaus, these thick walls often lead to shaft-like windows and a corresponding loss of daylight. Energy creation integrated into the building can more than offset a slightly higher energy loss without detrimentally affecting comfort. The Aktivhaus can avoid having mandatory large openings facing south and therefore avoid the risk of undesirable summer overheating from the passive use of solar energy. It can also use its envelope surfaces to maximise energy creation with active solar systems integrated into the building. In addition, the active use of environmental energy offers scope for new designs of building envelope, for example through the integration of wind and solar energy systems.

The Aktivhaus leaves more room for creativity. Freedoms in use and design increase. A building that in the end produces an excess of energy, in terms of its energy balance sheet, (i.e. an Energy-Plus building) is nothing too unusual. It is less oriented towards strict requirements, e.g. maximising the quality of the building envelope, than earlier buildings were. The critical factor will be to show an optimised energy balance between creation and consumption. However, this cannot be applied universally, because it will depend on the use and the density of development.

The approach both allows a realistic consideration of the building, its functions and its surroundings, and also opens up a field of possibilities, indeed opportunities to be grasped, of using the new energy creation technologies to develop new forms of expression for a type of construction that is fit for the future.

Emotions

Good Aktivhaus buildings react much more obviously to local circumstances. Therefore they promote a local, differentiated building form and concept suitable to the location. Aktivhaus buildings strengthen local identity and create an emotional bond through distinctiveness and uniqueness.

Good design appeals to the emotions and promotes bonding. Higher requirements for the quality of the design is a precondition for successful energy transition in buildings. This objective is paramount; the route to it is full of obstructions and opportunities. One obstruction is the difficulty of combining changed requirements and new technologies in a good building form. Intensive development and integrated design in dialogue are necessary here. However, uninhibited discussions about the unusual are fertile grounds for opportunities to develop something new: opportunities to find new forms of architectural expression that serve the objectives of sustainable and energy-efficient building, that use new materials or that combine in unusual ways, and create new forms by using new technologies intelligently.

If the opportunities are used, Aktivhaus buildings can be established as the standard of the future. This brings about acceptance and identification with this building concept and provides the basis for its successful propagation and establishment.

Energy balance

Since the various balancing systems differ greatly in their calculation methods, this section must begin with the clarification of some terms. Their explanations are based on a conventional energy balance for residential buildings in accordance with the German Energy Saving Ordinance. The following pages describe the current building energy standards, definitions and sustainability assessment systems in Germany and other German-speaking countries.

Developing the building energy balance

People in all corners of the earth have been building structures for shelter from the weather and danger for centuries. However, it is only in recent years that the energy consumed by buildings to achieve interior comfort has been considered and expressed in figures.

The final trigger was the first oil crisis in 1973, which was followed by another in 1979. As a reaction to the painful realisation that fossil resources would not last forever nor be available cheaply in the long term, many countries introduced legal instruments to be able to measure, compare and limit the energy consumption of buildings.

In Germany, for example, these developments revealed how the boundaries of consideration of energy consumption have shifted and the ultimate importance of total energy consumption. The Energy Saving Act (EnEG) was passed in 1976 by the German Parliament and requirements set out in DIN standards. The Thermal Insulation Ordinance (WschV), the German regulation on energy-saving thermal insulation for buildings, came into force in 1977 and for the first time defined minimum requirements for the building envelope as a device for reducing heat demand. Prior to this, from 1952, DIN 108 formulated requirements for energy losses from transmission and ventilation, which applied to individual components but not to the whole building envelope and

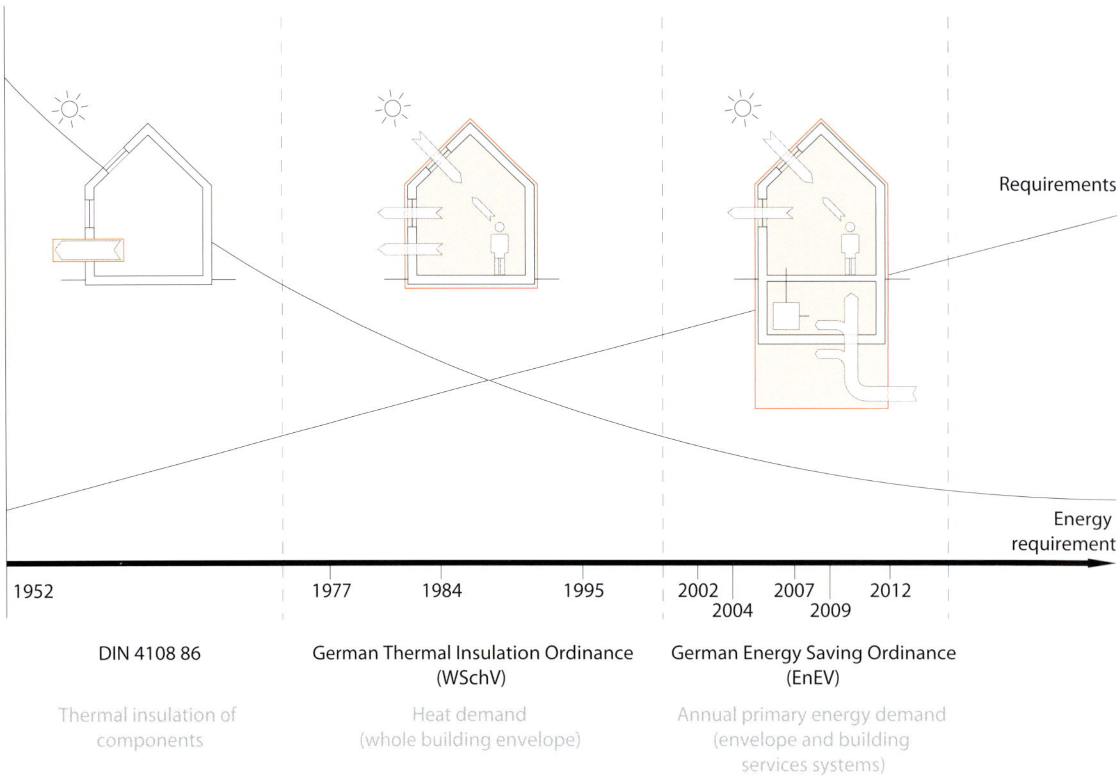

In addition to the reduction of energy demand, the diagram shows the expanded levels of consideration and the growing requirement profile.

were related to geographic location. While this allowed the calculation of the initially striking heating energy demand in temperate climates, the requirements on technical systems to cover the energy demand were stipulated separately in the Heating Appliance Ordinance (HeizAnlV). Amendments in 1984 and 1995 further sharpened the requirements on the insulating effect of building envelopes.

The Thermal Insulation Ordinance and the Heating Appliance Ordinance were finally replaced by the German Energy Saving Ordinance (EnEV) in 2002. This represented the first step in considering the building as a whole. The building services and the physical building elements were assessed together in terms of their overall effect on the building's energy efficiency. This more complex consideration of the issues meant further parameters became relevant.

The Thermal Insulation Ordinance used the average thermal transmittance, i.e. all the transmission losses through the wall, roof and windows, as the governing parameter. In the EnEV as well, the whole envelope is evaluated using the enveloping surface-related transmission loss (H_T'), albeit only as one of two main requirements. The energy balance is expanded by the addition of building services systems. The analysis considers not just the heat energy required to achieve the room temperatures, but also the losses incurred during creation, distribution and transfer, which are now important components of the building evaluation. The concept of final energy was introduced. The final energy balance describes all energy creation and distribution processes and the associated losses upstream involved in providing the energy. This approach allows the building operation to be holistically modelled and compared. The quality of the facade alone no longer determines the efficiency of a building, now the whole energy supply system is relevant. A building with a comparatively higher energy demand attributable to its envelope can compensate for this by having efficient technical systems. The primary energy balance, which assesses all the energy carriers on the basis of their effect on the climate, expands the options for optimisation by the use of renewable energy carriers and sources, such as solar radiation, environmental heat and biomass. The annual primary energy demand Q_p is the EnEV's second main requirement.

The EnEV has since been amended four times (2004, 2007, 2009 and 2014). It continues to be the basis of energy balances for buildings in Germany and allows properties to be compared in this respect. Over recent years, this has been the basis of various new building standards, which have widened the scope of the energy balance and increased the requirements. To understand these differences, it is first necessary to comprehend the composition and scope of an energy balance.

As well as the standards for residential buildings, DIN V 18599 was developed for preparing energy balances for the more complex non-residential buildings and allowed individual zones to be analysed according to usage profiles. It also took into account the energy used for cooling and lighting.

Principles of an energy balance

There are many reasons why different balancing systems cannot be compared to one another. The complex design of a computer model of a building's operation is based on different inputs and areas of consideration. There are also no rules to arrive at a uniform balance structure. Splitting an energy balance into its basic parameters reveals the differences and similarities.

An energy balance is always only a computer model and therefore a theoretical depiction of a building under standardised boundary conditions. For example, the standardised energy demand during operation is calculated but the actual energy consumption is not measured later. The actual value may deviate greatly from the theoretical results of the energy balance because it is influenced by parameters that cannot be represented in an energy balance, such as user behaviour. Faulty calibration of building technical services systems is often responsible for the actual consumption being way above the calculated values, especially at the start of building operation. Here it is expedient to fine-tune building services systems and install a good energy management or monitoring system for the building operation phase, but this is not specified in most cases.

Using the example of the EnEV, the following section shows in principle which approaches to balancing can be considered as the statutory minimum standard and which basic parameters are part of the balance.

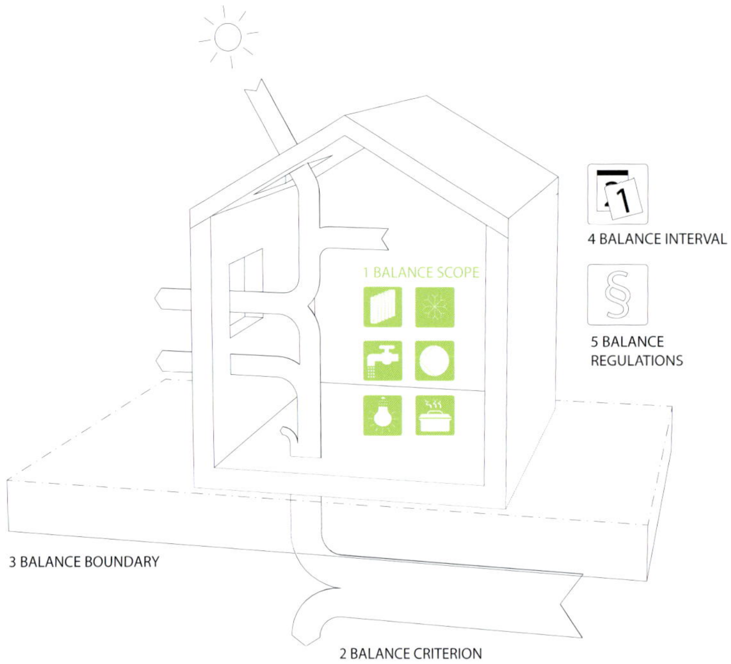

Balance scope

Energy is a conserved quantity – it is never just lost. However, it can leave the system in which the energy is used. This is commonly described as an energy loss. Strictly speaking, the energy still exists but is in another form and another place. Losses such as this can occur in a building, for example by heat flowing from the inside to the outside through ventilation and transmission. Energy losses therefore largely determine the energy demand. In this context, the energy demand is that quantity of energy used to keep the interior of the building at a comfortable temperature level (heating, cooling). A building can also accumulate energy. For example, the interior may heat up due to the presence of people and the waste heat from equipment. Solar radiation can enter through a window and carry heat energy into the room. In addition to these passive internal and solar gains, active technical components integrated into the building, such as photovoltaic panels forming part of the facade, can create energy.

The balance scope has the task of meaningfully differentiating the above-described complex systems of energy transfers, losses and gains for each of the uses. It circumscribes the extent of the assessment and prioritises the individual needs. Buildings consume a great deal of energy and therefore the EnEV addresses their operation and assesses every use of energy required to create comfortable conditions. In the field of residential buildings, this means the energy demand for heating, cooling, domestic hot water and auxiliary energy (e.g. for fans and pumps). In the field of non-residential buildings, the EnEV also considers the energy for lighting.

There are further sources of energy consumption in the operation of buildings, such as household appliances and domestic equipment, which are not taken into account in the official certificates for residential buildings, because they are too user-specific to be described using parameters. The incentive to save energy here is provided by a model of energy classes. Opening the field of view wider and leaving the building level, the assessor encounters further energy expenditures that are influenced only to a limited extent by the building and its position, but will gain in importance for increasing the efficiency of building operation.

A third expansion of the balance scope takes place when the building is considered over its whole life cycle. This adds further sources of energy consumption that go beyond operation and connect with the production, maintenance and demolition of a building. The development of some very good building operating concepts in recent years has caused the energy consumption for operating a building to shrink noticeably. The ratio of the embedded energy – the energy necessary for the manufacture of building materials and for the building works – to the operating energy approaches unity.

In principle, the balance scope can cover three areas: building operation, life cycle and user-dependent energy expenditure. In order to avoid highly complex, error-prone balancing systems and be able to evaluate specific areas, the balance scope defines the balance framework very narrowly. The EnEV considers parts of the building operation for residential buildings and an expanded area for non-residential buildings. These areas are shown against a coloured background in the diagram.

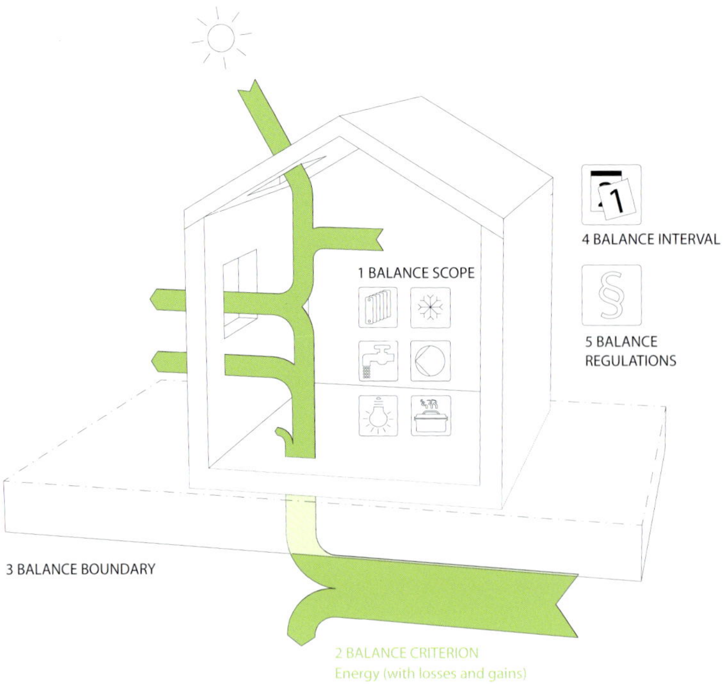

Balance criterion

The evaluation is determined by the balance parameters or the balance criterion. This in turn determines the calculation processes upon which the balance is based.

The complex calculation process of the EnEV considers the quality of the building envelope, all common forms of energy losses and gains, the efficiency of the building services components and the type of energy carrier used. The software uses this information to calculate the demands. The direction of balance is always opposite to the direction of energy flow. First, the quantity of useful energy required for the interior space, which depends on the quality of the envelope and volume of the building, is calculated. The calculation of end and primary energy is then performed, taking into account the technical requirements and selected energy carrier. The EnEV uses the resulting annual primary energy requirement Q_p as a requirement and a benchmark, and the specific transmission heat loss H_T' as a secondary condition. These values show how much energy escapes through the whole building envelope to the outside as heat.

The EnEV calculation focuses accordingly on the operation of the building. The relevant parameters are therefore operating energy values. Outside the area of operating energy and consequently extending beyond the evaluation in accordance with EnEV, natural resources are consumed and emissions released in the manufacture of the building materials as well as from the production, maintenance and demolition of a building. Because these processes have similar effects on our environment, they likewise provide figures for the energy balance evaluation. By specifying their units of measurement, these figures can be considered in the calculation and evaluated in terms of their energy (kWh) or efficiency (€/amount of energy).

Depending on the purpose of the evaluation and the comparison, useful energy, final energy and primary energy, emissions such as CO_2, material resources, energy costs or operating costs as well as embedded energy may make useful parameters. The focus of the content can always be directed by selecting the appropriate balance parameters.

Balancing methods, even those with the same balancing criteria, are usually only comparable with one another to a limited extent. As well as the balance criterion, all other balance parameters must also be identical, as must the calculation procedure. National preset parameters are a further influence on the outcome of the balance. For example, the primary energy factors differ from one another, depending on the make-up of a country's energy supply. A building with an electricity supply in Germany may have the same calculated final energy consumption as one in, for example, Switzerland or Norway, but will have a considerably higher primary energy demand due to the greater proportion of renewable energy carriers (e.g. hydroelectricity) used in those two countries.

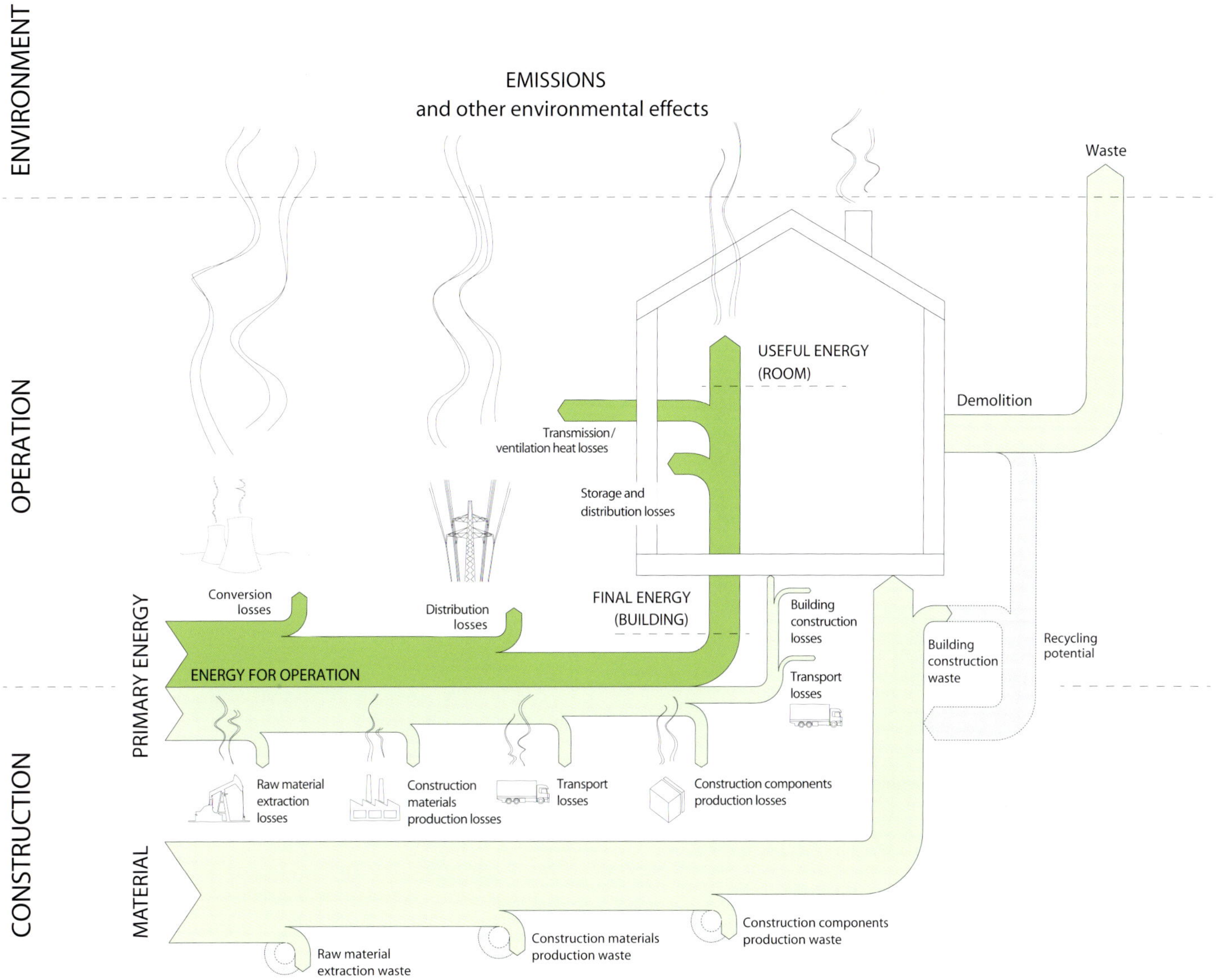

Potential balance criteria: a building generally consumes energy and materials and creates emissions as well as other effects on the environment. Accordingly, there are three areas in which, depending on the objective of the balance, the suitable point in the production and creation chain can be chosen as the criterion of the balance. The EnEV considers only the energy uses required to operate the building but the diagram shows the complete loss chain.

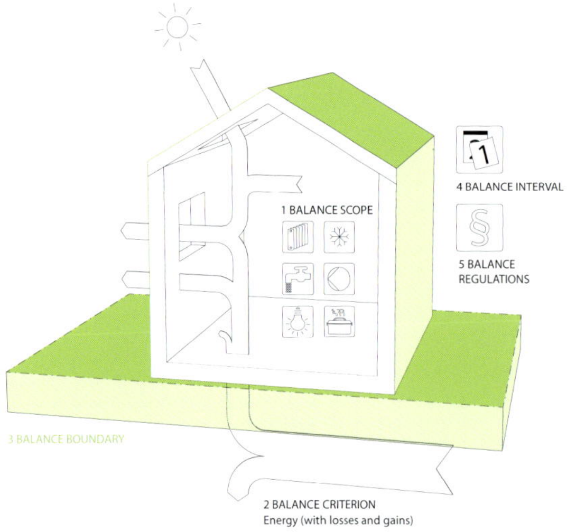

Balance boundaries

To make the balances comparable, spatial boundaries have to be set, as well as limits to content. These limits usually have no effect on the energy demand because the minimum boundary is the extent of the building and therefore the envelope – the main influence on the heat balance – is always included. Extending the boundary is mostly done to take into account of the location and aspects of energy creation that affect the balance. Normally any energy created by the building itself can be subtracted from the energy demand in a balance. Possible boundaries could be the building, the plot of land or even the district. If the purchase and sale of certificates is taken into account, then the balance boundary is invalidated in practice.

If the building is set as the balance boundary, then only the energy created directly in and on the building can be credited against the consumption. In the context of renewable energy, this is usually the use of solar energy in the building's technical equipment. In many cases, solar collectors are installed for the preparation of hot water, and photovoltaic modules for generating electricity. Another possibility is a combined heat and power plant inside the building, which, as its name suggests, creates heat and electricity. Above all in the area of renewable energy creation, a trend is emerging in which technologies are now made usable at an appropriate scale on buildings. For example, more and more small high-output wind turbines are being installed on roofs.

In most cases, the balance boundary is set as the area of the plot. A supplementary note along the lines of "in direct spatial conjunction" is usually enough to limit the balance to the area and infrastructure near the building. Not included in this consideration are large infrastructure items that, although they could be assigned to the grounds of the building, are operated as commercial plants and not used primarily for the building.

A balance boundary covering a city district is not often selected at the moment. Within this larger boundary, a wide spectrum of technical solutions can be included in any balance. In addition to the proliferation of, and often very effective yield from, photovoltaic arrays on public or shared areas, local heating networks involving various solutions based on combined heat and power plants are feasible, which are often more efficient on a district level than when supplying a single building. In addition, considering an area of this size offers the possibility of counteracting the climatically disadvantaged position of some buildings or exploiting time shifts in building use. It also offers the opportunity of linking listed buildings, which by their nature are difficult to improve with regard to their energy efficiency, with highly efficient new buildings.

In Germany, balance calculations have been able to take into account the electricity generated from photovoltaic panels near or on buildings only since the 2009 amendment to EnEV. The relevant provision is contained in section 5. It says the electricity from PV arrays near the building that are used primarily for the building can be subtracted from the final energy electricity demand. The calculation is done on a monthly basis. This type of system feeds power into the public grid only when a surplus exists on the site.

The German Renewable Energy Sources Act (EEG 2000) ensured that a huge number of photovoltaic panels were installed on roofs before 2009 by offering a feed-in tariff. This concept is based on feeding electricity into the public grid and not on self-use. This leads to lowering the primary energy factor in the case of the German fuel mix. Systems installed before 2009 were generally well designed and above all economically efficient, but were not set against the energy demand of a building in the calculations.

Likewise since 2009, systems for creating energy from renewable sources in Germany are regulated through the Renewable Energies Heat Act (EEWärmeG). This stipulates that part of the heating or cooling energy demand of new buildings is covered by renewable energy. How high this proportion is in individual cases depends on the selected technology and the energy carrier. This legislation also directs the focus back onto solar thermal panels in or on the building envelope.

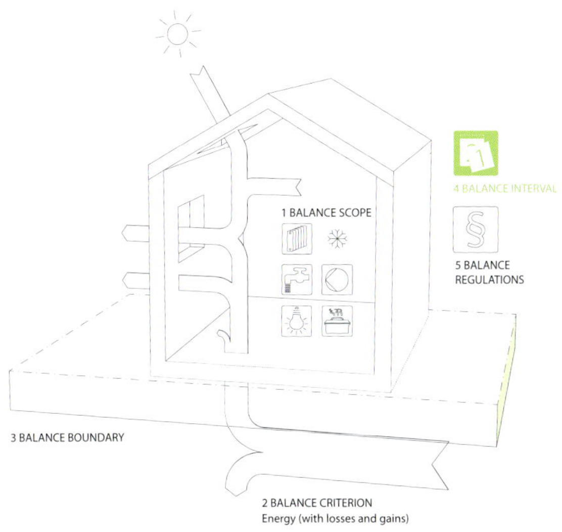

Balance interval

The described boundaries to the content and physical extent of a balance are also supplemented by a temporal component. The time horizon is particularly important for comparability, when credits for self-generated energies are taken into account. Itemisation against a timeline shows how close a balance is to the real picture.

Energy balances usually consider an annual mean or the annual total demand. They ignore seasonal differences. An annual balance is a good way of comparing different buildings with one another and ranking them in order of their efficiency.

If, in addition to demands, energy carriers are taken into account, an annual balance is not accurate enough. With the method of a credit based on the annual level, seasonal peaks of demand and creation are ignored. If a building creates energy, for example, through solar-active systems, the greatest part of the yield comes in summer and the transitional months. Conversely, a residential building at our latitudes has its highest energy demand in winter and the transitional months. This situation is better displayed in a monthly balance in which energy consumption and yield are expressed month by month. A monthly balance allows an approximate estimate to be made of how much of the created energy can be used within the building itself. The EnEV allows consideration by the month as well as by the year, but a monthly balance is necessary in order to credit locally created energy.

A daily balance can be produced if more accurate information is required about the performance of a single building. The monthly balance can be extended to consider day and night profiles. Specifically in the summer months, the monthly balance is too inaccurate for some purposes, for example for designing a small buffer store for peaks in load or yield. A daily or even an hourly balance in these circumstances offers more accurate information for sizing a buffer store to increase the amount of self-generated energy used by the building.

In theory, the time interval can be resolved down to the second. Then the energy balances are performed. These represent the transition from the parameter balance to building simulation. Because they are very time-consuming and depict only theoretical values, they are used only in exceptional circumstances, when the need is justified. When they are used, it is worthwhile monitoring the actual operation of the building and any optimisation measures.

This usually determines whether the very accurate representation (balancing hourly or by the second) is close to reality. They are also useful for more accurate planning and design of individual buildings (e.g. for energy recovery systems and storage buffers). However, if several buildings are to be compared with one another and their performances ranked, balances with greater time intervals (annual or monthly balance) are easier to deal with and usually accurate enough.

Using a monthly balance, the energy consumptions and the typical annual cycle can be evaluated on the basis of monthly mean values and used to develop a meaningful energy supply concept.

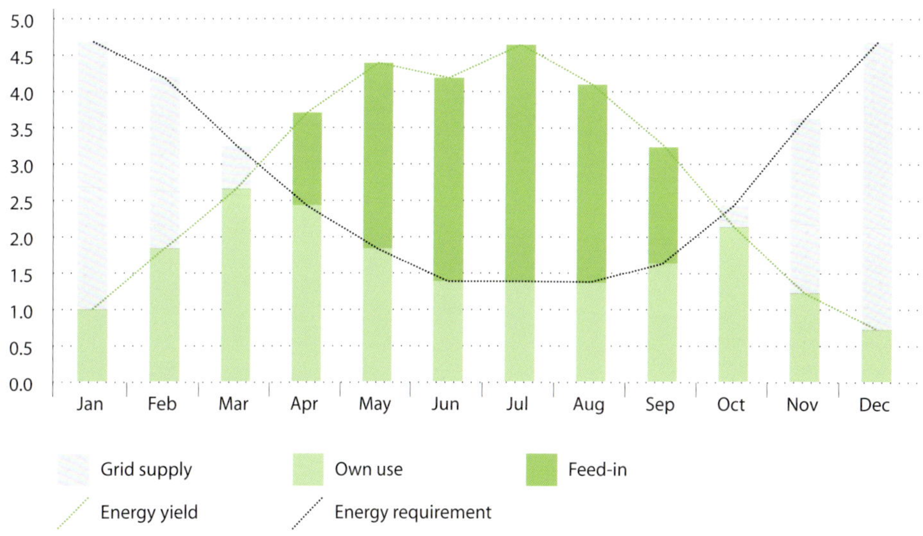

Solar Decathlon 2007, TU Darmstadt (DE), Darmstadt University of Technology, Department of Design and Energy Efficient Building

Energy production and consumption by the hour: the diagram shows the actual consumption of an Energy-Plus House designed and built by the students at the TU Darmstadt as an entry for the 2007 international university competition (image above). Afterwards it was mainly used as an office in Darmstadt and monitored to improve its operating performance. The energy consumption figures are very high, depending on the day. Use as a residence and an office also results in an energy demand at night. The very high energy yield from photovoltaics during the day (spring) was stored in an 8 kW buffer store and used to cover the night-time demand. Through this detailed monitoring and the resulting improvements in operating performance, self-use of the PV electricity increases and autonomous operation is ensured, even in spring.

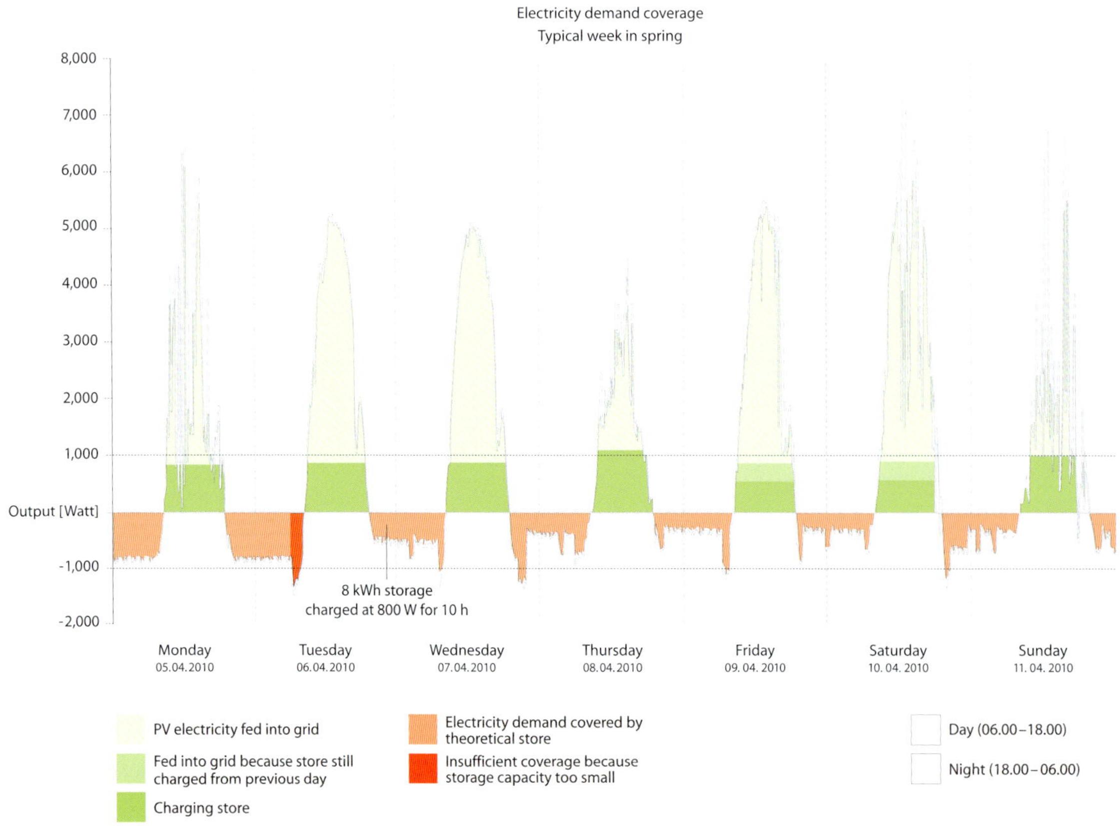

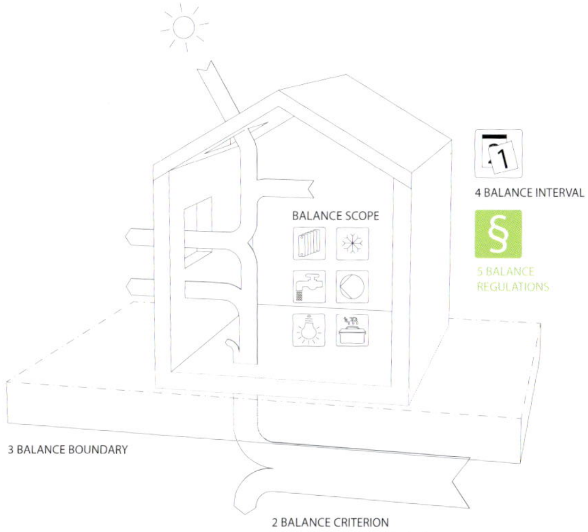

Balance regulations

After the above-mentioned boundary conditions and initial values of a balance have been determined, the actual calculation can be performed. The individual steps in the calculation are performed in accordance with the basic standards and technical rules. Because a balance can only ever be an approximation of reality, balances done using different calculation methods may give different results for the same building. However, they are credible if the parameters given in the standards and technical rules have been appropriately interpreted for the system. For the same reason however, it makes sense when comparing buildings to use the same method for the balances.

In Germany, the EnEV determines the overall method and refers to relevant DIN standards for the detailed calculations. The basic method for an EnEV calculation is the reference building method. On the basis of the values entered for the building being assessed, the method yields a standard thermal insulation value for the building envelope and a basic level of technological equipment for a computer-modelled reference building of the same dimensions, envelope area and use. The results for the proposed building are then compared with the designed values and the permissibility of the proposals assessed.

The detailed calculation method for residential properties is normally as specified in DIN 4108-6 and 4701-10. It is the older method of balancing and is relatively easy to perform. The result offers a good, comprehensive method of comparison. However, it is not very suitable for more complex models and investigations.

DIN V 18599 was developed in Germany for assessing the overall energy efficiency of buildings. It was produced in response to a request from the European Parliament, which has required a tool like this in all European Union Member States since 2006. This prestandard is an instrument for computer-aided assessment of complex building systems. Different use profiles can be modelled with their specific requirements and loads on the basis of a multizone model for a building. The pre-standard was first developed for non-residential buildings. More complex models of residential buildings can be assessed in a similar way using a single-zone model with this calculation method. The depth of detail and breadth of information are greater than those of the DIN 4108-6 method.

A balance in accordance with the Passive House Planning Package (PHPP) can be used to provide a detailed computer analysis of residential buildings that, for example, has also to take into account domestic electricity.

Static balances, which produce annual and monthly balances only, are of little help in creating realistic depictions of, for example, the thermal process within a building. The more accurate itemisation into hour and minute values, or into actual performance histories (load profiles), can be displayed only by dynamic simulations. A dynamic computer model can, for example, quantify and verify thermal processes and the way they develop dynamically in their physical locations. In contrast to a static calculation, the actual thermal air movements and energy flows in the space can be modelled. The accuracy of the results varies depending on the simulation software and the basic calculation model.

Secondary conditions

The framework for a building energy balance is defined by the adopted balance parameters. It is often the case that not all the parameters considered important can be taken into account. The computer models required for this could be too complex and prone to error and the required effort and cost are out of proportion to the result. Parameters that would be desirable but not of primary relevance to a particular system are therefore often included as secondary items in the requirements profile of a standard. Additional general requirements, such as the use of ecologically sound building materials, the avoidance of polluting emissions, the efficiency class of the domestic appliances to be used, or the stated objective of a highly economically efficient building can broaden the observation framework and raise a building's quality, without the need for further detailed balance methods. This naturally leads to the fulfilment of the secondary conditions not being verifiable in some areas.

An overview of building energy standards

A number of different building standards and evaluation methods have been developed in Germany to improve on the provisions of the established statutory framework (EnEV), and further reduce the energy demand and the environmental effects of operating buildings. The most important approaches in the German-speaking countries are explained in the following pages.

Efficiency House

BALANCE SCOPE

Incentive programmes were developed to achieve energy savings and emissions reductions stipulated at the national and international level. These sought to offer clients subsidies to adopt higher building energy standards than the statutory minimum requirements. The low-interest loans from the Kreditanstalt für Wiederaufbau (KfW) are one such incentive among others intended for private housing clients. Most often mentioned in this context is the Efficiency House. The programme currently has six efficiency categories, each with different terms and conditions of funding. In principle, the higher the energy standard, the higher the energy savings, and therefore the higher the financial support. The actual interest rates depend on the current state of the financial market, and therefore need to be checked in every case before a project starts.

Because public money funds the programme, the energy balance calculations for KfW Efficiency Houses must be in accordance with the EnEV. The increase in efficiency reflects the additional reduction in annual primary energy requirement and transmission heat losses compared with the minimum requirements of the EnEV.

Two categories are currently available for refurbishment projects: Efficiency House 100 and 115 (from the "Energy-efficient refurbishment" programme). The results are expressed as the ratio of the annual primary energy requirement to the minimum stipulated for a new building or for a reference new building under the EnEV, calculated as a percentage. An Efficiency House 100 has an annual primary energy requirement not exceeding the EnEV minimum requirement for a new building. An Efficiency House 115 must not exceed the EnEV requirement by more than 15%. The transmission heat losses for both standards must be 15% less than the primary energy demand. Both standards produce an efficiency improvement over the minimum requirements stipulated by the EnEV, according to which a refurbished building may have 140% of the energy demands of a reference building.

Furthermore, there is the category Efficiency House Monument, which was established explicitly for the rehabilitation of buildings that are listed as protected historic monuments.

Four further standards, Efficiency House 85, 70, 55 and 40, apply to new buildings ("Energy-efficient construction" programme). Refurbishment projects can also access these levels of funding, provided their requirements are achieved. This is possible usually up to Efficiency House 55 level.

As of April 2016, KfW will update the subsidies it offers for new buildings, focusing on those with a higher level of energy efficiency. The Efficiency House 70 for new construction will then meet the minimum legal requirements and be omitted from the subsidies. The introduction of the new Efficiency House 40 Plus standard is meant to also promote buildings that cover part of their energy demand with energy generated directly at the building.

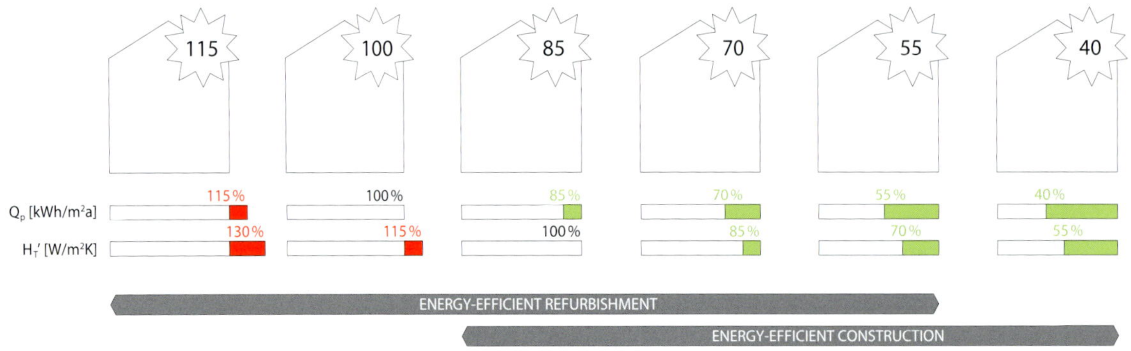

Overview of the KfW programme and the requirements of the Efficiency House categories

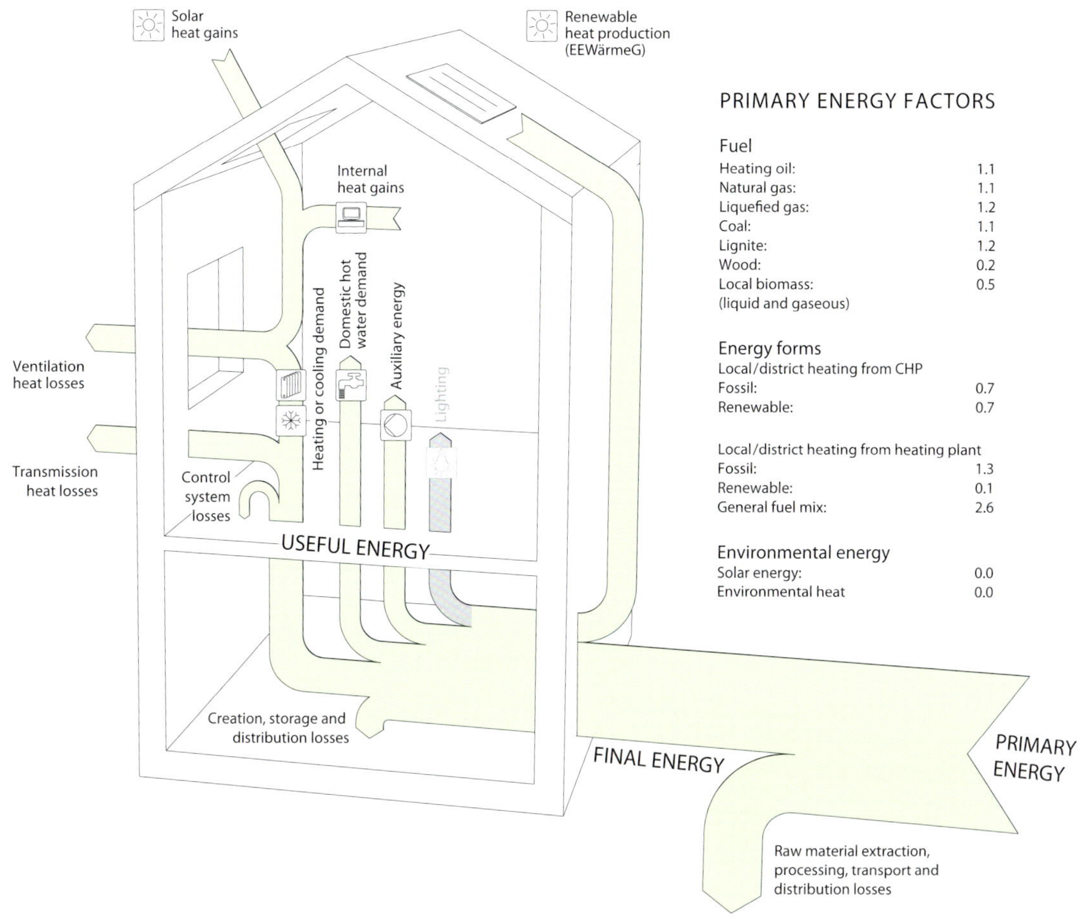

PRIMARY ENERGY FACTORS

Fuel	
Heating oil:	1.1
Natural gas:	1.1
Liquefied gas:	1.2
Coal:	1.1
Lignite:	1.2
Wood:	0.2
Local biomass: (liquid and gaseous)	0.5
Energy forms	
Local/district heating from CHP	
Fossil:	0.7
Renewable:	0.7
Local/district heating from heating plant	
Fossil:	1.3
Renewable:	0.1
General fuel mix:	2.6
Environmental energy	
Solar energy:	0.0
Environmental heat	0.0

The energy balance model for the Efficiency House corresponds to that of the EnEV. In order to achieve an Efficiency House standard, the outcome of the energy balance calculation must be under or exceed, as appropriate, the corresponding values for the reference new building.

Passivhaus

BALANCE SCOPE

The Passivhaus (Passive House) concept arose in 1987 as part of a research project by group of scientists. In the early 1990s, Dr Wolfgang Feist, founder of the Passivhaus Institut (Passive House Institute), established this system as a building concept and standard. Its main objective was to improve the heat balance of buildings by efficiently insulating the building envelope, and to minimise ventilation heat losses by ensuring airtightness and installing a heat recovery system. The result is the reduction of energy consumption for active climatic measures, hence the name Passivhaus. The Passivhaus is therefore a building that in principle does not require a traditional heating system and can be heated purely from the supply air.

To dispense with a traditional heating system completely when used for housing, a Passivhaus must retain heat. This is achieved first by adopting a very compact form and by avoiding cold bridges in the construction. Furthermore, the external walls, windows, roof and floor slab must have a very good standard of insulation. The design has to allow for insulation thicknesses in the walls

The community centre at Ludesch was designed to the Passivhaus standard, Architekten Hermann Kaufmann ZT GmbH, Schwarzach (AT)

and roof of than 20 cm (preferably 30 cm, and in certain circumstances up to 50 cm). Windows need to be triple-glazed, appropriately detailed and incorporate a thermal break. In order to achieve the Passivhaus standard, buildings must also be highly wind- and airtight to minimise ventilation heat losses. Fresh air must therefore be supplied through a mechanical ventilation system for ambient comfort. This must incorporate an effective heat recovery system to increase the overall efficiency of the ventilation system and further reduce losses. Fresh air is prewarmed, for example by passing it through an earth tube. The outdoor air is sucked in through a main inlet somewhere on the site and drawn underground into the building through a buried duct or a register. This raises or lowers the temperature of the air compared to that of the ground, which remains at a constant temperature all year round, and therefore saves some of the energy required to condition the supply air.

Internal gains from people, equipment and lighting heat up the space inside a Passivhaus, which retains the heat because of the high quality of its building envelope. Intentionally large windowed areas facing south provide further solar heat to the interior, while smaller-sized windows facing north reduce losses. A building optimised along these lines needs to be heated only in very low outdoor temperatures and therefore calls on externally supplied energy for heating for only a few days in the year. In many cases, spaces are heated by warm air. The passive building system reacts rather slowly and therefore further heating devices are required, depending on the energy-related qualities and internal zoning of the building, to ensure ambient comfort inside it. This is mainly worthwhile in rooms with uses that require them to be brought up to a higher temperature for short periods (e.g. bathrooms), rooms heated by waste air only, rooms with an exposed aspect, or rooms used only temporarily (offices, guest rooms). Correctly designed and built, the Passivhaus is a reliable building system. With such stringent demands on the airtightness of the building envelope, the building must be very precisely designed and built to ensure no defects occur during construction.

One means of verification and quality assurance is certification of the Passivhaus standard by the Passive House Institute in Darmstadt or by other certification bodies. Certification involves using various processes to check limits and other parameters set by the Institute and to confirm that the building fulfils all the criteria to be a quality-tested Passivhaus. In addition, the same criteria as the KfW's Efficiency House 55 or 40 apply for the evaluation and incentive funding of a Passivhaus.

The Passivhaus standard was first developed for housing in Germany. Use profiles for other climate zones and non-housing buildings of various types have been added in recent years: offices, care homes and schools, and since 2012, swimming baths and other similarly complex buildings. The main difference for non-residential buildings lies in weighting of the energy demand profile. In many use profiles, the electricity demand for office devices, equipment and lighting is determinant rather than the heating energy demand.

	EnEV	PHPP
Balance scope	Heating, cooling, auxiliary energy (residential building)	Heating, cooling, auxiliary energy, lighting, domestic appliances Aids
Balance parameter	H_T', Q_p	H_T', Q_p Q_h or P_h
Internal heat gains	5 W/m²	approx. 2.1 W/m² (with efficient domestic appliances)
Average room temperature	19 °C	20 °C
Solar gains	Flat rate 0.9	Calculated by software
Reference values	Energy reference area $A_N = 0.32 * V_e$	Heated living area
Balance interval	Month	Month
Secondary conditions	Cold bridges flat rate 0.05–0.15 W/m²K	No cold bridges (< 0.01 W/m²K)
	Airtightness n_{50} < 1.50 (with ventilation system) < 3.00 (without ventilation system)	< 0.60
		heat recovery n_{eff}, heat recovery 0.75
		Window U_W 0.85 W/m²K (installed) g-value > 50 %
		Opaque components U 0.15 W/m²K

Differences in the boundary conditions for the energy balance to EnEV and PHPP

A Passivhaus building is a highly optimised and therefore sensitive system. The energy balance must be based on a comprehensive calculation model to convert the design into a successful building. For this reason, the actual behaviour of the first Passivhaus designs was investigated using dynamic simulations capable of working with very short time intervals. The designs could then be optimised to achieve the high levels of comfort and efficiency, and ensure comfortable operating conditions. These simulations are complex and very time-consuming, and therefore too expensive to be performed for every Passivhaus project. As a result, the Passive House Institute produced its own energy balancing tool, the Passive House Planning Package (PHPP), based on the results of the simulations, which had been additionally verified by measurements. PHPP reduces the amount of input data for the calculation and displays the parameters prioritised on the basis of the simulations. The term "planning package" suggests the software should be used as a planning tool. Unlike the EnEV, which has the aim of comparing similar buildings and ensuring compliance with minimum requirements, the calculation according to PHPP supplies detailed statements about the building, specific consumption figures and the qualities of components. In addition to the energy balance and the U-value calculations, it is also possible, for example, to display the heat load, make predictions about summer comfort and design the necessary comfort ventilation. Using the results of the calculation, engineers can plan and design the building's technical equipment such as heating and hot water systems. Users of the package should keep in mind that the results may not meet the requirements of many national standards. The designer should make the client aware of these departures from standards, and obtain and record the client's agreement.

In the case of residential properties, the whole building is considered as one zone and evaluated using a monthly balance. The worksheets cover the various parameters individually, including:

- Design of the windows
- Design of the comfort ventilation
  - Calculation of the thermal balances, U-value calculation for all components including cold bridges
- Display of the heat load
- Predictions for summer comfort
- Design of heating and hot water preparation systems
- Verification for obtaining funding of Passivhaus buildings (e.g. by the KfW)
- Simplified verification in accordance with the EnEV

The primary energy demand, including the electricity required for lighting, household equipment and domestic appliances, must not exceed predefined maximum values. The scope of the balance produces a building concept that exceeds the statutory minimum requirements. Airtightness must be proven by a blower-door test on site.

Parameters of a Passivhaus:

Annual heating demand	< 15 kWh/m²a
Heat load	< 10 W/m²
Annual cooling demand	< 15 kWh/m²a
Annual primary energy demand	< 120 kWh/m²a
Over-temperature frequency	< 10 %
Ventilation with VHR	> 75 %
Electricity demand	< 0.45 Wh/m³

For the integration of renewable energy production, the Passivhaus concept was expanded in 2015 with classes:

- Passivhaus Classic PER ≤ 60 kWh/m²a
- Passivhaus Plus PER ≤ 45 kWh/m²a
- Passivhaus Premium PER ≤ 30 kWh/m²a

The requirements pertaining to heating demand remain unchanged. In place of the requirements for the primary energy demand, there is a requirement for the total demand of "renewable primary energy", which is determined using specific PER factors (PER = Primary Energy Renewable).

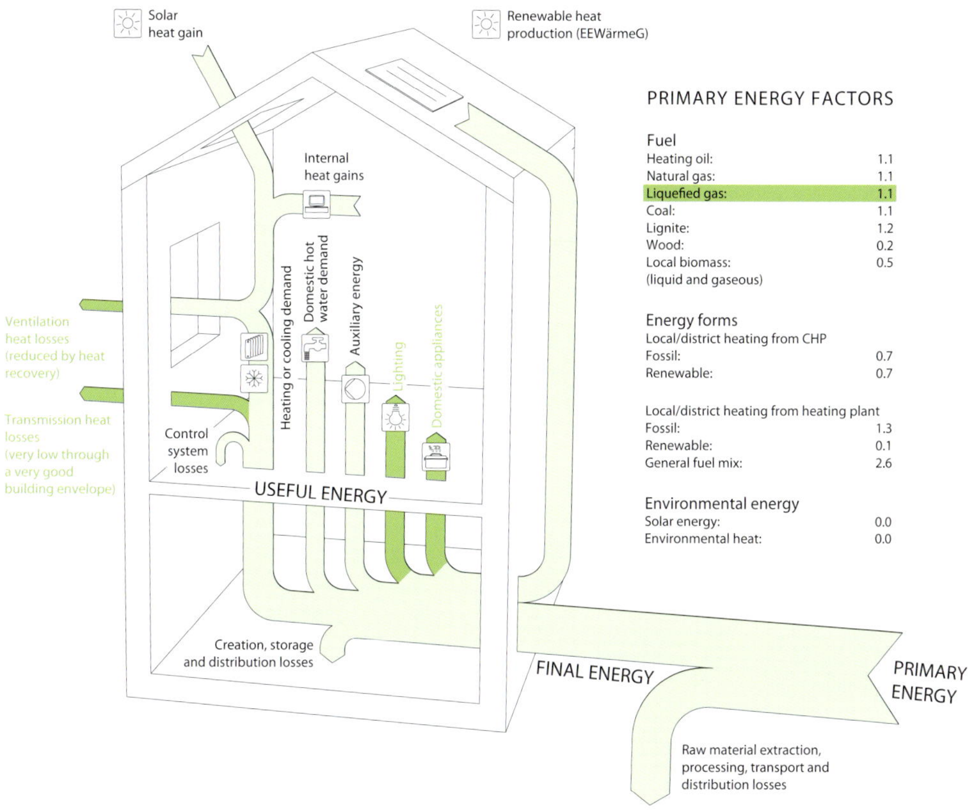

PRIMARY ENERGY FACTORS	
Fuel	
Heating oil:	1.1
Natural gas:	1.1
Liquefied gas:	1.1
Coal:	1.1
Lignite:	1.2
Wood:	0.2
Local biomass: (liquid and gaseous)	0.5
Energy forms	
Local/district heating from CHP	
Fossil:	0.7
Renewable:	0.7
Local/district heating from heating plant	
Fossil:	1.3
Renewable:	0.1
General fuel mix:	2.6
Environmental energy	
Solar energy:	0.0
Environmental heat:	0.0

Energy balance model for a Passivhaus. Ventilation and transmission losses are greatly reduced by the Passivhaus requirement profile compared to the EnEV values. The energy balance expands on the EnEV by taking into account the energy required for lighting and domestic appliances. Heating load and heating energy demand were added in 2010 as alternative ways of defining the objectives. Q_h ≤ 15 kWh/m²a remains as a mandatory requirement.

Modified or added criteria compared to the EnEV

Nearly zero-energy and zero-energy house

BALANCE SCOPE

Forty percent of the global final energy demand is required to condition the air in buildings. The development of compact systems for generating energy provided the stimulus for the idea of constructing buildings that would generate the energy they required to operate for themselves. As they no longer required any energy to be supplied from external sources, these buildings were called zero-energy buildings. The result of the energy balance calculation is, in theory, zero.

Climate change and the sharp decline in non-renewable sources of energy both led to political objectives being postulated on the international level to reduce energy consumption and emissions, and weaken dependency on imported energy. In this connection, an amendment to the EU's Energy Performance of Buildings Directive (EPBD) came into force in July 2010. The directive requires all new buildings in the EU from 2021 to be almost at the level of a zero-energy house (very low energy or nearly zero). This objective applies to public buildings from 2019. This is an ambitious objective for new buildings. In order to cut back the high energy consumption of older buildings, the same objective will apply to major refurbishments and extensions.

A number of different approaches were developed to define a zero-energy house on a conceptual level. The main problem is deciding the method of crediting the created energy against the energy consumption in order to arrive at as realistic a picture as possible.

A simple calculation setting the annual figures against one another does not mean that the building can actually operate without external energy. The building is more

With a good standard of insulation (MINERGIE-P) and energy generating technology, this apartment block in Dübendorf achieves a zero-energy balance over the year, kämpfen für architektur ag, Zurich (CH)

likely to create a surplus of energy at certain times, for example during the day in summer, which happens to equal the demand in winter. Practical seasonal stores with sufficient capacity to bridge this surplus and deficit have not yet been developed as single units for houses. The energy during periods of overproduction is therefore usually stored as electricity in the public grid and drawn from there again when needed. Such a building is a Net Zero-Energy House when judged on annual mean values, but not on a monthly basis.

Without doubt, the declared objective is a significant milestone on the way to solving the above problems in the long term and restructuring the present energy supply systems. However, the EPBD specifies only the direction. It does not make clear how seasonal imbalances should be handled. And the term "nearly zero-energy" has not yet been conclusively defined. The EPBD does not define the energy services to be considered or the relevant reference parameters either. All EU Member States must declare by 2015 how they intend to define the standard in detail and meet the objective.

For zero-energy buildings there is currently no basis in public law or an exact method of calculation for performing an energy balance and actually verifying this standard.

As its contribution to the process, the German Ministry of Transport, Building and Urban Development (BMVBS) published a definition for an an Efficiency House Plus as a model specification in 2011. This goes beyond the requirements for a nearly zero-energy house, because the building generates more energy than it consumes.

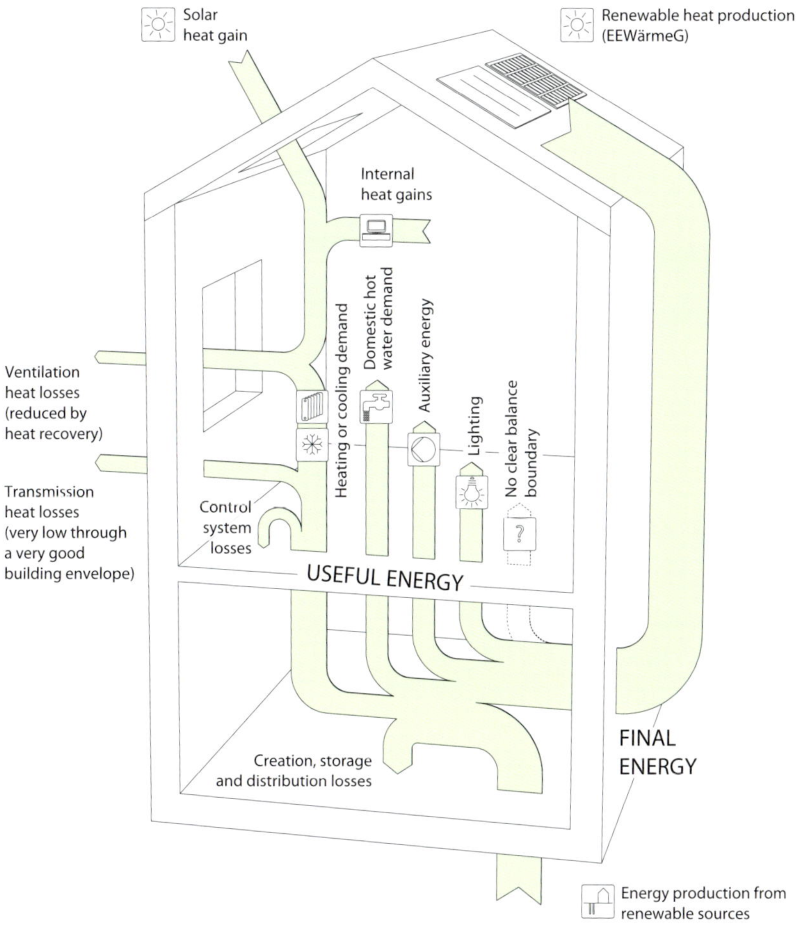

ENERGY DEMAND – ENERGY CREATION ~ 0

The model of the zero-energy house is not defined in detail. According to the energy-saving building guidelines, the overall balance must be more or less zero. This means the amount of energy created from renewable energy sources by the building must equal the energy used by the consumers. National standards have yet to define which consumers are included in the balance.

Efficiency House Plus

BALANCE SCOPE

Since the original guidelines for Energy-Plus buildings do not go beyond vague declarations of intent, the German Ministry for Transport, Building and Urban Development (BMVBS) published a first definition for the Efficiency House Plus with all the necessary parameters in 2011. This was the first Energy-Plus standard in any German-speaking country. The Efficiency House Plus standard is, however, still not a statutory requirement. Buildings of this standard have often been investigated and evaluated as model projects within the scope of a funding programme of the research initiative Zukunft Bau. In addition to the individual measures stipulated to obtain funding, which will be relevant in the future because of their degree of innovation, every building involved in the funding programme is monitored for 24 to 30 months. As well as consumption data, the energy supplied to the building is recorded and evaluated in detail. Measurements inside the buildings allow conclusions to be drawn about indoor comfort in relation to outdoor climate. Monitoring offers the client the basis for an energy management system and exposes weak points, so that measures can then be introduced to optimise the building operation. The evaluation and comparison of all the monitored projects also provide generally applicable findings about the new technologies used, and information for updating and further developing building standards.

The energy balance for an Efficiency House Plus is an extended version of an EnEV verification calculation in accordance with DIN V 18599 using a monthly balance and assuming the average standard climate for Germany. As a future-oriented standard, it also takes into account energy consumed for purposes other than just the operation of the building, and the air-conditioning of the interior. The electricity required for domestic appliances and processes is included in the balance scope. Since this is not currently included in the EnEV calculation process,

House P is an Efficiency House Plus by calculation alone because both the annual primary energy and final energy demand are less than zero, ee concept GmbH, Darmstadt (DE)

a flat rate of 20 kWh/m²a is assumed, with a maximum limit of 2,500 kWh per housing unit. The flat rate of 20 kWh/m²a is composed of 3 kWh/m²a for lighting, 10 kWh/m²a for domestic appliances, 3 kWh/m²a for cooking and 4 kWh/m²a for other energy consumers. The quantity calculated for the project is distributed equally to the previously calculated monthly end energy demands. The demands are given separately for each energy carrier to allow the individual primary energy factors to be correctly applied in a later consideration of primary energy. The monthly energy yield from energy generating technologies (such as photovoltaics) is then set against the monthly demand.

The balance boundary is the site. Unlike the EnEV, all renewable energy created on the site is included in the balance (on-site generation). If more than one building stands on the site, the calculated total amount of energy is distributed among the buildings in proportion to their usable floor areas. A credit for or the calculated amount of own-use electricity counts in full, but only up to the level of demand. The remaining energy yield after deduction is considered as having been fed into the grid. The reduced final energy quantities are then multiplied by the relevant primary energy factors. Deviating from the EnEV procedure, the primary energy factors are used in a similar way to DIN V 18599. The electricity fed into the grid is multiplied by the primary energy factors specified by the program and then deducted from the annual balance. The result must have a negative annual primary energy demand and a negative annual energy demand in order to achieve the standard.

In fact, almost all buildings are credited for electricity generated from photovoltaics – that is, during the summer – and therefore receive energy from external sources in the winter. In terms of primary energy, the positive balance is relatively easy to achieve, because the excess electricity is highly weighted. The positive final energy balance is therefore the more onerous requirement to fulfil. Heat pumps are often used to achieve this; systems based on combined heat and power, on the other hand, do not lead directly to a clear result in the balance. All systems that have losses on site and do not capture environmental energy make achieving certification more difficult.

The Efficiency House Plus and its limiting values in overview:

Annual primary energy demand $Q_p < 0$ kWh/m²a

Annual final energy demand $Q_e < 0$ kWh/m²a

All other conditions of the EnEV, such as summer heat protection, must be observed.

Secondary requirements:

Appliances with the highest efficiency label must be used (label A++ or better). A smart meter for evaluating the building operation and for determining the level of own-use of the created electricity must be installed in the building.

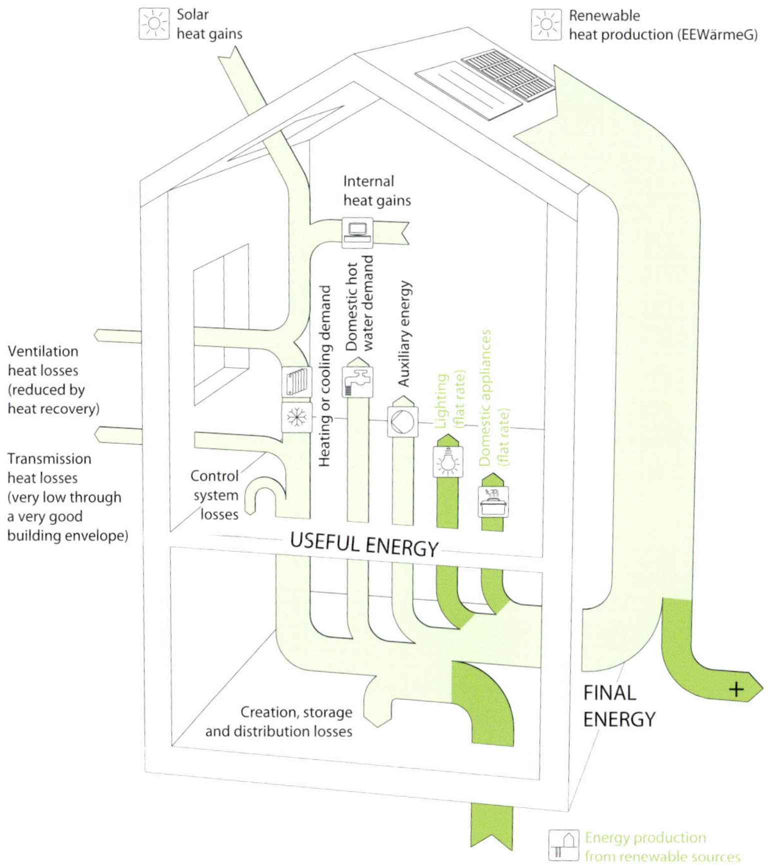

PRIMARY ENERGY FACTORS

Fuel	
Heating oil:	1.1
Natural gas:	1.1
Liquefied gas:	1.1
Coal:	1.1
Lignite:	1.2
Wood:	0.2
Local biomass: (liquid and gaseous)	0.5
Energy forms	
Local/district heating from CHP	
Fossil:	0.7
Renewable:	0.0
Local/district heating from heating plant	
Fossil:	1.3
Renewable:	0.1
General fuel mix:	2.4
Replacement electricity fuel mix:	2.8
Environmental energy	
Solar energy:	0.0
Environmental heat:	0.0

Energy balance model of the Efficiency House Plus. Flat rate quantities of energy for lighting and domestic appliances are entered into the balance according to the BMVBS. The balance results in a negative figure, expressed as the quantity of energy produced in excess of the fulfilled demand. The excess energy is usually fed into the public grid. The Efficiency House Plus provides the foundation for increasing the renewable fraction of the fuel mix in Germany and improving its primary energy factor in the long term.

Modified or added criteria compared to the EnEV

Active House

BALANCE SCOPE

In 2010, a number of companies and interested parties from the international construction industry came together under the name Active House Alliance in Copenhagen. This network marked the starting point for the development of the Active House, a housing construction standard that directs its attention beyond energy efficiency standards to indoor climate qualities and the environmental effects caused by the building. The building model is currently a specification for new buildings and refurbishments with detailed definitions of the requirements relating to energy, indoor climate and the environment. The standard is conceived as an open-source model. Discussions among subject matter experts from all over the world take place on a specially designed website at www.activehouse.info. The outcomes of these debates about the various subject areas and the results of workshops and meetings are brought together centrally and used to update and further develop the model of the Active House. The originators intend to develop a universal standard based on the latest science and practical experience.

Main categories of the Active House model

The Active House has a high efficiency standard, extending beyond the statutory requirements. Because it is an international model, the relevant country's national regulatory framework applies to the requirement levels and energy balancing methods in each case. The nationally recognised conversion factors for primary energy and emissions also apply. In Germany, for example, these requirements are defined in the EnEV. The parameter is specified as the annual primary energy demand. This comprises the energy demand for operating the building (heating, cooling, ventilation, hot water preparation), domestic appliances, lighting and the credit for own-use energy created from regenerative sources. Thus the Active House extends the balance scope of the EnEV. The supply concept of the building is based solely on renewable energy sources. These can either be exploited by technology on the building or on the site, or obtained from a public grid. The technical systems must be selected on the basis of their economic viability. If they are to be attached to the building, they must integrate into the architecture. In addition to meeting the purely technical requirements in the calculations, the building should have a building management system designed for user-friendly operation. Various checks, certificates and tests are demanded for quality assurance and validation purposes during the design and construction process.

The Active House should offer a healthy living climate and therefore sets requirements for the parameters

The VELUX LichtAktiv Haus was refurbished following the principles of the Active House guidelines.

influencing comfort and health inside the building. The objective is to create a good indoor climate, easily influenced by the user. The parameters for the categories as daylight factor, operating temperatures in summer and winter, air quality, has to be calculated and evaluated. The evaluation shall be based on dynamic calculation tools for the main rooms and allows for an adoptable method for indoor comfort levels. In addition the noise and acoustic levels shall be evaluated. In the area of indoor climate, evaluation of the interactions of the building with the interior takes place in parallel with investigation of its interactions with the outdoor environment. The motive here is to examine the environmental compatibility on an ecological level and the integration of the building into the cultural context. In relation to environmental compatibility, this means avoiding pollutants, contributing to biodiversity, incorporating a high content of recycled building materials and designing a recyclable building. A life cycle impact assessment of all the important building components (external walls, roofs, ceilings, foundations, windows, doors, internal walls and main technical components) forms the basis of the evaluation of the building's environmental effects. The period of observation for the energy balance is currently 50 years. The evaluation covers all common categories of influences on the environment. Reduction in the consumption of fresh water can be added to the influences on the environment, reductions in consumption and the maximisation of energy created from renewable sources. Easy-to-clean surfaces and the use of greywater and rainwater can reduce the consumption of fresh water. Moreover Active House evaluate qualitative parameters like outlook, low emitting building materials, architectural design solutions, roads and landscape, infrastructure and building management. The result of the design and realisation of an Active House is not a single number obtained by applying various energy balancing principles and considering the wider scope of observation. Therefore, the results are categorised and displayed on a radar chart, which presents the project qualities in a clear way and allows comparison with other projects.

Aktivhaus energy classes

According to the results of the energy balance the building is classified as:

Final energy demand

1: ≤ 40 kWh/m²a

2: ≤ 60 kWh/m²a

3: ≤ 80 kWh/m²a

4: ≤ 120 kWh/m²a

All domestic appliances should meet the highest efficiency standard.

Primary energy balance (including energy creation)

1: ≤ 0 kWh/m²a for the building

2: 0–15 kWh/m²a for the building

3: 15–30 kWh/m²a for the building

4: ≤ 30 kWh/m²a for the building

1: 100 % or more of the energy is produced on the site or in the energy system
2: ≥ 75 % of the energy is produced on the site or in the energy system
3: ≥ 50 % of the energy is produced on the site or in the energy system
4: ≥ 25 % of the energy is produced on the site or in the energy system

The radar chart shows the performance of a project and all the areas examined in the Active House Radar. The numbers of points scored in each category plot out an area. The dimensions of the area visualise the building quality. The shape reveals the building's strengths and weaknesses. The energy balance method for the Active House complies with the applicable national model. In Germany, for example, the operating energy balance complies with the EnEV. This energy balance model can be found in the section on the Efficiency House.

MINERGIE® standard

Swiss researchers were working on the MINERGIE® label in parallel with the developments in Germany. This voluntary building standard has been the yardstick for efficient buildings since the middle of the 1990s, and has more than one level.

MINERGIE® (basic standard)

BALANCE SCOPE

Various energy parameters are specified in the basic standard for a MINERGIE building, depending on the building's use. The building category determines the energy demands considered, which may include space heating, hot water preparation and the electricity for mechanical ventilation. The energy required for cooling, humidifying and dehumidifying is also taken into account if the equipment is installed. A residential building may not exceed an energy demand of 38 kWh/m²a. The reference value is the energy reference area, which in Switzerland is defined as the gross floor area. The energy parameter is in principle the total amount of energy supplied to the building for thermal conditioning during the course of a year. The level of observation is the final energy. Each type of energy making up the total supplied energy is weighted within the energy parameter to reflect the availability of the energy source. For example, the weighting of electricity generated from fossil fuels is 2.0, while wood attracts a factor of 0.7 in the energy parameter. Solar energy has a weighting factor of zero, and therefore the energy parameter calculation does not count the regenerative part.

This procedure is similar to the consideration of primary energy. The difference in figures, for example, between the primary energy demand of this standard and the 120 kWh/m²a of the Passivhaus standard arises because the MINERGIE system takes into account only the energy required for space conditioning. In a Passivhaus building, domestic electricity usually represents 70% to 90% of the primary energy demand.

A second requirement of the MINERGIE standard relates to the heating energy demand. This must be 10% below the limiting value for new buildings given in SIA 380/1. SIA 380/1 is a standard published by the Swiss Society of Engineers and Architects (SIA) and regulates the thermal energy requirements in buildings. A secondary requirement is the recommendation to use highly efficient appliances in order to achieve the MINERGIE standard. In addition, the building should have a ventilation system with heat recovery to ensure comfort.

The multistorey housing block Kraftwerk B was built as a MINERGIE-P building with an additional Eco quality seal, grab architekten ag, Altendorf (CH)

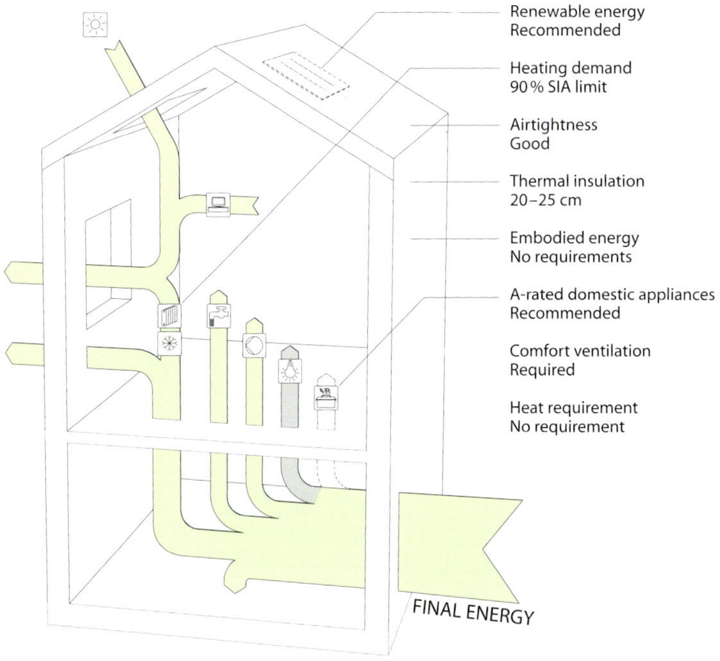

MINERGIE performance value
Heat: 38 kWh/m²a

Weighting factors in MINERGIE energy performance value

Sun, environmental heat, geothermal	0
Biomass (wood, biogas, sewage gas)	0.7
District heating (at least 50 % renewable energy, waste heat, CHP)	0.6
Fossil energy carriers (oil, gas)	1.0
Electricity	2.0

MINERGIE-P®

BALANCE SCOPE

Compared to the basic standard, the MINERGIE-P® concept is an additionally optimised building system notable for its further reduced energy consumption. The required weighted energy parameter for a residential building is 30 kWh/m²a. The heating demand must be less than 40 % of the limiting value given by the SIA. This very low energy consumption assumes, like the Passivhaus standard, a highly optimised and sensitive building system that must satisfy further requirements to provide a comfortable ambiance and fault-free operation. Areas such as thermal comfort in summer, airtightness of the building envelope and the integration of comfort ventilation must always be allowed for in the design.

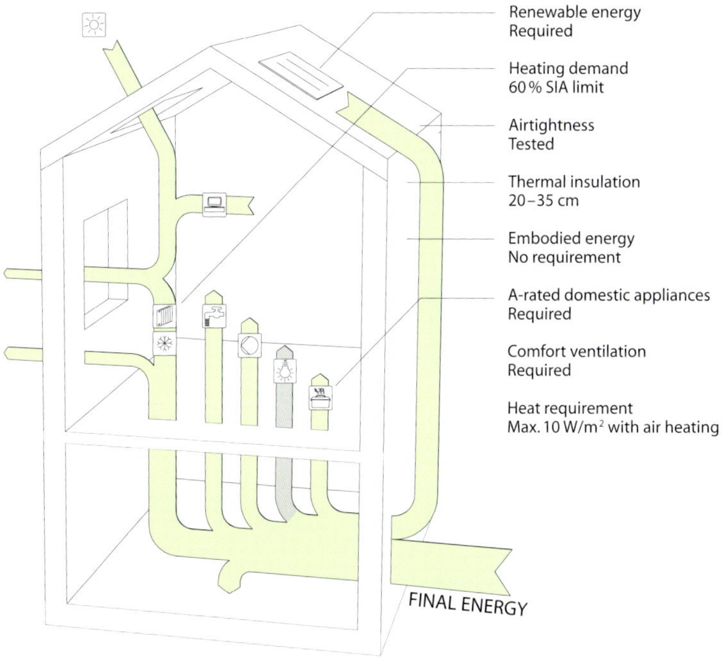

MINERGIE performance value
Heat: 30 kWh/m²a

MINERGIE-A®

BALANCE SCOPE

In addition to demand reduction, the MINERGIE-A® concept regulates the energy supplied to the building. By incorporating this requirement, Switzerland is reacting to the EU Energy Performance of Buildings Directive (2010/31/EU) and the nearly zero energy building. To achieve the MINERGIE-A standard, a building must first fulfil the requirements for a lower heating demand in accordance with the MINERGIE basic standard. In addition, it must achieve an energy parameter of 0 kWh/m²a or less. This is commonly done by covering the demand using renewable energies. Biomass is permissible for supplying heat if the heating plant is linked hydraulically to the household heating system. Suitable combinations include wood heating with solar thermal collectors for generating heat, if both technologies charge a thermal store and at least 50 % of the annual energy comes from the collectors. As with the Efficiency House Plus label, heat pumps are often used and their electricity demand covered from regenerative sources.

The concept's energy parameter is primarily an energy balance of demand and generation and as such requires a moderate amount of insulation compared to MINERGIE-P, providing the building has an efficient renewable energy supply. Other limits apply as in the basic standard. The zero balance of the standard means other expenditures of energy are considered as well as the operation of the building. Embodied energy is also brought into the overall energy assessment. An upper limit of 50 kWh/m²a is set as an upper limit for the building construction (building envelope, interior parts and building services equipment. The calculation assumes a building life of 60 years.

MINERGIE-ECO®

The ECO suffix is not a standard in its own right. It is a supplement to the three main MINERGIE standards. This also means only a MINERGIE, MINERGIE-P or MINERGIE-A building can be called an ECO house. The ECO standard seeks to create a healthy and comfortable living ambiance and reduce a building's effects on the environment. In addition to the operating energy balance, the evaluation also examines the aspects of daylight provision, sound insulation, indoor climate, building ecology, the embodied energy used to manufacture the materials, and the energy expended in the construction process, such as in the construction of the building and its demolition.

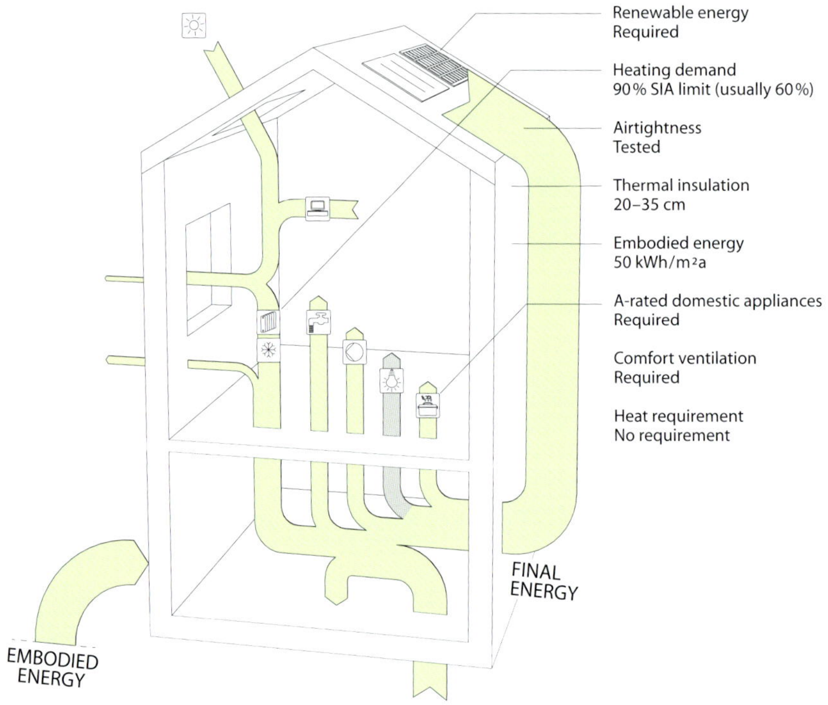

MINERGIE performance value
Heat: 0 kWh/m²a (biomass 15 kWh/m²a)

Overview and summary of the MINERGIE standards

Italy

CasaClima

The building label CasaClima (in English, ClimateHouse) was established in 2002 in the province of South Tyrol as part of a programme of work on energy-efficient buildings in the German-speaking part of Italy. The CasaClima Agency was founded in the autonomous province of Bolzano in 2006. This body is responsible for publications and the further development and certification of the standard.

In principle, there are three different classes of CasaClima buildings, which are based on their energy demand (heating and hot water):

CasaClima B (5-litre house):
heating energy demand less than 50 kWh/m^{2}a

CasaClima A (3-litre house):
heating energy demand less than 30 kWh/m^{2}a

CasaClima Gold (1-litre house):
heating energy demand less than 10 kWh/m^{2}a

The standard unit of measurement is the heated gross floor area: the sum of the heated areas in each storey.

In addition to the qualitative requirements, a CasaClima house must satisfy further demands. The design must ensure a compact form, a very good thermally insulating and airtight building envelope, the passive use of solar energy and optimised building services systems. Considering these parameters in the design and ensuring careful construction provide a good foundation for an efficient building.

The CasaClima initiative has established an additional label, CasaClima Nature, which goes beyond the consideration of energy alone to include the careful use of resources and the energy used in the manufacture of the building materials. CasaClima Nature sets out ground rules, for example, for avoiding fossil fuels, synthetic insulation, pollutants and tropical woods. Using a points system, the building is graded into one of the CasaClima classes.

Climate House A
3-litre consumption

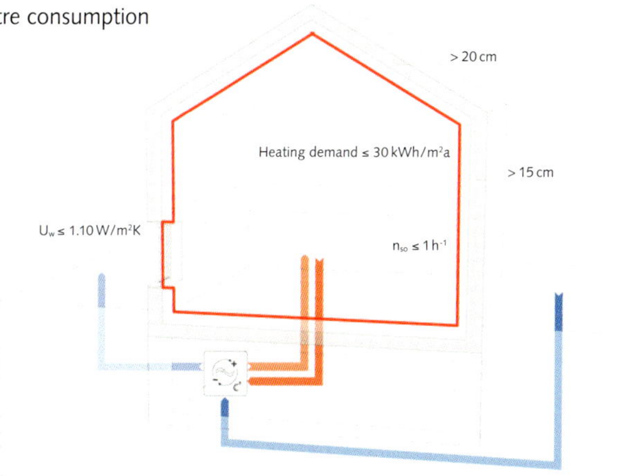

Climate House B
5-litre consumption

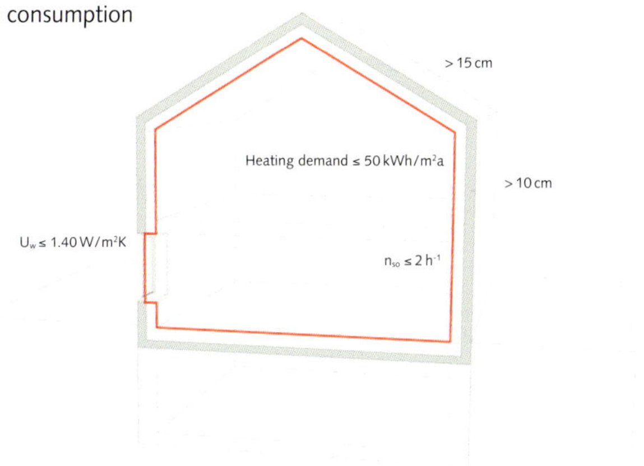

Climate House Gold
1-litre consumption

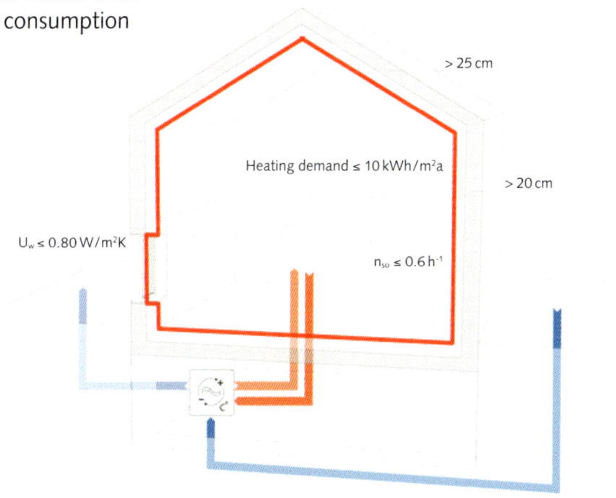

Beyond energy

All the building energy standards described on the previous pages have the primary aim of ensuring that buildings have an efficient supply of energy. The focus lies mainly on the consideration of operating energy, which is justifiable when viewed against the background of the huge energy consumption of buildings and the variety of available means of regulating it. The comparability of the numerical results from the various standards is possible only with difficulty because of differences in input parameters, such as national primary energy factors, and in their calculation methods.

Life cycle considerations

Several standards go beyond purely energy balances and extend their scope of consideration. Some energy balances include the energy expended in the construction of the building and the manufacture of the materials, and examine the environmental effects of the building. These aspects will gain further importance in future in the context of reducing operating energy.

Although embodied energy is already taken into account in some existing standards, enhanced standards and benchmarks backed by political will and statute are necessary to ensure the environmental effects of the construction and demolition of buildings are included in the calculations. Life cycle assessment (LCA) is the current tool of choice here. This method of calculation is based on standard ISO 14040 and considers the environmental effects of a building and its recycling potential over its entire life cycle. Life cycle assessment uses an inventory-based approach to analyse inputs and outputs, and reports on the effects in a variety of categories. The evaluation usually covers the categories of global warming potential (GWP), ozone depletion potential (ODP), photochemical ozone creation potential (POCP), acidification potential (AP), eutrophication potential (EP) and primary energy input (PEI). As a rule, no priorities apply to individual categories, because the consequences of the different environmental categories are not comparable scientifically or in their effects. A comparison based on numerical results alone is not very enlightening either. For this reason, buildings are often evaluated using more than one comparable LCA. Comparing numerical results against those of a reference building provides a better benchmark for evaluation and ranking. Current methods supplemented by an LCA allow a building to be evaluated from construction, through operation and finally to demolition.

2,000-watt society

In addition to the building and, most relevantly, its positioning on the site, the user's energy consumption also affects the global environment. No energy balance standard is available for judging the latter. It is very difficult to obtain a meaningful picture of user energy consumption from household electricity, because of very divergent styles of behaviour and the resulting consumption patterns. Energy expenditure on transport, personal consumption and the like contributes to the world's rising energy consumption. The model of the 2,000-watt society was developed in Switzerland to examine this issue. The idea is not about retrospectively establishing the energy demand of the user, but developing a forward-looking model to allow global energy objectives agreed by representatives of the world's governments to be achieved. One of the main objectives is the plan by the Intergovernmental Panel on Climate Change (IPCC) to reduce primary energy consumption and greenhouse gas emissions per head.

The 2,000-watt society model is a vision of a future where everyone in the world has a daily limit of 1 tonne of emissions per head and 2,000 watts of energy. According to the IPCC, this should keep the planet's climate-related temperature rise down to 2 kelvin. The 2,000-watt limit includes the energy-consuming areas of living, transport, nutrition, consumption and infrastructure. Living standards play a crucial role in achieving this objective. As well as the use of efficient household appliances, the 2,000-watt society model also encourages an adjustment in user behaviour. 2,000 watts equate to a primary energy demand of about 17,500 kWh per year. This target figure corresponds to the historical global mean in 2005. In relating to 2005, it concentrates less on achieving a reduction of the total primary energy demand and more on striving for its equal distribution between developed and developing nations to counteract a steep rise in energy consumption similar to what occurred after 1950. The model takes into account efficiency improvements by high consumers, while allowing a margin for development for previously disadvantaged populations.

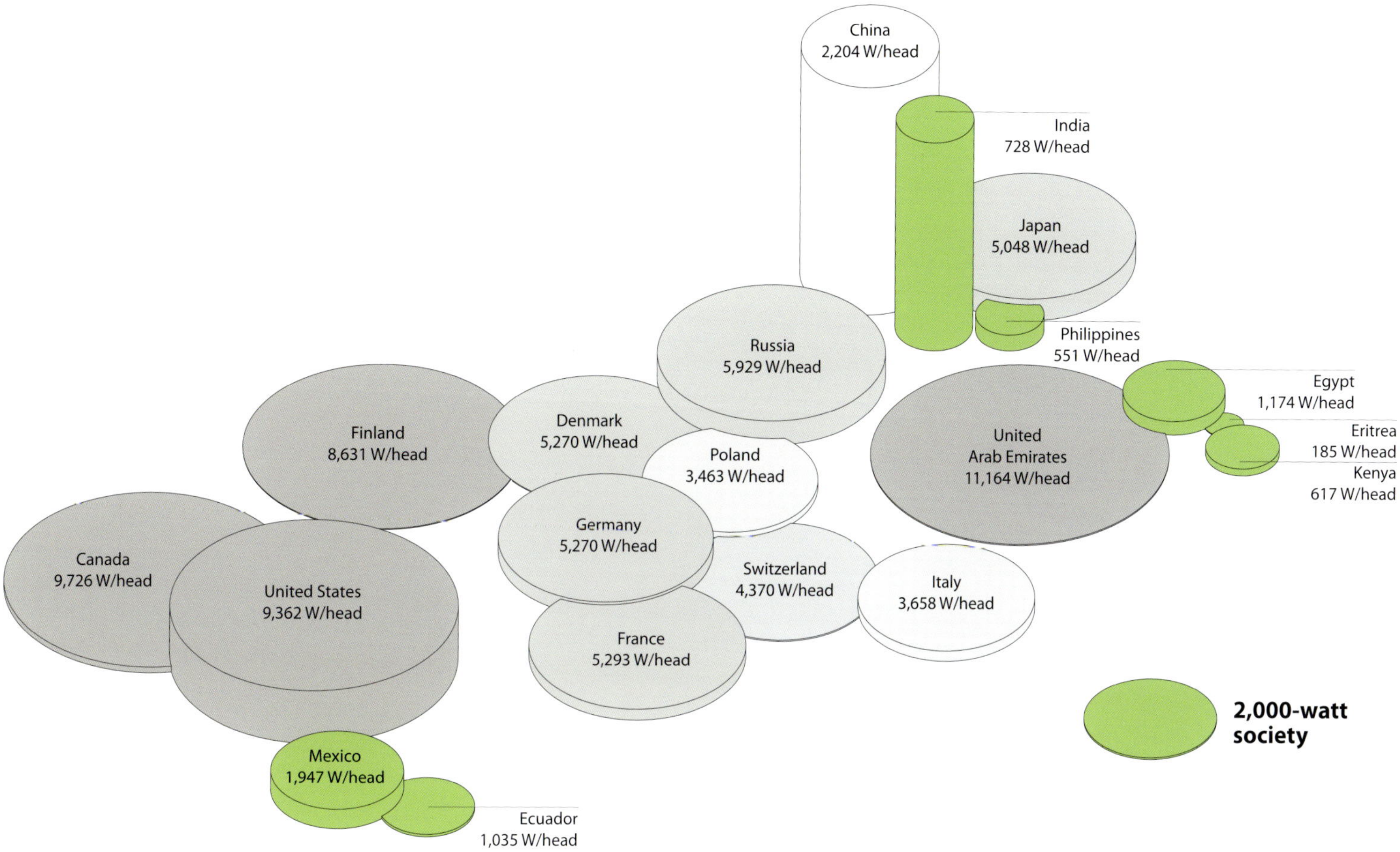

Primary energy consumption per capita in various countries. The area of the circles indicates the relative size of a country's per capita consumption, and the height of the columns represents the country's population. Highly developed countries are generally a long way from the target of the 2,000-watt society.

Other energy balance fields

The energy balances and building standards introduced on the previous pages demonstrate the diverse spectrum of approaches, ranging from considerations of heating energy, through increasingly comprehensive analyses of building-related characteristics to a complete evaluation of the living situation of humankind. Each country has developed evaluation tools and standards, all of which could be useful in driving forward the design and construction of buildings fit for the future. However, the field of building evaluation itself will also undergo significant changes in the years to come. Global and national climate protection objectives, such as energy consumption profiles conditioned by the ongoing need to make savings, will also foster this process.

The diagram shows the areas of consumption influenced by buildings, and the possible parameters for inclusion in an energy balance. The elements highlighted in colour depict the areas addressed by current national statutory requirements – with the rest indicating the future fields of development. Architects and engineers who already take into account the principles of holistic project design can create buildings today that will withstand a future building evaluation.

Possible areas for inclusion in a building energy balance. Shown in green are the areas addressed by the regulatory framework of the EnEV. Reducing the energy required to operate a building will focus increasing attention on life cycle and the energy expended on everyday living, and eventually on the way energy balances are performed.

Sustainability evaluation

Other certification systems to evaluate the sustainability of buildings have been established that go beyond the energy-related parameters and evaluation methods considered in this book. Energy considerations form only a part of these systems, which generally seek to ensure that national standards are at least met and that renewable energy use is maximised.

The first such system for the evaluation of the sustainability of buildings BREEAM (Building Research Establishment's Environmental Assessment Method) was developed in 1990 by the Building Research Establishment (BRE) in the United Kingdom. The further development and adaption for a wide range of uses quickly led to it being adopted outside the UK. Today, BREEAM is the world's most popular certification system. This was followed by other evaluation and certification systems, including HQE (Haute Qualité Environmentale) in France in 1996 and LEED (Leadership in Energy and Environmental Design) from USGBC (U. S. Green Building Council) in the USA in 1998. LEED is an established certification system mainly in the USA but also on the international level.

The German Sustainable Building Council, Deutsche Gesellschaft für Nachhaltiges Bauen e.V. (DGNB), developed its sustainable building certification scheme, the Deutsche Gütesiegel für Nachhaltiges Bauen, in Germany in 2008. Using about 50 criteria, the system evaluates a building in terms of its ecology, economy, social, cultural and functional aspects, technology, processes and location over the whole of its life cycle. In other words, the design and construction phases as well as the building and its operation. The level of fulfilment of each criterion leads to an overall score and the award, depending on performance, of a bronze, silver or gold certificate. The system does not evaluate individual measures but rather the overall performance of a building.

The DGNB started with the certification of offices and administration buildings. Since then, various other usage profiles have been added, for example, residential and educational buildings, hotels and industrial buildings. In some usage profiles, the criteria have been amended to allow certification of existing and refurbished buildings.

Sustainability certification DGNB

Ecology
- Life cycle assessment – environmental effects from emissions
- Risks for the local environment
- Environmentally compatible material extraction
- Life cycle assessment – primary energy
- Drinking water demand and waste water volume
- Land consumption

Economy
- Building-related costs over life cycle
- Flexibility and capability of changing use
- Marketability

Social and functional aspects
- Thermal comfort
- Interior air quality
- Acoustic comfort
- Visual comfort
- Possibilities of user intervention
- External space qualities
- Security and risk of malfunction
- Barrier-free building
- Public accessibility
- Bicycle-friendliness
- Urban design and building concept processes
- Art in architecture
- Layout qualities

Technical services
- Fire protection
- Sound insulation
- Heat and moisture protection technical quality of the building envelope
- Ability to adapt the technical systems
- Ease of cleaning and maintaining the building
- Ease of demolition and clearance

Process
- Quality of the project preparation
- Integrated design
- Evidence of optimisation and complexity of the design approach
- Sustainability aspects ensured in the tender and award process
- Creation of the conditions for optimum use and management
- Site/construction process
- Quality of the finished building
- Problem-free building commissioning

Location
- Microlocation
- Image and status of location and quarter
- Transport connections
- Near to facilities relevant to the building's use

Assessment areas as part of sustainability certification. The topics show the complexity of the certification system, which goes far beyond purely considering matters of energy efficiency. From the total of six areas of evaluation, five are relevant for the building assessment.

Aktivhaus design

How do I design an Aktivhaus? This chapter explores the question. Beyond the fundamental internal and external framework of conditions applying to every building project in whatever context, it shows the fundamental design strategies and presents examples to illustrate the design process.

Residential buildings provide living space. In order to create high-quality space in which to live, living space that will continue to please the resident over the long term, the designer must provide optimum thermal conditions and good, clean air for comfort inside the building. In addition to the comfort requirements arising from the needs of the users, their habits and activities, and the quality of the interior space, the climatic situation sets important preconditions for the design of the building.

As the interlocutor between the indoor and outdoor conditions, the building has to satisfy many different requirements. The building envelope can achieve some of these goals. In most cases, however, achieving the desired comfort in the areas of heating, cooling and ventilation requires technical solutions. In the context of global climate change and declining fossil fuel resources, attention now turns to saving operating energy and increasing the proportion of energy supplied from renewable sources.

The outer envelope of a building acts as the interlocutor between the indoor and outdoor worlds. Over the course of a year, external climatic conditions often fail to correspond with the requirements set for the interior of the building. Within a relatively small geometric space, the skin must insulate and seal the building to ensure the inside remains comfortable, without excessive expenditure of energy (the use of technology).

Fundamental requirements of the building project

External framework conditions and internal requirements applicable to the building provide the preconditions for the development of the building energy concept. These two aspects have to be thoroughly explored when beginning a project. At this point, the requirements defining the project are identified and the most suitable strategies to fulfil them derived.

Interior requirements

The building use, any project-specific constraints and conditions, and the generally applicable comfort criteria define the internal requirements. Subjective requirements are project-specific and can arise from the layout of the building and imposed conditions (such as in refurbishments) or from specific requirements and wishes of the client.

Ambient comfort

Ambient comfort criteria are generally applicable and are subject to many technical documents, from design regulations to DIN standards. The sensitivities of the human body also supply a subjective definition of ambient comfort. People perceive disturbances to their comfort caused by heat, cold, odours, noise and glare, through their skin, nose, ears and eyes. The building is designed to help alleviate these disturbances, by acting as an intermediary between the inside and outside environments, between requirements and circumstances.

The requirements of users for a defined use remain the same over time and are generally independent of geographical location, but this is not the case for the measures necessary to fulfil these requirements. During the evenings in the transitional months either side of winter, residential buildings may have to be heated to provide a comfortable temperature, whereas cooling may be necessary in office buildings with high internal loads, prolonged use during the day or high solar irradiation.

The following criteria are generally decisive in any space for the wellbeing of the users:

- Thermal comfort
- Hygienic comfort
- Visual comfort
- Acoustic comfort

The criteria and their indicators are discussed individually in the following sections. Because thermal comfort is usually determinant in the development of an energy concept for buildings, this is then followed by detailed descriptions of the relevant points for a building.

It is well known that people can perceive the same indoor climate differently. While some may feel as if they

Thermal comfort	User	Clothing Type of activity Number of occupants Period of presence
	Surroundings	Weather/climate Outside air temperature
	Room	Temperature Air speed Air humidity
Hygienic comfort	User	Type of activity Period of presence
	Surroundings	Outdoor air quality Emissions-creating industry in the surroundings
	Room	Air humidity Emissions/pollutants Air quality (CO_2) Odours
Visual comfort	User	Type of activity Period of presence
	Surroundings	Proportion of diffused sunlight Proportion of direct sunlight Reflective surfaces in the surroundings
	Room	Daylight quotient Illuminance Luminance distribution Glare Colour rendering Visual relationship with outside world
Acoustic comfort	User	Type of activity Number of occupants Period of presence
	Surroundings	Ambient noise
	Room	Qualities of surfaces Reverberation time Airborne sound insulation Footfall sound insulation

Ambient comfort criteria and indicators for buildings

are freezing, others find the same temperature comfortable. However, specific statutory provisions apply to ambient comfort in buildings, and there are guidance values that represent a comfortable climate for the majority of people. DIN 1946-2 says thermal comfort prevails when a person is satisfied with the temperature, humidity and movement of air, and does not wish it to be warmer or colder, nor the air to be drier or moister. This is generally the case when the body is in equilibrium with the room, in other words the body is in heat balance; of course – as described – there can be subjective differences of perception.

User and use

The human organism creates body heat from a process of combustion. The body gives off heat through convection, radiation, perspiration and breathing, raising the temperature of its environment to some extent. Physical movements can accelerate these processes and increase the amount of heat given off. A person gives off an average of 80 watts per hour while resting, and 210 watts per hour during moderate work. The heat given off by the human body depends on the ambient temperature as well as the level of activity. As the temperature of the surroundings rises, the amount of heat given off falls.

In addition to the mean room air temperature, radiation temperature and level of activity, the amount of body heat given off also depends on further factors such as the speed of air movement, the relative humidity, and the type and condition of the person's clothing.

Room temperature (radiation temperature, operative temperature)

Generally, for a Central European climate, a normally clothed, seated person would find a temperature of 20 to 22 °C pleasant in winter and 22 to 24 °C in summer. But the temperature people find comfortable depends not only on body heat and air temperature, but also on the surface temperature of the surrounding areas (walls, ceiling, floor and heated surfaces) because they exchange heat with the body by radiation. If they are too hot, the resulting effect can be just as uncomfortable as cold surfaces such as poorly insulated windows. Because half the heat given off by the human body is by radiation, the presence of inadequately insulated surfaces increases the heat lost from the body and produces an uncomfortable cooling effect.

Surface temperatures should therefore not be below 18 °C and, for continuously perceived comfort, not more than 2 to 3 K below room temperature. In order to create an even, comfortable climate in the room, the temperature difference between the individual surfaces or building components must not be greater than 5 K. Consequently, this reference value is relevant for the planning and design of panel heating systems.

The temperature perceived by the human body is a combination of the air and the radiation temperature and is called the resultant or operative temperature. The reference values for the operative temperature must be viewed in conjunction with the outdoor temperature and the season. Too low an indoor temperature during the summer months is just as uncomfortable as too high an

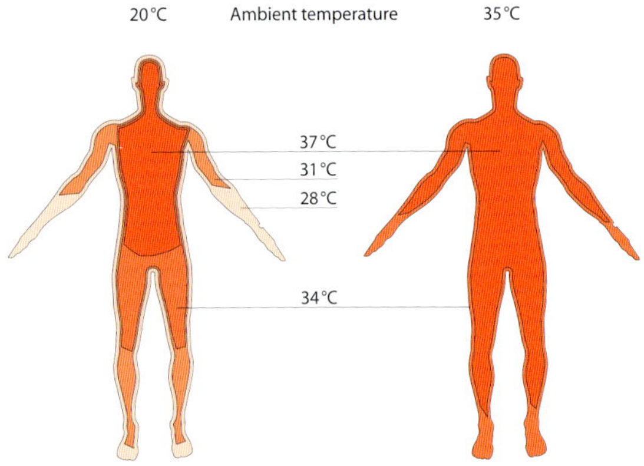

The way the human body manages its heat is influenced by the ambient temperature. In cool temperatures, the body loses heat and cools. This happens first to the arms and legs. In very hot temperatures, the body cannot give off heat and therefore overheats. A comfortable ambient temperature is one in which the body achieves an optimal relationship between the heat stored and the heat given off.

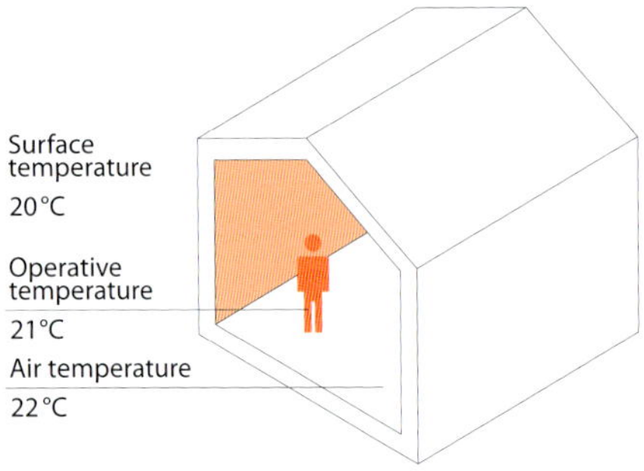

indoor temperature in winter. On the other hand, temperatures slightly outside the comfort limits are still perceived as pleasant in summer if the difference between the indoor and outdoor temperatures is sufficiently great. This is because the body adapts to the season. The measures necessary for active cooling to work in these zones bordering the comfort limits would be technically very complex and would have only a relatively small effect.

Relative humidity

A further factor influencing comfort is the relative humidity of the indoor air. The human body regulates its core body temperature by radiation and perspiration. For this reason, the relative humidity of the air directly influences the feeling of wellbeing too. The absolute humidity of the air is the quantity of water in g/m³ that the air can absorb. This depends greatly on the air temperature. The relative humidity expresses the saturation of the air as a percentage. Very warm air can take up a lot of moisture; cold air, on the other hand, cannot. If warm air is cooled, the relative humidity rises. Moist, warm air is perceived as being close. High humidity in summer prevents the body from regulating and lowering its temperature by evaporation. If cold outdoor air is heated, the relative humidity falls sharply. Dryness leads to the body emitting excessive amounts of moisture and the mucus membranes and eyes drying out. People perceive both extremes as unpleasant.

For greater comfort inside a building, the relative humidity should not exceed 70%. The recommended lower limit for relative humidity is 30%. Relative humidity is affected by the heating system, or more precisely by the type of heat transfer, and by the choice of materials used in the building's interior. Materials that store moisture and can release it again (such as loam) help in a natural way to smooth out humidity peaks. Reducing the rate of air change to the hygienically required minimum so that too much moisture cannot enter or escape the building is crucial. As a rule, simple passive measures like these can achieve a high level of comfort. Active measures, however, are necessary to be able to control humidity fully. Rooms that have to fulfil high requirements because of their type of use have suitable mechanical humidifying and/or dehumidifying equipment fitted.

Improving the energy properties of the building envelope generally increases comfort because the measures reduce temperature differences. A good concept takes into account individual requirements and conditions as well as external influences. However, it is worthwhile examining the user's desired parameters. Significant savings can be made on the costs of the system and its operation if the user is willing to tolerate short periods during which these desired values are not quite reached or are slightly exceeded. Values considered by calculation to be optimum do not necessarily provide the best possible environment for people's comfort. Some means of adjustment to suit the seasons and weather conditions or the ability to create special areas in which the user can influence the indoor climate conditions (thermostat, solar screening, opening windows) are helpful in this respect.

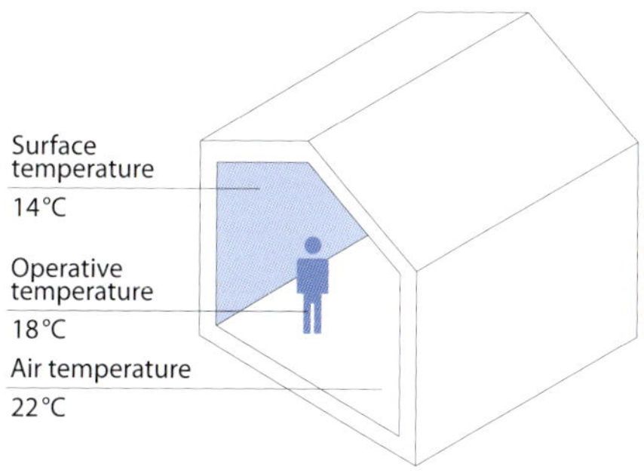

People perceive the operative temperature of a room from a mixture of the temperature of its surfaces and the air temperature. The smaller the difference between the two temperatures, the better the comfort of the room.

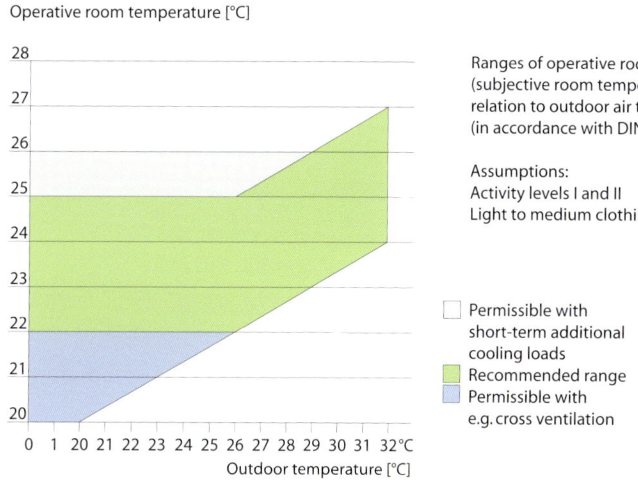

Ranges of operative room temperatures (subjective room temperatures) shown in relation to outdoor air temperature (in accordance with DIN 1946 Part 2)

Assumptions:
Activity levels I and II
Light to medium clothing

Requirements dependent on use

The type of use of a building determines the important boundary conditions for its design. These affect the spatial layout, the room sequence, which also depends directly on use, and the energy supply concept. Different building uses result in different user behaviours and consumption profiles, which affect the energy demands for heating, cooling, ventilation, hot water and lighting.

Comparing the energy consumption of a new and a refurbished residential building with a non-residential building shows them to have different demand profiles. In the case of housing, the energy used for heating continues to be a significant factor at our latitudes. It is relatively simple to save energy. Concepts such as Passivhaus exert a huge influence here and demonstrate how this can be done economically. In non-residential buildings, electricity consumption is considerable. Identifying specifically where this energy is expended and implementing new energy-saving concepts will be key to reducing this figure in the future.

Comparison based on consumption per square metre shows hardly any difference between the two types of use. Only the absolute figures (kWh/a) show a tremendous difference. The project section of this book describes the three projects in detail.

In residential buildings, very little of their demand arises simply from their type of use. In most cases, the needs of the user exert the greatest effect on the requirement profile. All users have their own preferred settings and different levels of understanding of energy savings. Individual advice about possible measures that directly influence user behaviour is a good first step. A change in user behaviour can save up to 15% of a household's energy consumption.

Not only spatial requirements, such as a mandatory sequence of usage units within a building, but also technical standards covering aspects such as fire protection and sound insulation can mean that not every type of energy efficiency improvement measure can be integrated into the design. In this context, the task is to find a suitable solution for each proposed building. As described in the principles of this book, deliberately choosing an open definition of an Aktivhaus points the right way ahead.

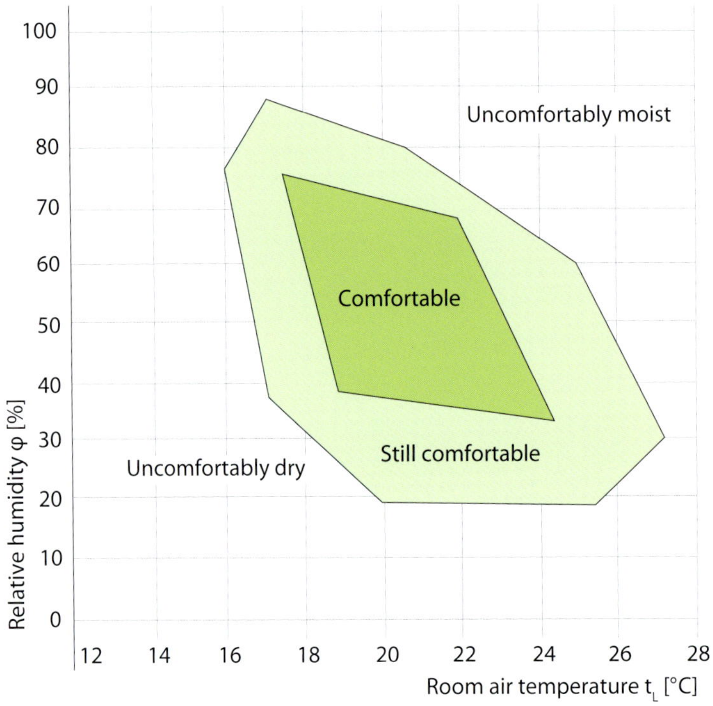

The comfort window defines the zone that most people in Central Europe perceive as comfortable. The graph shows how the relationship between room air temperature and relative humidity is crucial to the subjective perception of comfort.

Comparison of final energy demands of residential and non-residential buildings

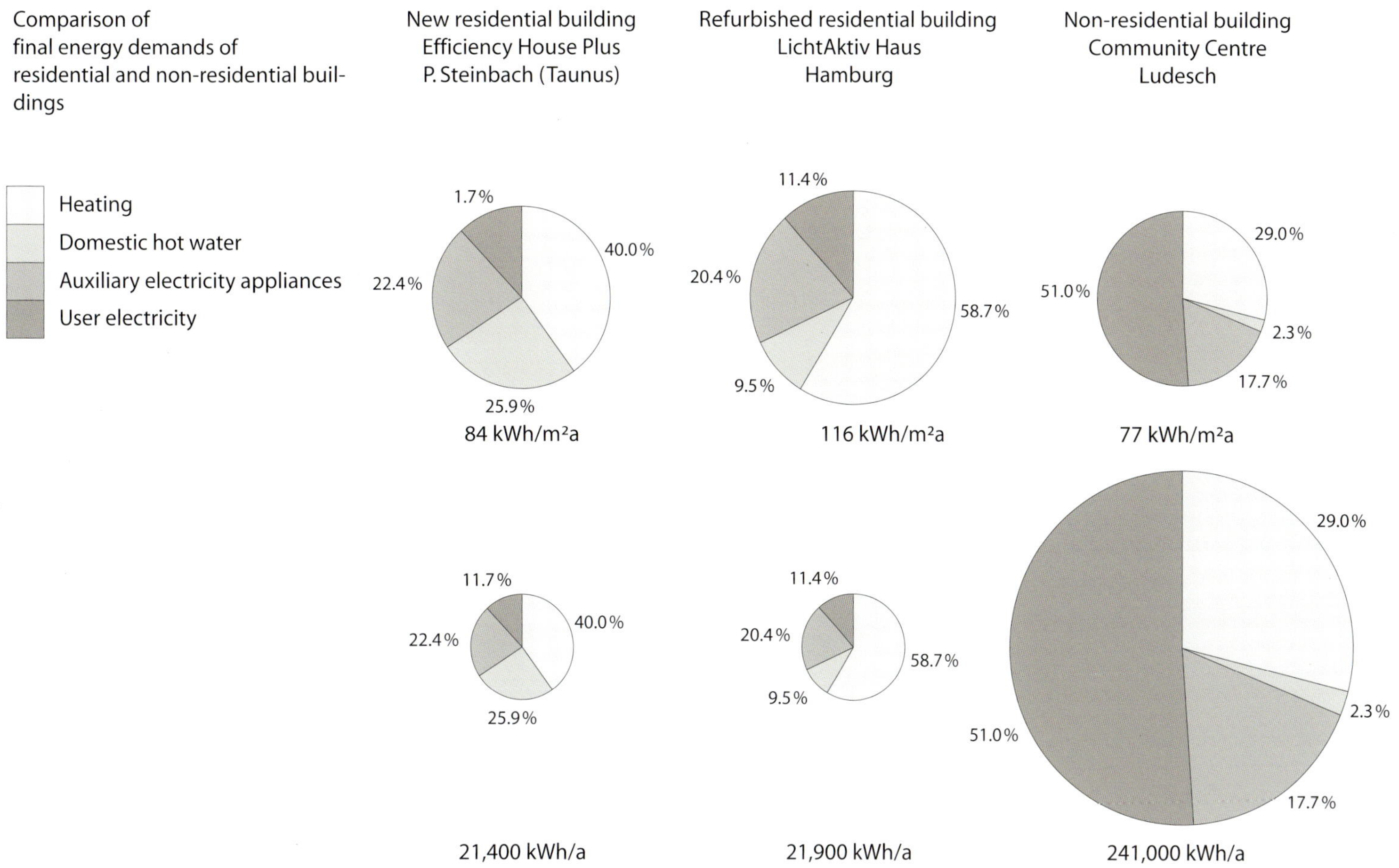

Efficiency House Plus P., ee concept GmbH, Darmstadt (DE)

LichtAktiv Haus, TU Darmstadt, Ostermann Architekten, Hamburg (DE)

Community Centre Ludesch, Architekten Hermann Kaufmann ZT GmbH, Schwarzach (AT)

External boundary conditions

The external boundary conditions are mainly influenced by the climate prevailing at the location. Macroclimate describes large-scale climatic effects extending over more than 500 km. Microclimate, on the other hand, defines the climate at a clearly delineated location (city, between buildings, a particular site). An exact analysis in advance of the design is important to developing a building concept suitable for the specific climate zone.

Climate

The climate of a location describes the typical boundary conditions such as average solar radiation, precipitation, average temperatures, seasonal differences, day lengths and prevailing winds. Climate should not be confused with weather. The latter is always only a snapshot, whereas climate describes a continuous state. Very strong environmental influences can change the climate. The change takes place at first unnoticed over many years. Nevertheless, we find ourselves currently in a phase of climate change characterised mainly by a warming of the atmosphere and more frequent extreme weather events.

Climate zones

The spherical shape of the earth and its inclined axis lead to regions of different solar radiation and temperatures. The distribution and effect on the atmosphere of the land and water masses give rise to further regional climatic peculiarities. These regions are called climate zones and given the names:

- Polar zone
- Temperate zone
- Subtropical zone
- Tropical zone

The extent of the zones is defined by latitude. The further the zones are away from the equator, the greater their seasonal nature and fluctuations.

POLAR ZONE

The polar regions of the earth are found in the northern polar circle, the Arctic, and in the opposite polar circle, the Antarctic. They are described as polar deserts because temperatures are below or at zero all year round, which permits little or no plant growth. Even in the warmest month, the temperature is continuously below 10 °C. The daily temperature differences are also very small. The long hours of daylight in summer and the continuous darkness in winter lead to wide annual temperature differences for locations deep inside continental land masses (e.g. in Siberia). The intensity of solar radiation in these regions is very weak because of the shallow angle of solar incidence and the filtering effect of the earth's atmosphere. A major proportion of the radiation is

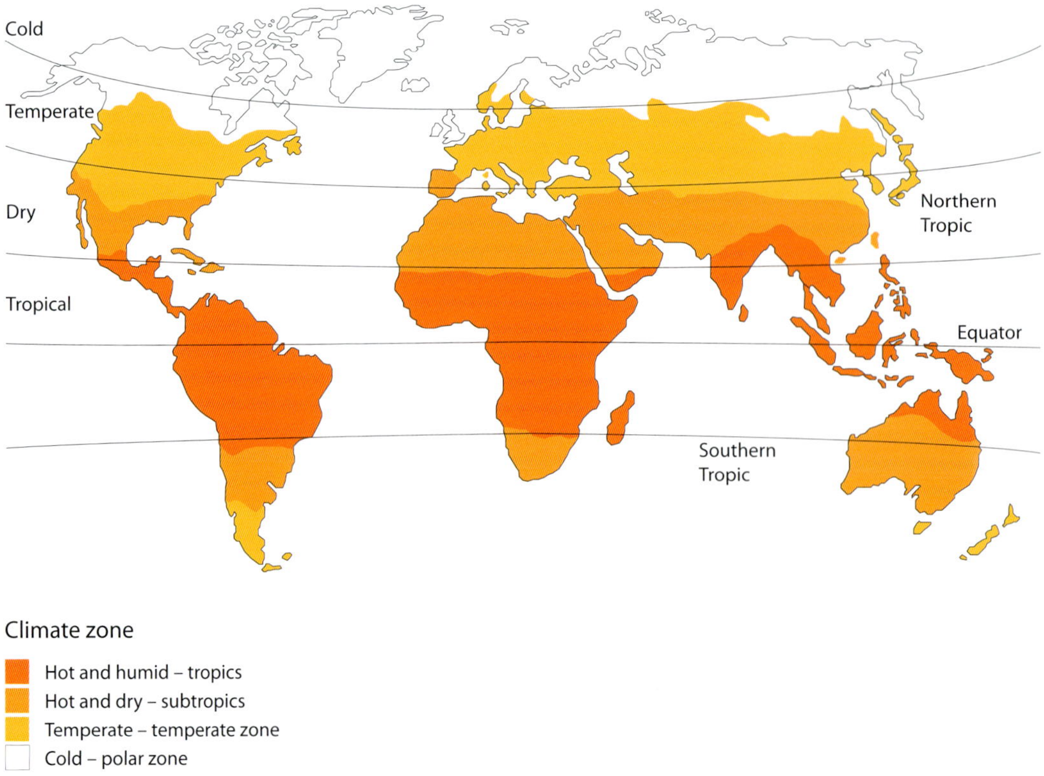

The earth's climate zones

reflected by the ice masses. Long periods of frost penetrating deep into the soil layers reinforce the already dry climate.

TEMPERATE ZONE

The temperate zone borders the polar circle and is characterised by its moderate climate. The temperate zone extends to about the 40th circle of latitude. It includes various climate characteristics: the western maritime climate, the warm summer continental climate, the semi-continental climate, the cool continental climate and the eastern maritime climate. The zone can be divided into cold, cool and warm-temperate climates. This heterogeneity also shows in the intensity of radiation. For example, the often cloudy skies over Central Europe provide a high proportion of diffuse solar radiation, whereas the transition areas leading to the tropics show a higher proportion of direct solar radiation.

The temperate zone also has noticeable temperature differences over the day and year. The temperature differences over the year are most obvious and have a range of 18 to 20 K in Central Europe. These very pronounced seasonal differences result in complex building requirements. Seasonal differences become less in locations closer to the equator. The day length varies according to the season. In summer, it can be up to 16 hours between sunrise and sunset, whereas in winter it can be as short as 8 hours. Because the amounts of precipitation are low and distributed evenly over the year (in Central Europe e.g. about 600–1,000 mm per annum), the weather can be described as changeable. Humidity varies between 60% and 80% in a medium to high range.

SUBTROPICAL ZONES

The subtropical zones lie between the tropical zone and the temperate zones, i.e. between latitudes of 25° and 40° north and south. They are characterised mainly by their very warm summers and mild winters. Solar radiation is at its most intense in the summer. This leads to high air temperatures during the day. During the night, in contrast, temperatures can fall to medium or low levels. The day-night fluctuation is on average 20 K and the annual fluctuation is small.

In summer, in addition to the high temperatures, the climate is very dry. The relative humidity is 10%–50%. This is accompanied by very low average annual precipitation (approx. 0–250 mm per annum). Rainfall occurs only seldom and in the form of brief, heavy downpours.

The dust content of the air is very high because of the large areas of desert in the subtropical zones. Wind varies and can be very strong in some areas. In the desert regions, this can lead to sandstorms. The subtropical zones are thinly populated because of their rather unfavourable climatic conditions.

TROPICAL ZONES

The tropical zones are found either side of the equator. Solar radiation is intense, but is reduced and diffused by the mostly cloudy skies. In spite of this, the amounts of solar radiation are high. Seasonal weather effects are almost completely absent. The highest daytime air temperature in an average year is approximately 30 °C, the night-time air temperature approximately 25 °C. The day-night difference is therefore small, but is still more than the seasonal fluctuations. Day length of 10.5 to 13.5 hours is likewise relatively constant. High amounts of precipitation (approx. 1,200–2,000 mm per annum) contribute to the high fertility of the land. The resulting sultriness is reflected in the high relative humidity of 60%–100%.

Winds are relatively light. In the rainy seasons, however, they can occur as tropical storms or even cyclones.

Autochthonous building

Handed down over the centuries, autochthonous building has developed in many regions of the earth into climatically optimised types of construction. They demonstrate that, even with limited technology, an optimum living environment can be created for people and their needs. With the arrival of generally available and inexpensive energy, types of construction optimised for specific locations developed into an international architecture, which in turn was optimised for each location by the use of technical building equipment. The price for this was increased energy consumption for building operation. User satisfaction is not bound to rise as a result. *Sick building syndrome* occurs more often in these types of buildings and productivity declines, as does people's acceptance of the built environment. These old building traditions could provide information about passive strategies that could be integrated into contemporary concepts.

	BUILDING	ZONING	PRECIPITATION
POLAR ZONE	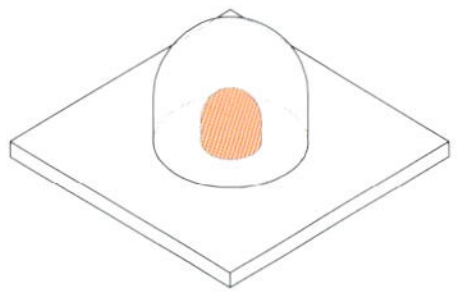	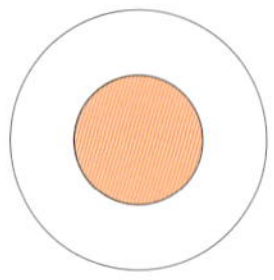	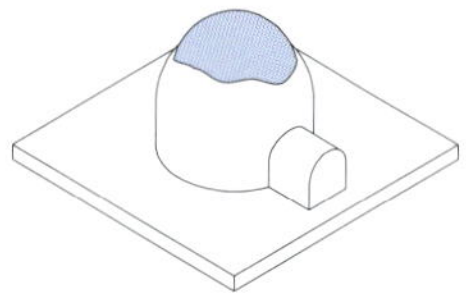
REQUIREMENTS	Protection from cold (all year round)	Protection from cold (all year round)	Heavy snowfall
CONSTRUCTION MEASURES	Very compact volume Very good insulation Low facade permeability	e.g. onion principle, create buffers to protect warm zones from cooling	Loadbearing capacity of structure
TEMPERATE ZONE	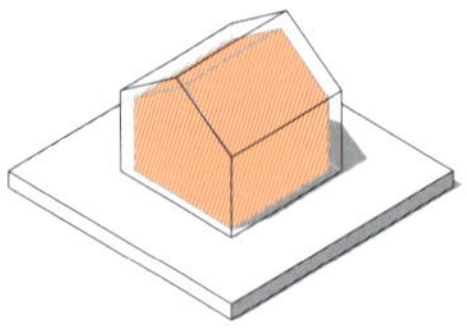		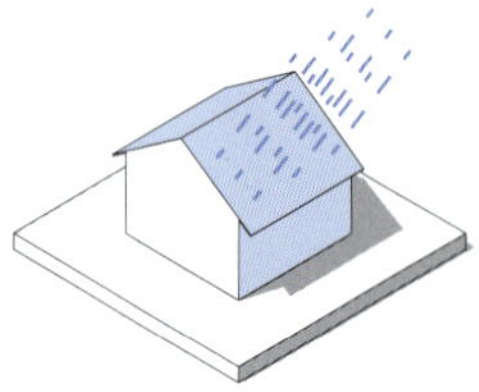
REQUIREMENTS	Protection from winter cooling Protection from summer heat	Protection from winter cooling	Localised protection from frequent heavy precipitation
CONSTRUCTION MEASURES	Compact volume Good insulation High airtightness	Main usage areas in the south to use passive heat	Protection of building (e.g. by roof overhang) Facade protection on the weather side
SUBTROPICAL ZONES	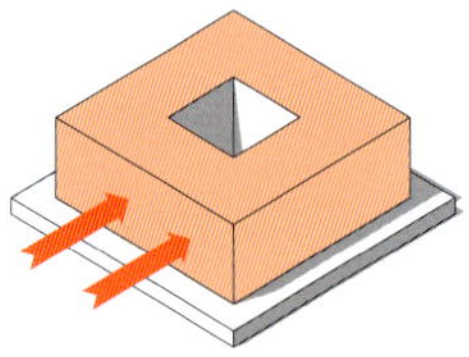		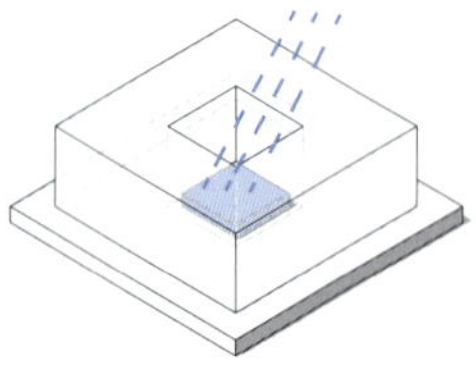
REQUIREMENTS	Protection from strong heat	Protection from excessive heat	Low rainfall in desert regions
CONSTRUCTION MEASURES	Shape building to create shade	Arrange good shade for main use areas (e.g. pergolas)	Collect precipitation and water
TROPICAL ZONES	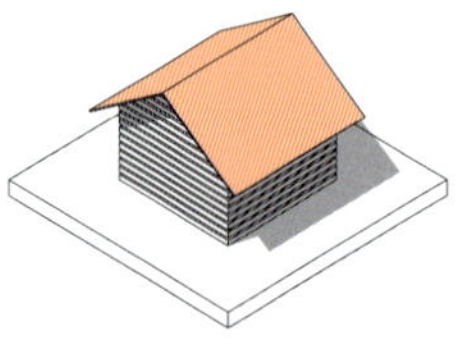		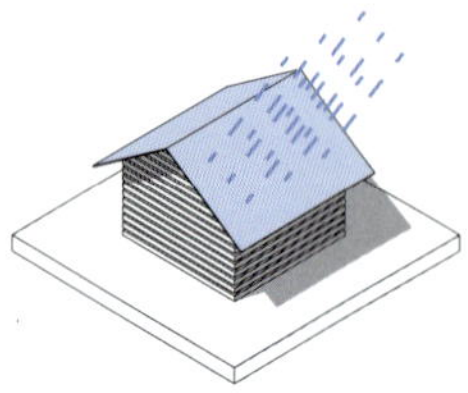
REQUIREMENTS	Protection from heat and moisture	Protection from heat and moisture	Protection from high precipitatio and humidity
CONSTRUCTION MEASURES	Create shade with building form and orientation (roof shape)	Well-shaded open areas with good through ventilation (usable almost all year round)	Provide good rainwater drainage

AIR	SUN	GROUND
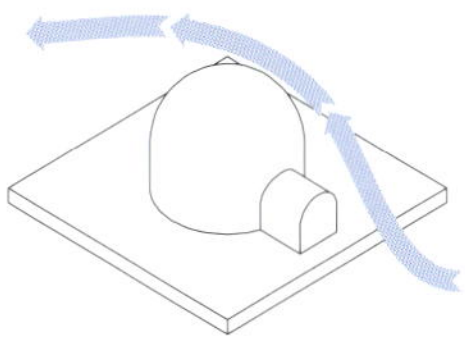	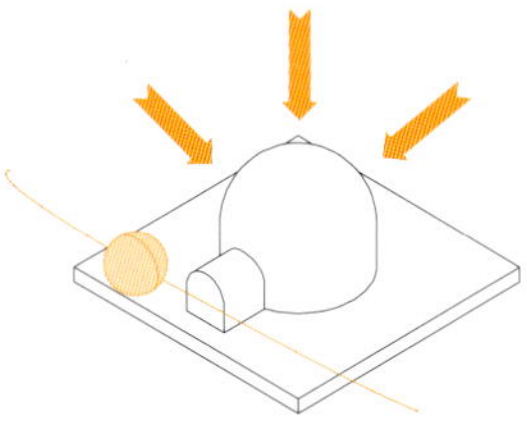	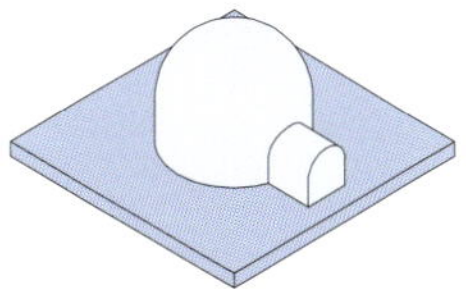
Gales and storms in the cold season	Moderate solar radiation / high reflection	Protection from ground freezing
Eliminate windward faces (wind redirection) Requires porch in entrance area	Open to low sun path in summer (no shading necessary) Heat-absorbing surfaces	Avoid founding directly on soil
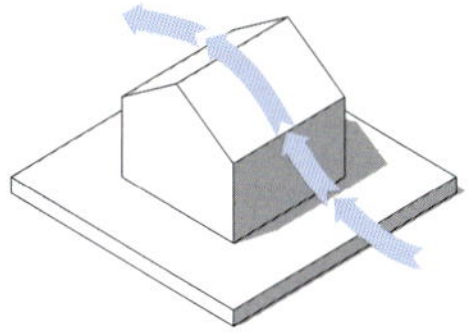	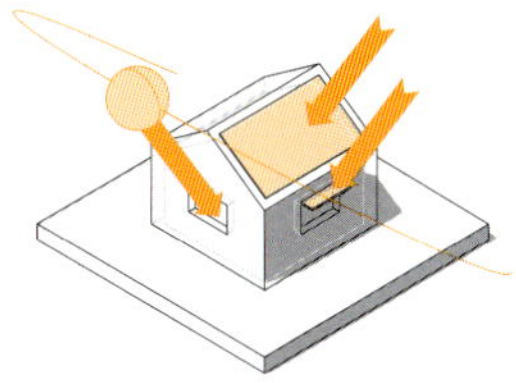	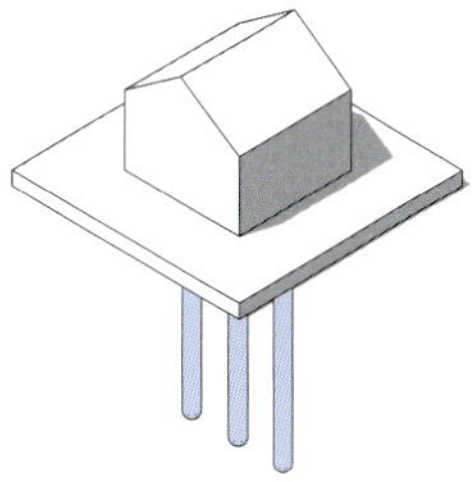
Higher requirements not necessary	Insulation to prevent excessive cooling and overheating	Higher requirements not necessary
Observe basic principles (avoid causing wind turbulence, use prevailing summer wind direction for cooling)	Use solar radiation for passive heat in winter Suitable solar shading and thermally active envelope surfaces in summer as appropriate	Ground freezing and thermal properties of the soils are not critical (can be used for heat generation)
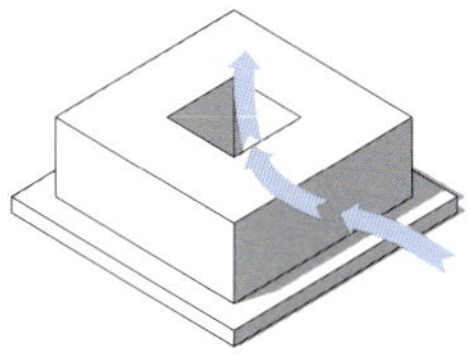	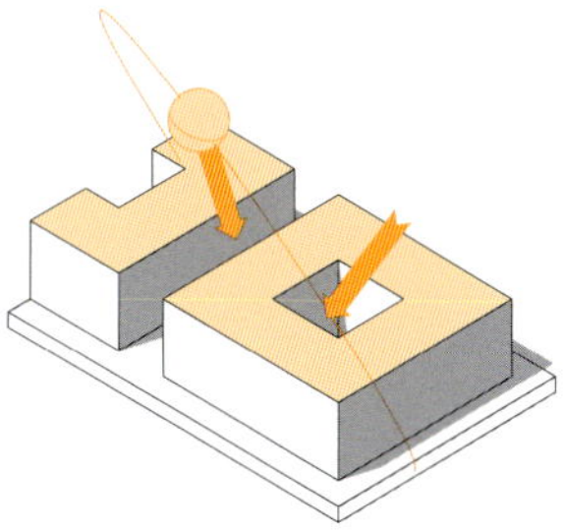	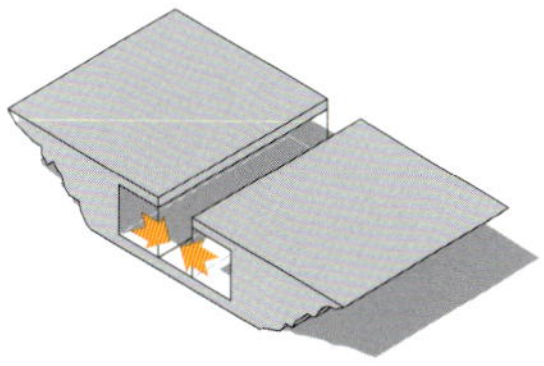
Sandstorms are common in desert regions, otherwise medium wind speeds	High proportion of direct solar radiation striking the building (almost all year round)	Dry, mainly sandy soils
Use porch in main wind direction for passive cooling and constant good through ventilation	Provide buildings and open areas with shade or solar shading, integrate thermal stores into the construction (e.g. soil)	Exploit constant ground temperatures, where possible (e.g. earth houses, or earth tubes for ventilation)
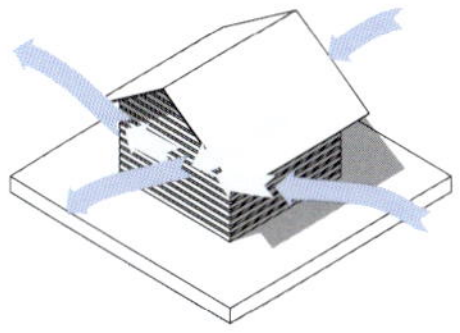	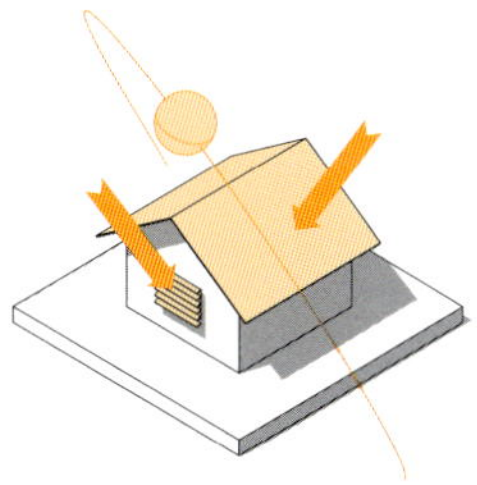	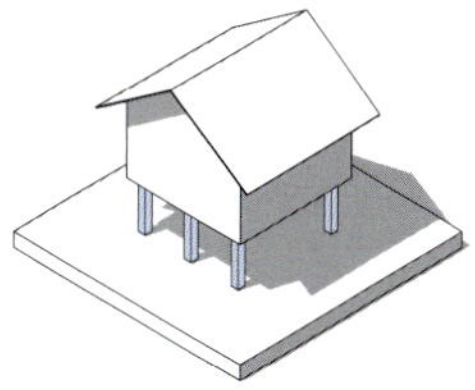
Protection from moisture in the interior	Protection from direct radiation from the east-west travelling sun	Protection from heavy rainfall (and animals)
Continuous through ventilation of the interior to cool or remove heat and moisture using supply air	Building shaded by e.g. roof overhang Shading of interior by solar screening	Building on piles is worthwhile in regions subject to monsoon-like rainfall

Requirements and the resulting building measures in the various climate zones

Microclimatic analysis

Every building design should begin with an analysis of the parameters for the actual location and the specific use. The analysis of the climatic conditions of the site must extend beyond the principal characteristics of the climate zone, which represent only a macroclimatic definition. For the actual design, it is important to know the microclimate at the location and its effects on the building, and to estimate its potential for providing energy. The microclimatic conditions can be strongly influenced by the tectonics of the surrounding buildings and therefore may deviate from the characteristic features of a climate zone. Steep hillsides can lead to winds, which in turn can affect average temperatures. Any climatic consideration of the area around the proposed building should therefore take place on various levels and go beyond the overall climate zone to consider the urban context and the actual site itself.

The analysis of the microclimate takes place after the survey of the geographical or tectonic situation. The following points should be taken into account:

- The amount of solar radiation at the site and the shade situation
- The amount of rainfall on the site and infiltration capacity
- Main wind direction and strength, and a frequency distribution analysis of wind direction and strength, possible wind funnelling and channelling by surrounding buildings and geographical features
- Surrounding green space and planted land quantified in area and height
- Analysis of the condition of the ground in relation to soil and groundwater

An analysis on various levels is necessary to assess the local climate. The urban area (1), for example, gives information about fresh air corridors and green spaces in the city. The characteristic building typology prevalent in the neighbourhood (2) is indicative of the amount of impervious ground and surfacing materials, whereas the block (3) and the immediate surroundings of the building (4) provide information on the actual climate and usable potential.

SUN

The sun is the driving force for practically all renewable energy sources and fossil energy carriers. It offers free daylight for lighting and energy. Solar radiation at a location can be extracted from climate data sets from various sources. The mean global [horizontal] solar radiation values are adequate for preliminary design and analysis. This indicates how much energy (kWh) provided by the sun falls on a horizontal surface of 1 m^2 in area in an average year. Mean global [horizontal] solar radiation in Germany, for example, is approximately 1,000 kWh/m^2a. Each location is different and the value increases from north to south. The mean global [horizontal] solar radiation, the efficiency of the photovoltaic (PV) panels and their angle of inclination can be used to calculate, for example, the yield and therefore to gauge whether a solar system would be worthwhile.

A survey of the shading of the proposed site by adjacent buildings, bushes and trees, and topographic features also needs to be performed. A simple massing model is adequate to simulate the path of the sun at any time of the year. It is sufficient to consider the four seasons (represented by the situation on the 21 March, 21 June, 21 September, 21 December). Instead of expensive simulations, simple spatial studies can be used with a polar radiation diagram in a similar way. A polar radiation diagram is a projection in plan of the solar conditions related to the four cardinal directions over a specific time period (year, month, week, day). The diagram can be used to calculate the shading and insolation conditions of the building in its proposed position and orientation.

WATER

Rainwater can be used on a site (e.g. as greywater or for cooling) but its use is not without risk. To ensure the building can be used over the long term without risk of damage, the designer must calculate the amount of rainwater falling on the site using the annual precipitation records. The values are expressed in millimetres per annum and range, for example, in Germany between approximately 500 mm and 1,200 mm (e.g. Lüdenscheid 1,203 mm, Halle 521 mm). The calculation should take into account how much if any infiltration capacity is available on the site to accept heavy precipitation and whether enough water falls on the site to be collected and used as greywater.

A flooding calculation needs to be performed too. If the site being considered is in an area prone to flooding, the analysis should also consider whether building there is sensible or possible. If building is considered sensible, then engineers must decide on the flood protection measures to be included in the design and construction of the building.

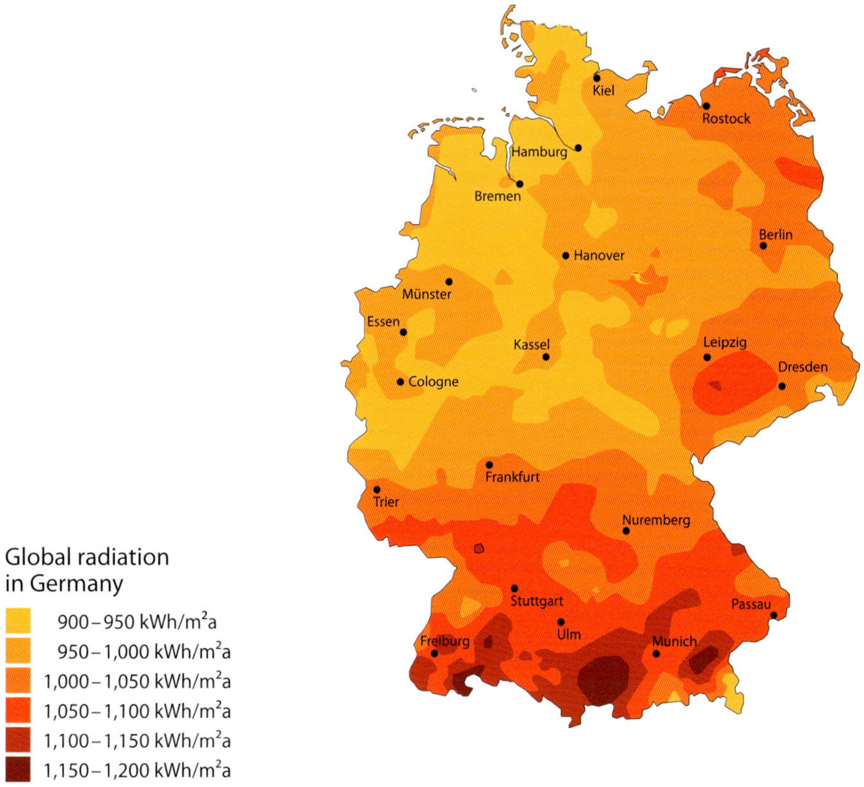

Distribution of the mean [horizontal] global radiation in Germany

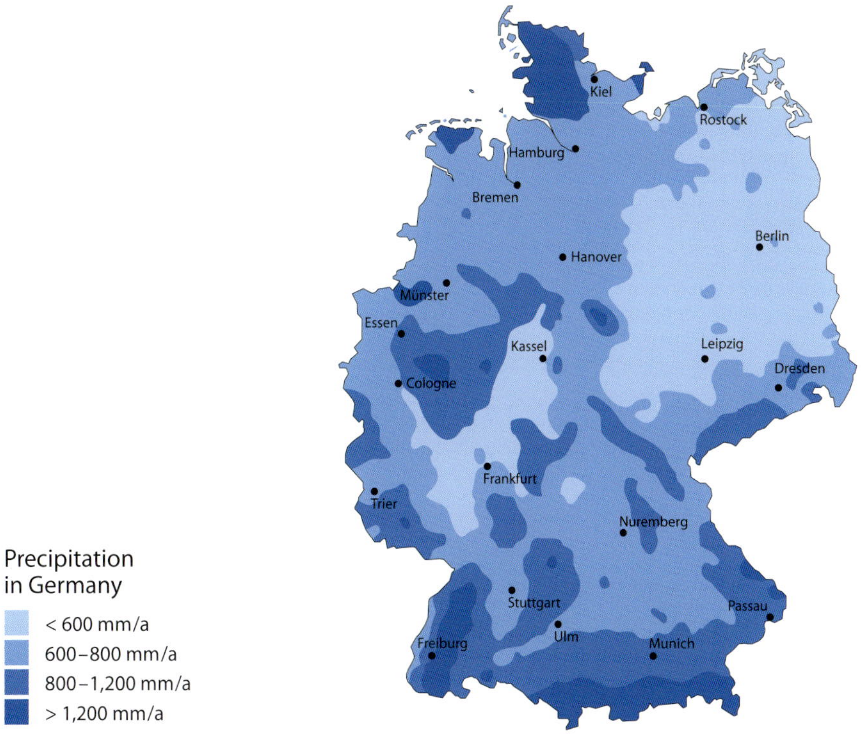

Distribution of annual precipitation in Germany

AIR

Strong winds can lead to uncomfortable conditions, but they can also generate energy. An appropriate analysis should therefore consider wind speeds and direction, their frequencies and distribution over the year. Climate data sets can supply this information.

The surrounding built environment and the local topography need to be considered too. Buildings can divert, channel and strengthen wind. These effects can be estimated using a suitable wind rose for the site and a map of the surrounding area. Wind modelling provides detailed results from a massing model or wind tunnel tests.

Wind roses for Frankfurt am Main. The diagrams show the prevailing wind conditions for each season. The wind roses show wind speeds (m/s) and frequency distributions (%) by wind direction. Their information can be used as a basis for developing the building form and for the active use of wind energy.

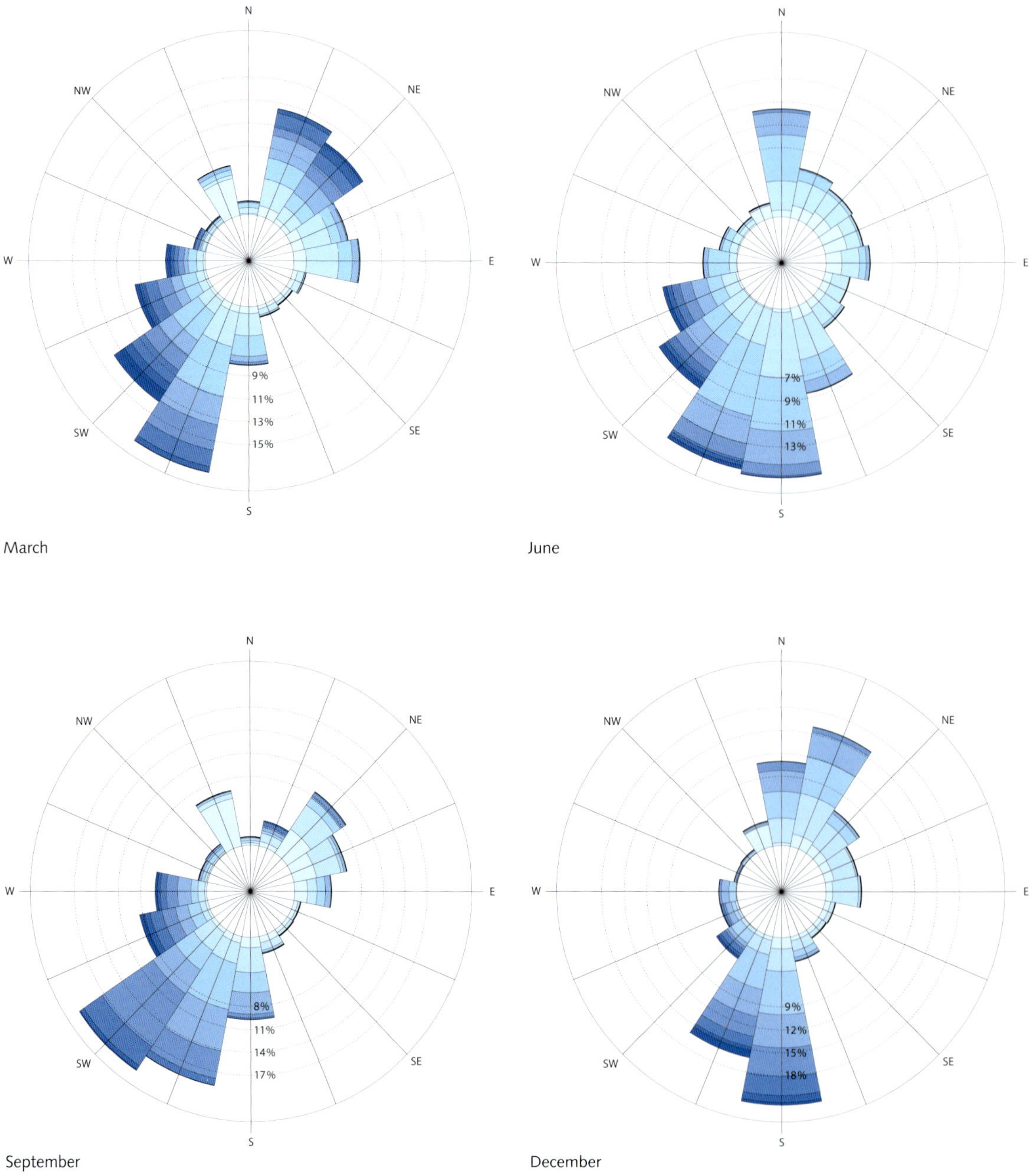

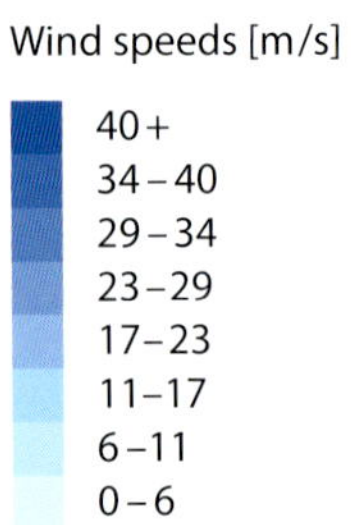

FLORA / FAUNA

Green space and groups of trees filter and cool the air, exerting a positive effect on the microclimate. They can, however, create unwanted shade.

An environmental assessment based on a study of the local flora and fauna is essential, particularly for sites in densely developed inner-urban areas. Green space has a positive effect on the climate. If there are insufficient green spaces in the area, then some attempt to compensate should be made in order to keep down the dust in the air and avoid creating heat islands.

GROUND CONDITIONS

The soil type on and under the site can present problems as well as potential advantages. Loose, sandy ground requires great care in designing the foundations. Energy can be extracted and stored in the form of heat from the ground and any groundwater it contains. Soils reports provide information about how suitable the ground on the site is for energy extraction and storage, about foundation options and any contamination present.

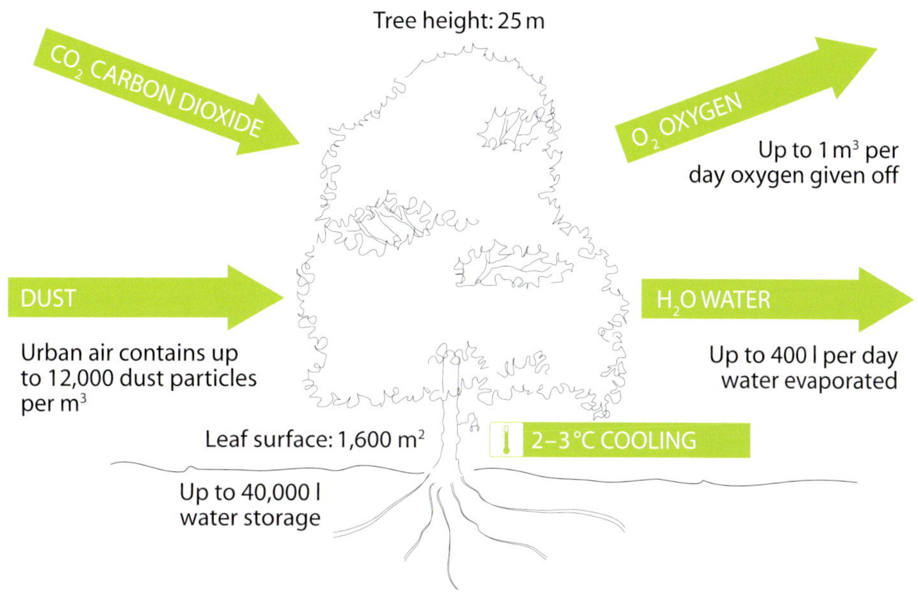

Climate balance of a tree. Green features can influence the local climate.

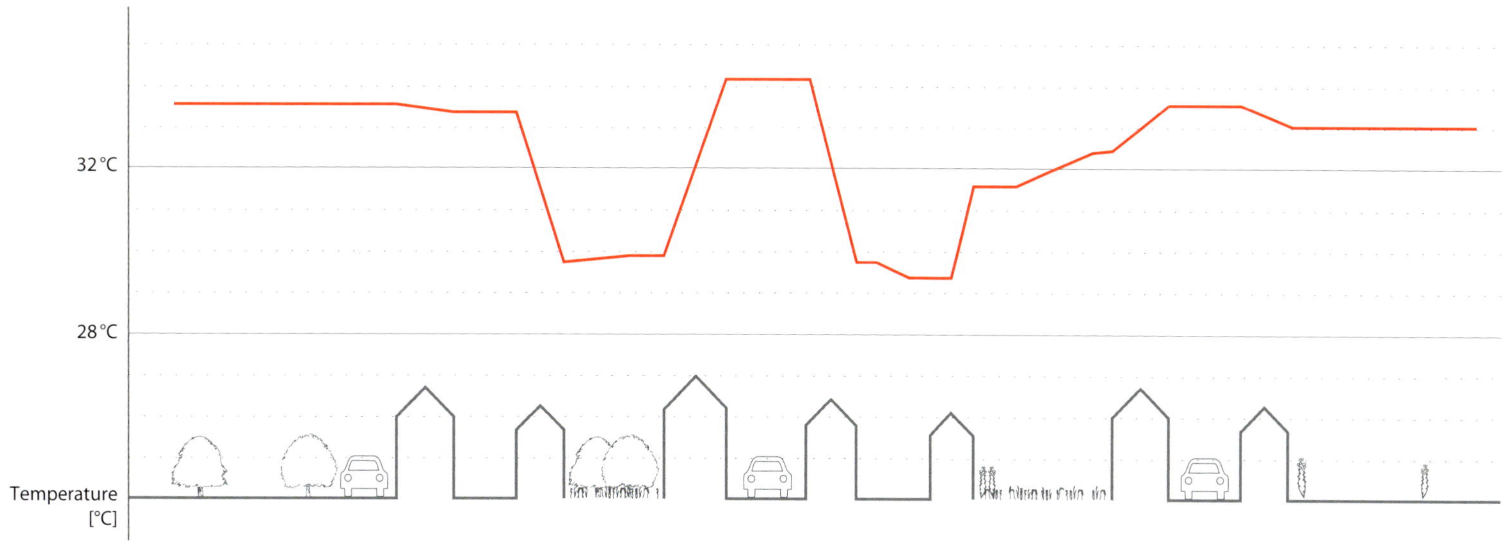

Some effects of planted areas on the urban climate, for example, on daytime temperatures.

Development of a conceptual idea

An objective can be developed as a first conceptual idea for the building following the completion of the analyses and assessments of the internal and external boundary conditions.

The concept should be seen as a contextual solution based on general good practice that will be optimised for the specific location as the design proceeds. As well as the position of the building on the site, the designer should identify the volume and its essential features, and the desired building energy standard.

This objective for the energy characteristics agreed with the client is influenced by the renewable energy sources available for active exploitation on site, and by the passive building design measures worthy of implementation. The building design can be developed along the resulting directions. An Aktivhaus demands new design forms, which may deviate from the established strategies required for the Passivhaus concept. The building envelope takes on a new task with activated, energy-generating roof and facade surfaces. The Aktivhaus examines not only consumption itself, but also the balance between consumption and the regenerative energy created by the building environment and the site. This calls for new design strategies and creates new building forms. A Passivhaus has large glazed areas facing south to make passive use of solar radiation in winter and consequently also has shading elements designed to prevent overheating in summer. An Aktivhaus, on the other hand, limits the glazed areas in the south to the area reasonably required for comfort, natural lighting and the atmosphere inside the building and uses opaque and transparent building envelope surfaces to create energy.

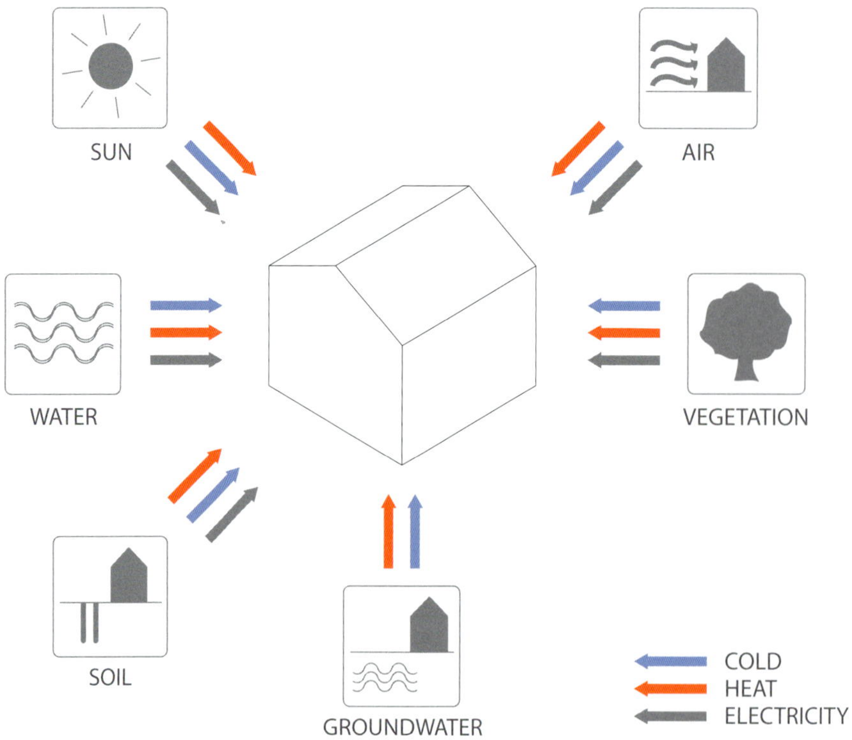

Potential usable energy sources on the site

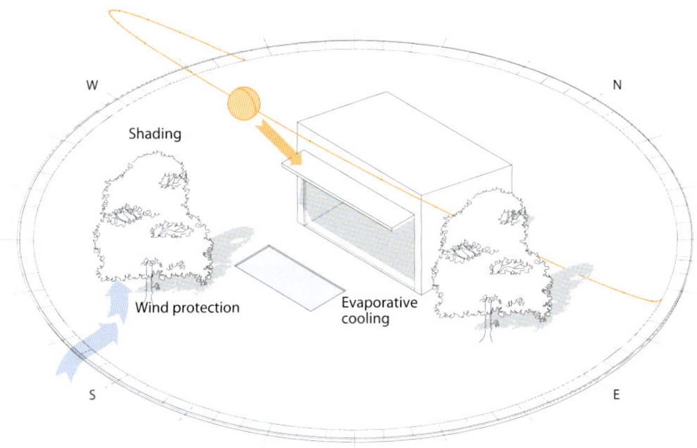

Passivhaus in summer

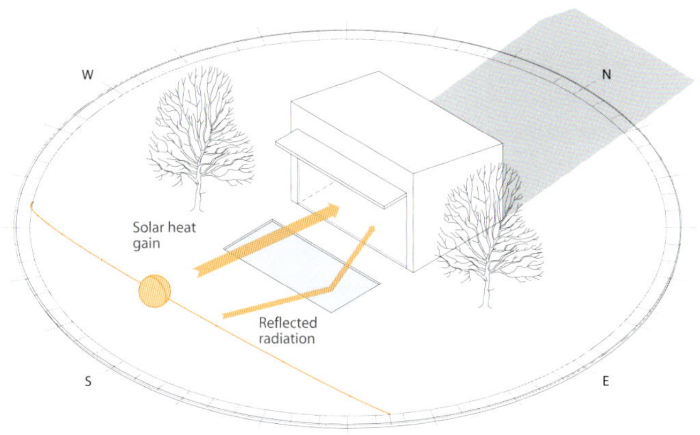

Passivhaus in winter

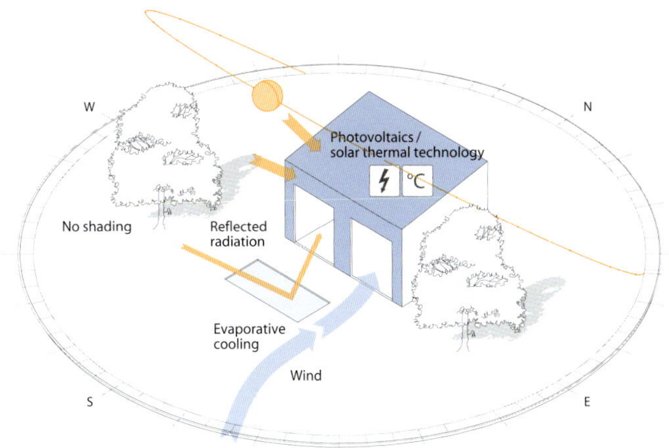

Aktivhaus in summer

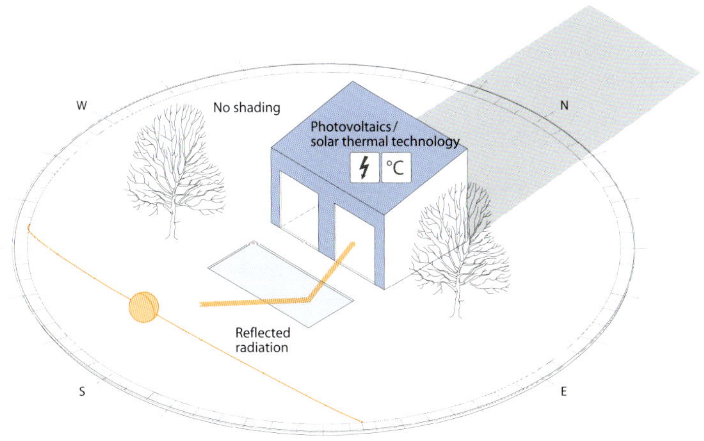

Aktivhaus in winter

Paradigm change in design strategy. While a Passivhaus (top) has large glazed areas facing south in order to make direct use of the heat in the interior, the Aktivhaus (bottom) has the optimum ratio of opaque and transparent surfaces for the optimum integration of active technologies. The opaque areas should be unshaded if possible, while shading is required for the transparent areas.

Design strategies

An efficient and yet robust building concept can be developed only on the basis of a design strategy that takes into account all the relevant requirements and boundary conditions. The resulting situation is unavoidably highly complex and can be resolved only in a design team of specialists coordinated by the architect. In order to appraise the specialist designers' concepts and integrate them into the planning and design of the building, the architect has no choice but to broaden his or her knowledge of energy concepts, building technical services and their integration into the building's architecture. However, the architect should not become a specialist designer. Specialist designers must be brought into the preliminary design process at an early stage. Valuable advice in the beginning can influence the design and simplify the integration of the technology required for specific location. The integrated approach to design process is gaining in popularity. It ensures an optimum outcome, a lower incidence of error, and often construction savings.

For non-residential buildings in particular, it makes sense to involve not only the specialist designers and the client but also the building operator in the design. After successful completion and handover of the building, the designer is usually no longer involved. But it is precisely at this interface where problems arise with the fine-tuning of the systems.

Every design for a modern building must attempt to achieve a balance between active and passive measures. All the worthwhile passive measures should be fully exploited before they are supplemented with active technologies. Passive means that the energy demand of a building is reduced as much as possible by the design, construction and the choice of materials. Where this is insufficient or inappropriate, generated energy, in other words active technology, should supplement the basic passive provision. The supplementary systems should be highly efficient and use renewable energy.

This interplay of passive and active components provides all five services to a building: heating, cooling, ventilation, lighting and electricity.

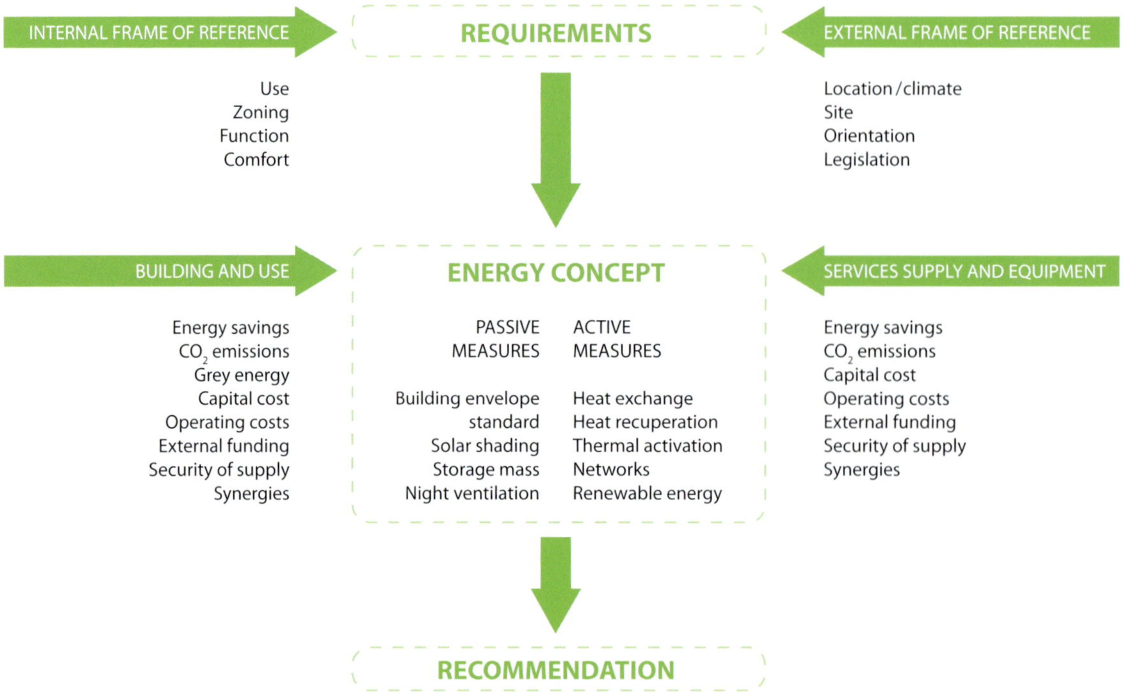

Strategic process for recommending a building energy concept on the basis of the specific use and site

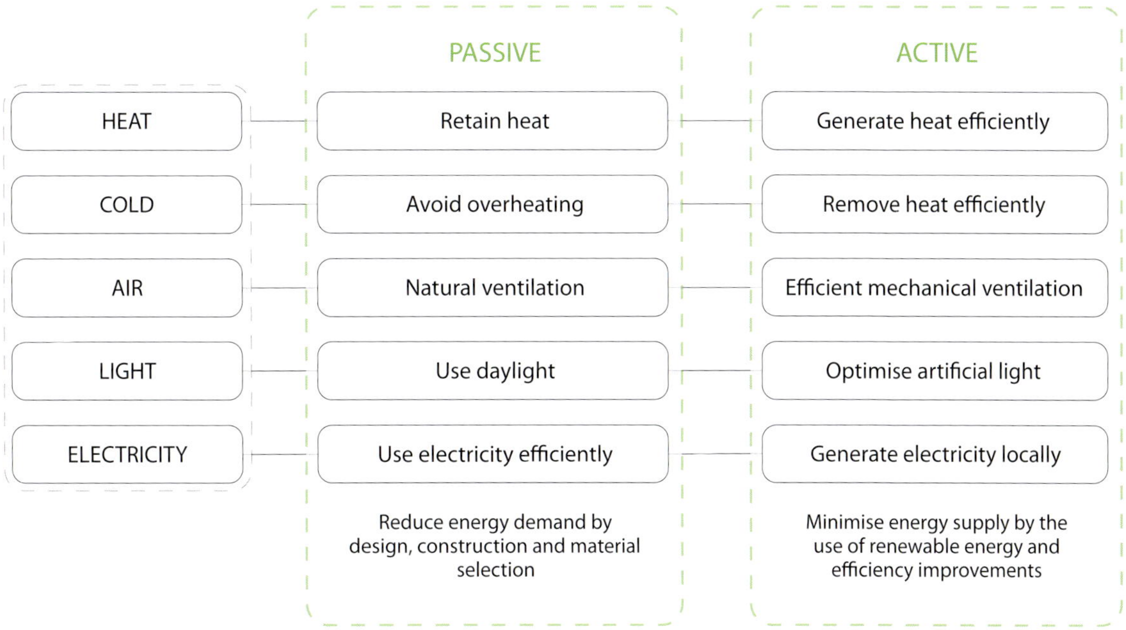

Interplay of passive and active. The building concept should first seek to optimise the services using passive components. Only then should the active systems come into consideration. Intelligent and robust systems arise from a combination of passive and active measures, a melding of structure and technology into one system.

Building design

The building form develops not just from consideration of urban design, function and form; it also depends on the local climatic conditions specific to the site and the energy benchmarks. The design of an energy-efficient building follows generally applicable principles. To what level they can be implemented in any one context must be judged against the background of the specific requirement profile.

The cheapest kilowatt hour is the one that is not consumed. Therefore the building design should concentrate on saving energy. First, by reducing the energy transmitted through the building envelope, which comprises largely of the floor, roof, walls and windows. Next, the area of the envelope, the energy-transmitting surfaces, should be minimised. The parameter for this is the A/V ratio, the relationship of the surface area (A) to the volume (V). With better insulated building envelopes, the effect of minimising the surface area has less effect on reducing the energy consumption.

The Aktivhaus represents a paradigm change not only in relation to the level of insulation, but also with respect to energy creation. With an Aktivhaus, the designer is no longer forced down the optimum path of the lowest A/V ratio, but can choose a ratio appropriate to the project. As the building envelope is reduced in area, it offers less space for energy generation.

The site's own potential usable energy has been mentioned earlier. Many of these energy sources (such as the heat contained in the ground, in groundwater and in surface water) have no direct effect on the building form and appearance. One exception is the use of the sun as an energy source. The orientation and shape of the building can make the difference between a good or poor yield. Typical spatial arrangements vary according to climate zone, largely dependent on whether the design seeks to protect the occupants from the sun or is open to the sun in order to make direct and active use of it. In Germany, for example, the latter is the case. The intensity of the radiation falling upon a surface varies according to its inclination and angle towards the south. Software packages can calculate the approximate amount of radiation. The input required includes the orientation and inclination of the surface, distribution of radiation compared with the local mean global radiation, and the parameters of the selected technology (efficiency etc.). As the building cannot face in the optimum direction on all sites, perhaps due to the surrounding buildings, this calculation is helpful in optimising the design.

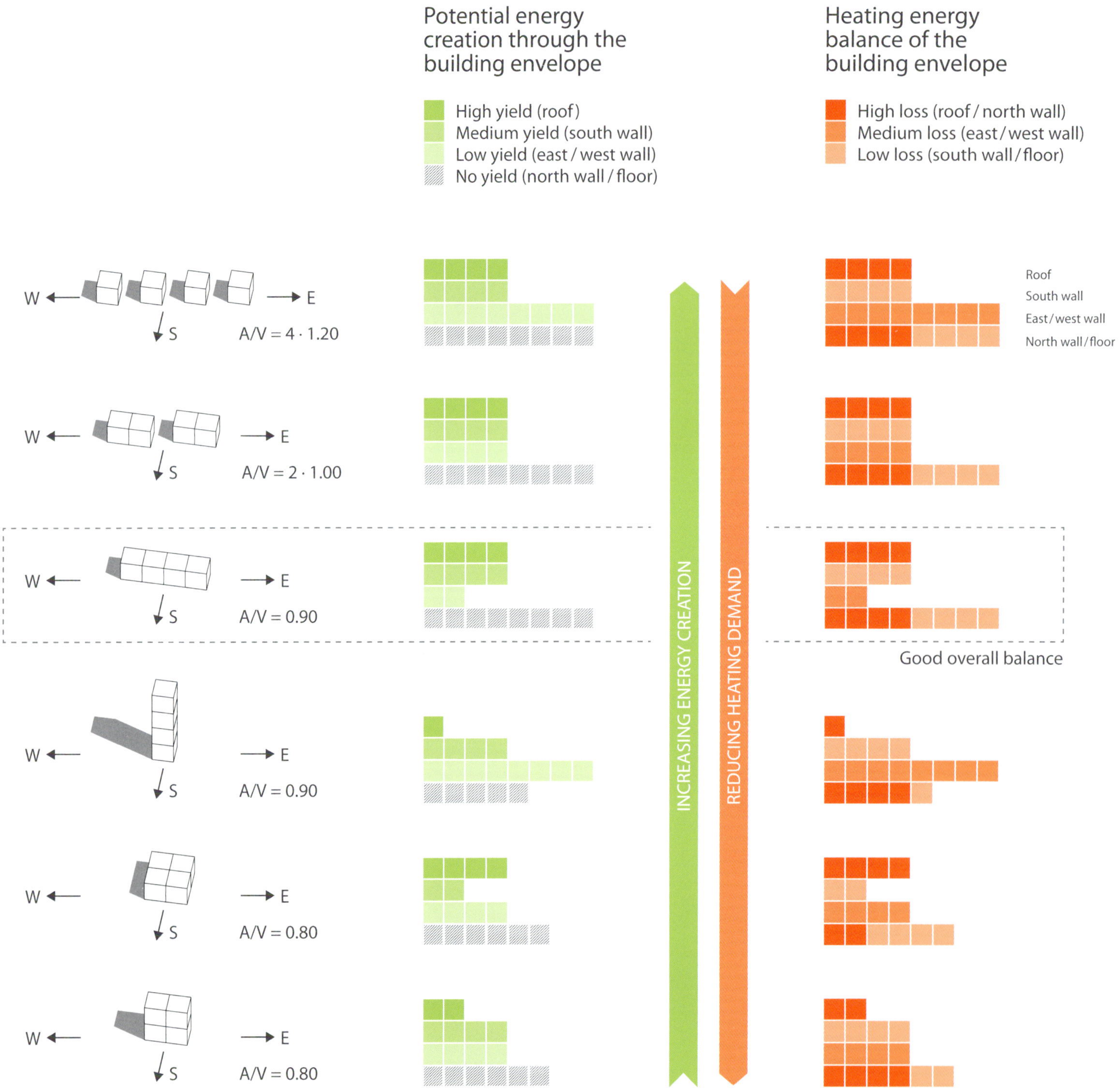

The compactness of buildings is an important parameter in optimising their energy losses. The smaller the area of the envelope (A) relative to the volume (V), the better the loss performance. Aktivhaus buildings, however, generate energy from their facades. Therefore, minimising the envelope area alone is not enough to fulfil the designer's goals. The diagram shows the relationship between energy losses and gains over the envelope calculated for different degrees of compactness for the same floor area. A medium degree of compactness produces the best result. Other parameters, such as the size of the roof area and its orientation to the south, are more important to energy creation than just the area of the envelope. Two equally compact buildings can have different potentials for creating energy, depending on whether they have been optimised in orientation and building form.

Wind can also influence the building form. The orientation of the building must not lead to unpleasant channelling of the wind. The building form should not offer any large surfaces to the prevailing winds because this could lead to continuous cooling of the facade and complete cooling of the whole building in winter.

The usage requirements of the individual rooms determine the design of the interior, which leads to the development of the building and plan layout. The rooms are arranged according to various zoning concepts (onion principle, linear zoning, horizontal zoning) to ensure that they gain appropriately for their use from their orientation and their position in the sequence of rooms. Warm, well-lit rooms are generally placed in the south, while storage spaces, for example, which are better cool and dark, should face north. These north-facing rooms also contribute to the buffer effect against the heat losses from the north facade.

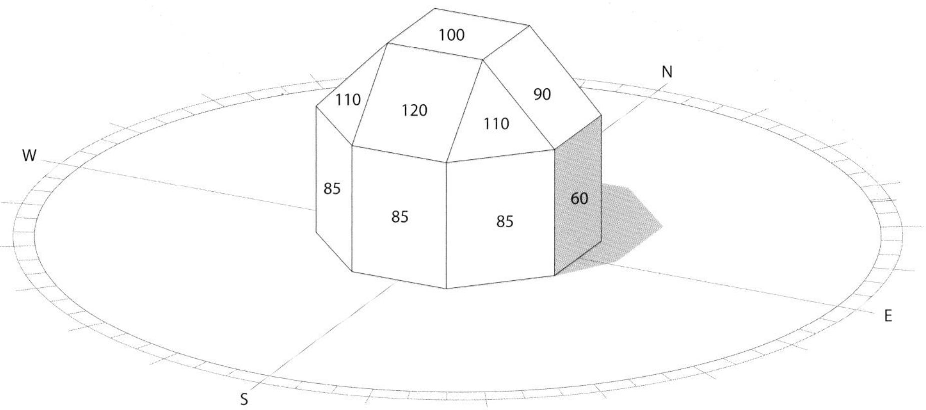

Influence of orientation on the yield of a photovoltaic system in Germany

Energy yields from active solar technologies integrated into the building depend on the orientation and inclination of the building surface. Rough guidance values are a help at the beginning of a project. In the later detailed design, however, these early estimates must be checked by calculations based on the actual systems.

They show that insolation on a south-facing wall is 15% less than on a flat roof. Seasonal distribution is another important factor: the angle of the sun means a system integrated into a facade creates more energy per day in the transition months and in winter than a flat roof system. Active facade surfaces can therefore fulfil the building's needs, depending on the objective for creating the energy (maximum yield over the year versus evenness of the yield or relevance of winter).

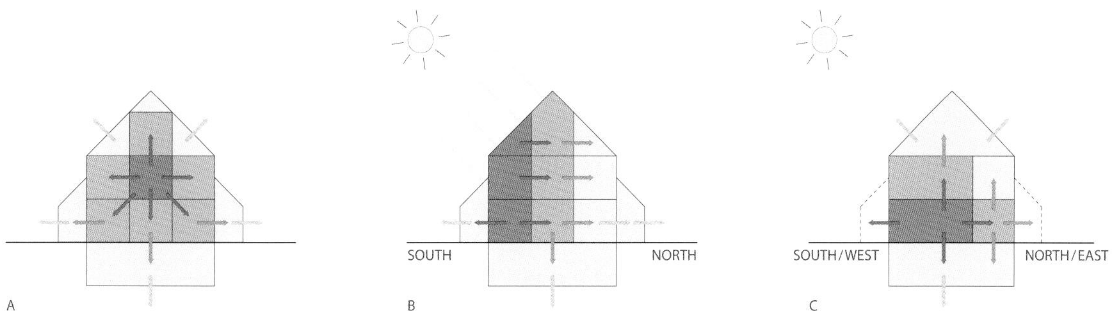

Example of zoning in a building. The orientation of individual rooms to a specific compass direction may be advantageous in terms of energy, depending on their use and requirements. The diagram shows some zoning types: onion principle (A), vertical (B) and horizontal (C) zoning.

Building envelope design

The primary function of the building envelope is to ensure comfort in the interior by providing adequate insulation, a high proportion of natural light inside through an appropriate proportion of window area, and by offering suitable solar shading to eliminate overheating.

The processes in the design dealing with these topics, along with some possible solutions, are shown later in this chapter (see p. 86 ff).

After the building form has been designed to take into account climatic and energy considerations, the next step is to decide on the quality of the building envelope. Materials and type of construction have a central role to play in this. They are not only key to reducing the embodied energy in a building, but also influence the microclimate on the site. Highly reflective surfaces, for example, can increase the reflection of solar radiation and lead to local overheating of outdoor spaces in the absence of adequate cooling influences in highly built-up areas. The same effect can also arise in summer from heat-absorbing, intensively insolated, solid surfaces. Planted building envelopes contribute to cooling in summer due to evaporation. This has a positive effect on both the microclimate at the site and the building itself because cooling of the external facade surface reduces the transmission of heat into the building through the facade.

Surface temperatures of different roof forms

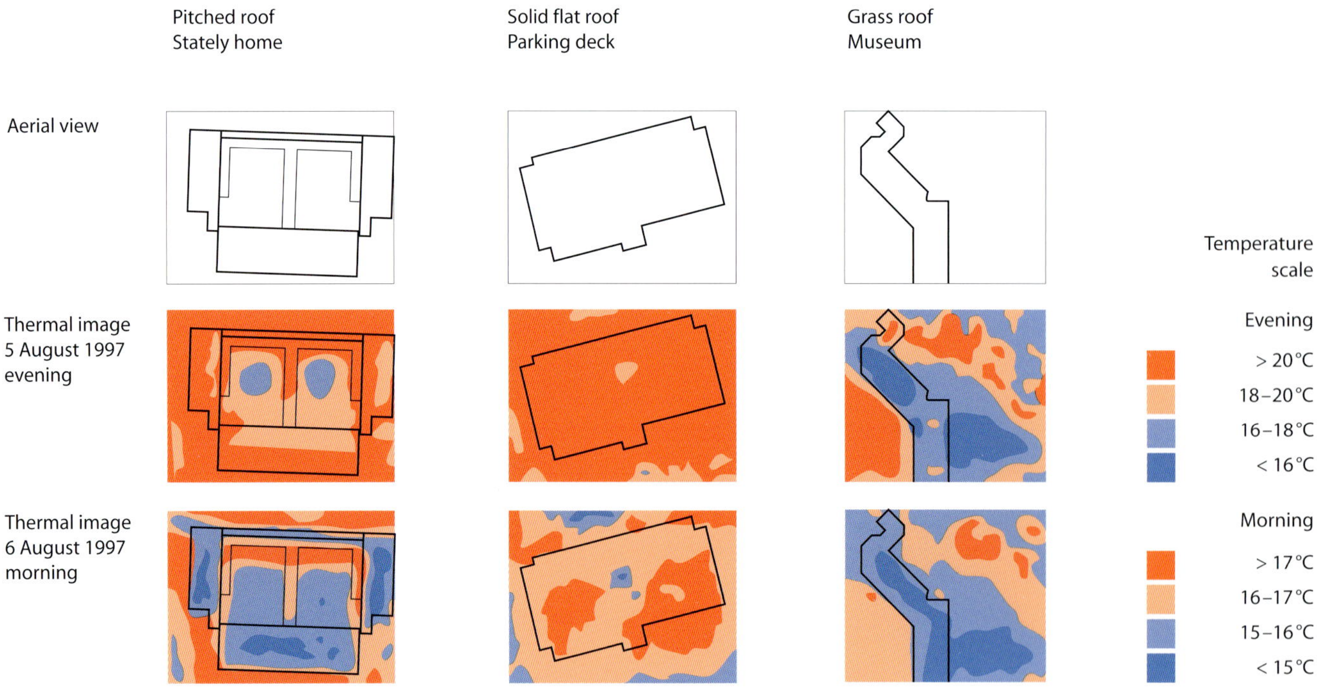

The diagram shows differences in the external cooling behaviour of roof surfaces overnight in summer. The grassed roof is able to dissipate most heat by evaporation from the plant surfaces and thus contributes to ensuring the interior does not overheat. The form and solid construction of the flat roof means that it achieves the smallest cooling effect.

Energy supply

The parameters of a forward-looking energy supply are explained in detail in the "Toolkit" chapter (see p. 120 ff). The objective in principle is to design as simple and robust a technical services supply system as possible, and base it on as few as possible different energy sources, technologies and transmission media. This reduces the frequency of system failures and avoids high maintenance costs. In addition to the conventional components of the building services, Aktivhaus buildings also have systems for generating energy. An efficient control and regulation system is therefore worthwhile. These systems are described in the chapter on "Control and regulation" (see p. 154 ff).

An innovative building should attempt to cover its whole energy demand and therefore in addition to the creation of heating energy and hot water, it should consider reducing the consumption of power by household appliances and, in housing projects, by lighting, even if the statutory energy balance principles do not currently require it. Increasing the efficiency of these energy services is now the focus of attention. Only highly efficient equipment and energy-saving lighting systems, such as LED (light-emitting diodes), should be used. The Efficiency House Plus national standard already incorporates this as a secondary requirement for household appliances.

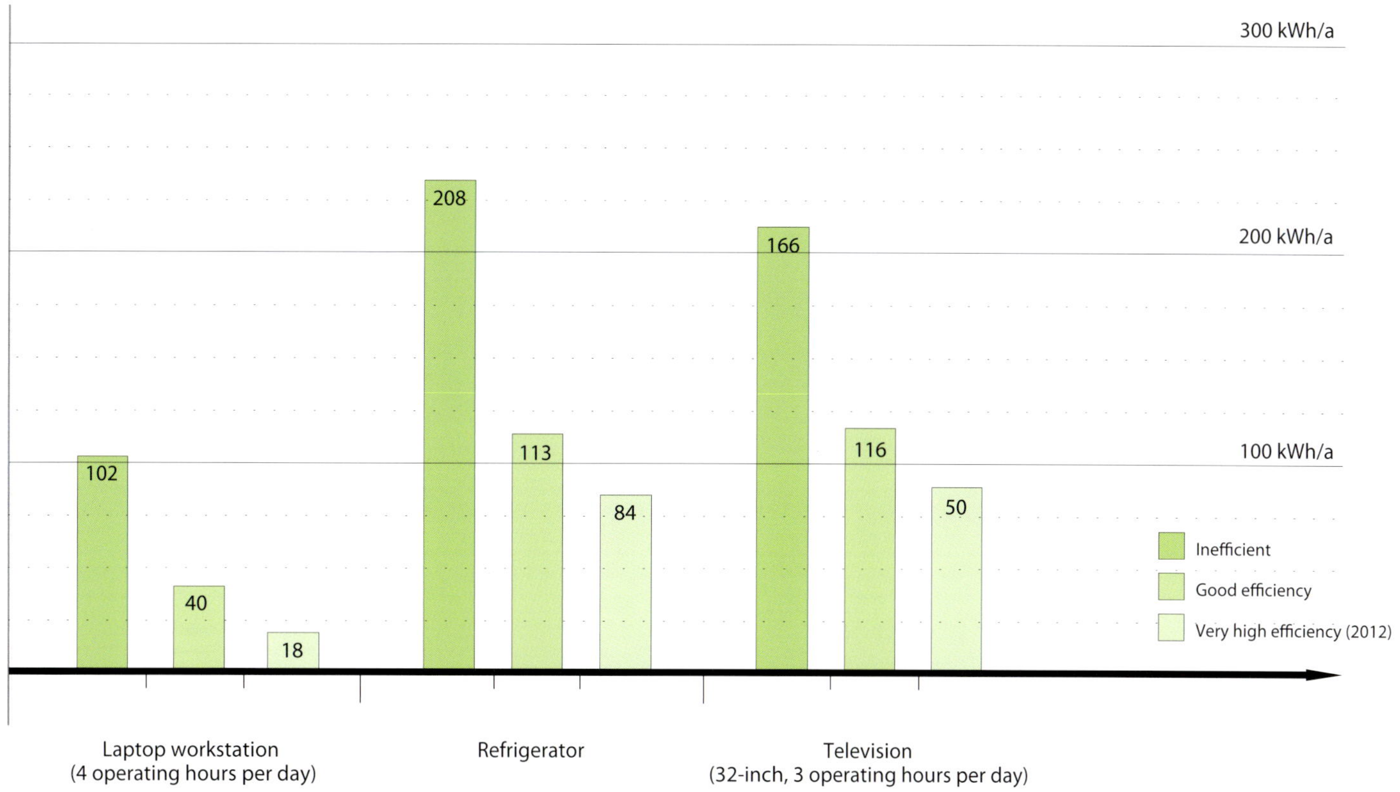

The diagram shows how much the efficiency of domestic electrical equipment has risen. This development can reduce the enormous proportion of household electricity over the long term. It is important, however, to avoid any rebound effect.

Examples of integrated design

The following two example projects, one a new building, the other a refurbishment, illustrate the process and the result of an integrated design process. Design considerations and contradictions are explained by way of these actual examples. Both examples come from the field of small-scale residential projects. Design approaches and innovative design solutions are, however, readily transferable to other usage types.

New building

The contribution by Darmstadt University of Technology to the international university competition Solar Decathlon 2007 provides the new building example. The objective of the competition was for students to develop and build a prototype for a future-oriented residential building that would create more energy than it consumes. Because the competition took place in Washington, D.C., USA, the building concept had to be designed for this location as well as for Darmstadt, even though the climatic characteristics of the two locations are different.

Development of a conceptual idea

The rules of the competition set out the energy objective of a positive energy balance for heating, cooling, hot water, auxiliary electricity, lighting, all domestic appliances and transport. It was decided very early in the process to adopt a Passivhaus type of construction in order to approach the objective. For the building design, building envelope quality and the energy supply, a passive building design approach was used initially to optimise the complete system. Active technologies were then adopted to the extent necessary to ensure comfort and to reach the very precisely defined Energy-Plus objective for the building. The focus was on integrating the technology harmoniously into the building concept and achieving a pleasing architectural appearance. The usage profile was defined precisely in the competition rules too. The building had a footprint of 54 m² and was designed for two people. Architecturally, the small building had to be based on the idea of a simple, flexible housing solution with smooth transitions within the building and between internal and external space.

To achieve these ambitious objectives and for quality assurance reasons, extensive modelling and energy balance calculations accompanied the design process, which allowed increasingly reliable statements to be made on energy and heat flows within the building, and between the building and its environment. This was the basis for implementing the ideas and harmonising the components.

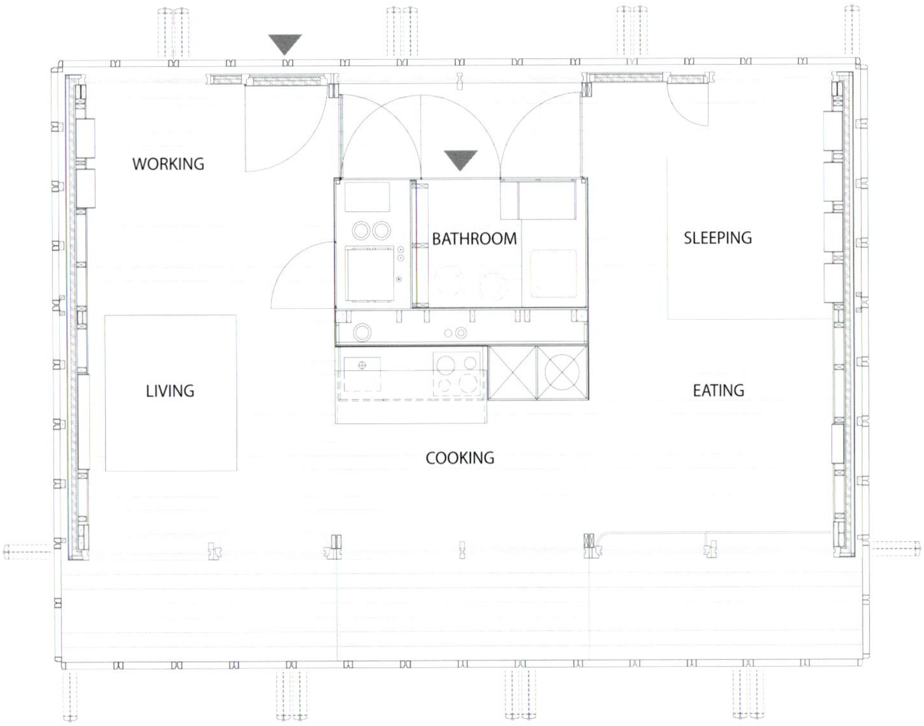

Solar Decathlon 2007, Team Germany, TU Darmstadt (not to scale)

Passive measures

BUILDING DESIGN

Internally: a plan layout zoned according to temperature

In order to reduce heat losses, the layout of the building was constructed using a model based on zones of different temperature, which allows a dynamic usage pattern depending on the season. In essence, there are three concentric zones surrounded and partially regulated by envelope surfaces of various designs. The external facade envelops the entire building including a veranda as a sheltered exterior area in the south. The facade shades the veranda in summer, which offers an external space for enjoying leisure time outside. The veranda can be used as an extension of the interior space by opening the south facade in favourable weather conditions. In the transitional months, the facade provides shelter from the wind and rain so that conditions are still pleasant enough to sit out in the open air on warmer days. In winter, the facade can be closed to provide a buffer space. The house is able to react dynamically to the outside circumstances and its usable floor area grows and shrinks according to the climatic conditions.

The second layer takes the form of the thermal envelope and encloses the heated interior of the building. A high insulation standard reduces heat losses and creates a comfortable climate.

The third zone, the inner core space, is the warmest area. The technical services room and the sanitary area lie at the centre in the heated volume and as far away as possible from the outer facade. Some means of rapid space heating is desirable in the area of the sanitary rooms to create a comfortable climate.

Passivhaus tends to be a sluggish system because it relies on internal and external solar gain for its heat. Early in the design, it was decided to increase user comfort in the warm building core by installing an additional surface heating system. As the design progressed, this became specifically an underfloor heating system with only a small number of loops activated efficiently by a solar-thermal system or a heat pump.

Externally: a compact building volume to optimise the envelope surface

The strict competition rules and the tight room schedule left very little scope for variation in the design of the building. Because of the stipulated maximum height of the building, the optimum of a roof inclined to the south for solar use would be possible only at the expense of the usability of the interior space. Therefore a simple cubic form was adopted. To check the feasibility of the design, a rough energy balance between the consumption and the energy potentially created by activating the flat roof and the facade surfaces facing west, south and east was carried out. The result showed that the estimated annual yield for the flat roof would be only about 10% less than that of a roof with a 30° slope. The greater area of the flat roof, all of which would be activated, more than compensated for this. The comparison of consumption and creation confirms the positive energy balance and confirmed the design decision.

A generally compact structure was chosen to reduce energy losses through the envelope surfaces and the building was optimised as much as the project rules would allow.

A/V ratio

Vacuum insulation installed in the facade. The insulation boards were installed in two crosswise layers to reduce cold bridging at the battens as much as possible. Continuous compressible foam strips around the boards seal to a certain extent and create a clamping effect to keep them in place.

BUILDING ENVELOPE DESIGN

Thermal insulation and airtightness of the building envelope

The compact form and an envelope reduced as much as possible in area should have the highest standard of insulation to cut the energy demand for heating and cooling to a minimum.

The objective of the design team was to achieve a very high insulation value within a slender wall thickness. The very tight limits on floor areas in the competition would have further reduced the usable floor area using conventional insulation. Therefore the opaque components were insulated using vacuum insulation panels (VIP). These have a porous core made from fumed silica, from which all the air is evacuated within an enveloping layer of aluminium foil. The vacuum in this insulation material means it has ten times the insulation performance of conventional insulation. The high insulating effect of the VIP enabled a 6 cm thick insulation layer (2 × 3 cm VIP panels) to achieve a U-value below 0.1 W/m²K.

The design requires much more careful detailing than conventional systems to ensure that the delicate insulation panels are installed without cold bridges. The intermediate battens of the two insulation layers are offset from one another to reduce the number of unavoidable weak points where the battens cross. The battens are made from a wood-like recycled material manufactured from compacted polyurethane. The use of these and insulating boards edged with compressible foam sealing strips reduces cold bridging and the associated losses. The consideration of this building set out here clearly indicates that the design of a particularly energy-efficient building can progress quickly from a conceptual statement of objectives to issues of detail that require new solutions and unconventional ways of thinking.

The fully glazed south facade and the largely glazed north facade define the main elevations of the building. To achieve a high standard of insulation despite the large proportion of glass, quadruple glazing was used for the north facade, with triple glazing for the south. The quadruple glazing was a prototype developed specially for the project. Increasing the number of panes increases the number of insulating glazing cavities to reduce energy losses further. On the other hand, more glass increases reflection and reduces permeability to light, thus reducing the amount of passive energy from solar radiation entering the room. Therefore quadruple glazing is used only in the north. Opaque surfaces insulated with VIP, like the other opaque surfaces, supplement the transparent parts of the north facade to reduce its transmission losses further. The aesthetics and energy qualities of several variants were examined and compared. The designers performed rough energy balances before deciding on whether to pursue these variants further.

Triple glazing in the south facade more fully exploits the solar heat energy available from the sun in winter. Modelling of the situation here showed that the greater energy gained by the room compensates for and may even exceed the higher transmission losses of the triple glazing compared with the quadruple glazing. The timber frames of the glazing were manufactured in oak with a recycled compacted polyurethane core to increase the insulating effect.

The quality of the building envelope shows in the high insulation and airtightness standards achieved. The more airtight the building, in other words the less air escaping from inside the building with the windows closed, the less heat or energy is lost.

Unavoidable cold bridges that cannot be mitigated by appropriate detailing detrimentally affect the otherwise good quality of the envelope. They are minimised in the design to greatest possible extent. For transport, the relocatable building divides into three modules, at the four corners of which are steel members running up through the base of the structure and out through the roof. These butt connections are used to suspend the modules on a cross beam for lifting by a crane. Weak points like these can also occur in other projects for different reasons. What is important here is to assess priorities. Achieving the optimum energy performance everywhere may not provide the correct solution for the project. In the case of this project, it could not have been transported to Washington had it not been for this arrangement. However, the building and its envelope still achieve the Passivhaus standard for Washington, D.C.

Attaching the facade cladding

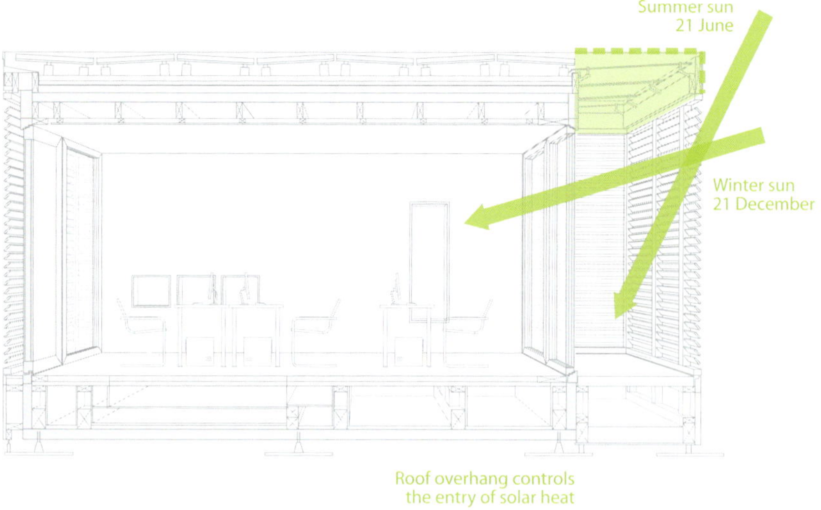

Control of passive solar gain

The quality of the building envelope not only reduces energy losses; appropriately designed glazed areas can also increase the direct use of passive heat. This is becoming more and more important in buildings that are highly airtight and insulated. Transparent elements make up the whole of the south facade to maximise this passive energy gain on cold winter days. The fixed horizontal roof cantilever over the south facade blocks the sun at its high position in summer, and stops the entry of much of the heat from direct radiation. In contrast, the low winter sun reaches well into the interior through the window with the slats open and provides passive heat. To counter the risk of overheating in summer further, articulating slats on the outer building envelope regulate the entry of solar radiation and the associated development of heat.

ENERGY SUPPLY

Latent heat storage in lightweight construction

In parallel with these considerations, the designers also had to think about ambient comfort in the interior space. The right choice of materials and type of interior surfaces can help regulate temperatures by passive means and therefore contribute to the energy supplied to the building.

Storage masses can play a considerable role in achieving a comfortable room climate. In contrast to masonry or concrete, timber is limited in its ability to store heat by its material properties. However, phase change materials (PCM) working passively in the ceilings and opaque walls of the interior fitting out store latent heat to avoid rapid overheating in summer and raise the level of comfort of the indoor climate. A PCM changes from solid to liquid at its transition temperature. At this change of state, the material can absorb a great deal of heat energy and store it temporarily. This means: if the temperature rises in the range of this change of state, the latent heat store can take up a large amount of heat energy and then release it again when the temperature falls. By this transfer of the positive properties from solid to lightweight construction, this physical effect significantly helps to save energy and reduce the weight of the structure. Many different raw materials exhibit phase change and can be used in buildings in various situations and components. The PCM in the Solar Decathlon 2007 is composed of microscopic plastic spheres with a paraffin wax storage medium core incorporated in plasterboard. The heat storage capacity of the 1.5 cm board is the equivalent of a 9 cm thick concrete ceiling or a 12 cm thick brick wall. The charging or melting temperature of PCM can be selected to suit the specific conditions of use. A temperature of 23 °C was chosen for the Solar Decathlon project in order to remain within the very tight temperature range of 22–24 °C specified in the competition rules.

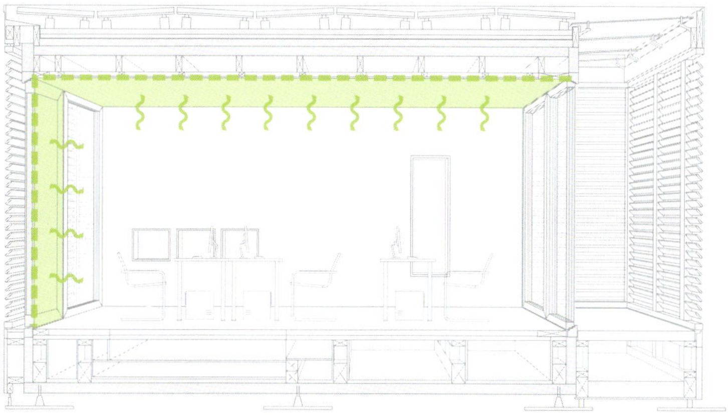

PCM acting as thermal storage mass inside the building

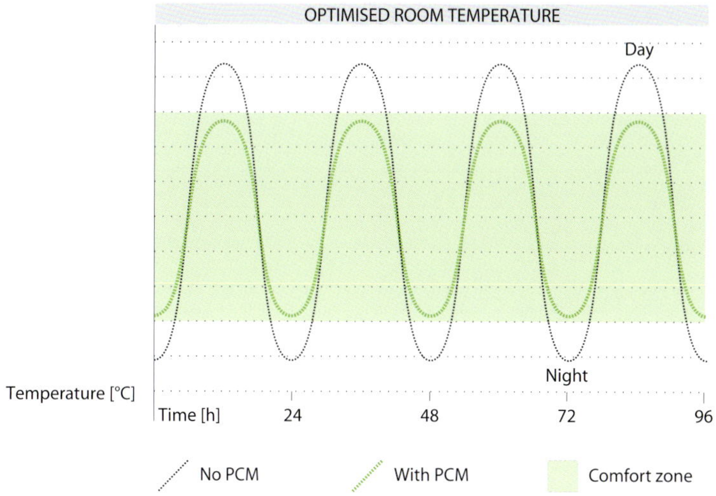

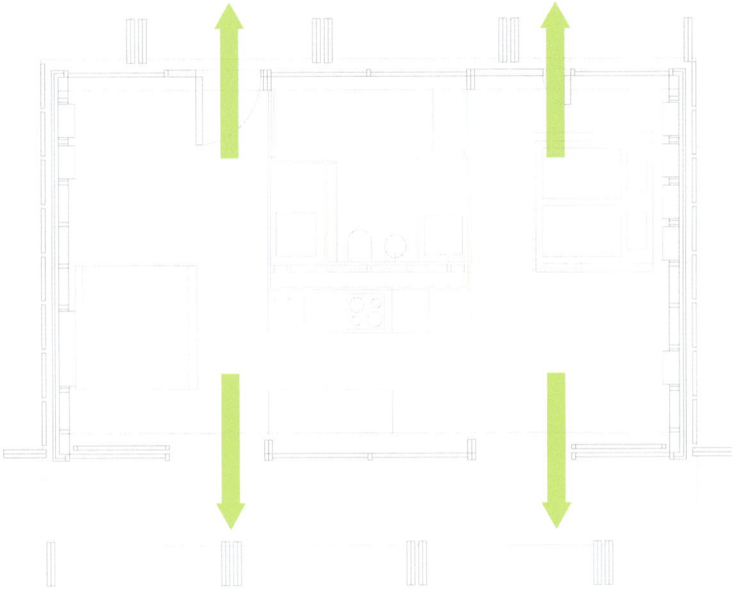

Cross ventilation to "discharge"

Night ventilation for cooling the thermal storage mass

Repeatedly reactivating the latent heat storage masses requires an effective means of charging and discharging them. Direct solar radiation is best for charging, while night cooling by natural cross ventilation is good for discharging.

Melted by charging during the day, the encapsulated PCM turns solid overnight by the effect of the cool surfaces enclosing the room or cool night air. It returns the stored heat to the room. If the heat is not required, it is transported outside by ventilation air flows. On the following day, the PCM can take up heat again and contribute to a comfortable indoor climate. If this cooling does not take place or outdoor temperatures remain above the charging temperature of the PCM for a prolonged period, the daytime cooling effect is limited.

Efficient use of the system and powerful ventilation air flows, especially during prolonged periods of high temperatures, rely on the building having opening windows on opposite facade surfaces.

Supplementary passive night cooling system by PV modules

The building designers looked for ways of reducing temperatures without having to use active cooling systems during long periods of high temperatures. Engineers from different disciplines developed an innovative cooling system comprising a mixture of passive measures backed up with active ones by linking various parts of systems in the building together. The system is based on thermally active building components using capillary tube mats in the ceilings. The double floor of the building contains a water tank. During summer nights, a pump draws water from the tank up to the roof and sprays it on the roof skin or the PV elements. This achieves adiabatic and atmospheric cooling at the same time. The cooled water flows back into the tank. By day the water flows through the capillary tube mats. During the day, the tank water serves as a heat sink for the warmed water flowing out of the capillary tube mats. This has a cooling effect in the interior of the building. In addition, the charged PCM cools and changes state. As a result, it can again take up heat energy from the room and contribute once more to cooling. The effect of the PCM multiplies through this combination of simple technology and contributes by passive means with little consumption of energy (for the water circulation pump) to cooling.

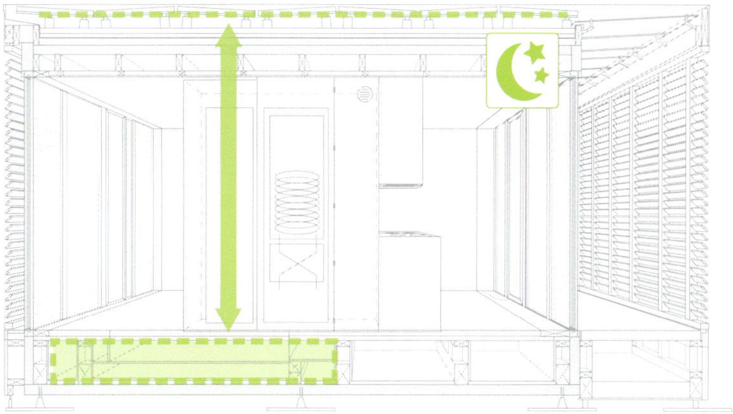

Store cooling during night

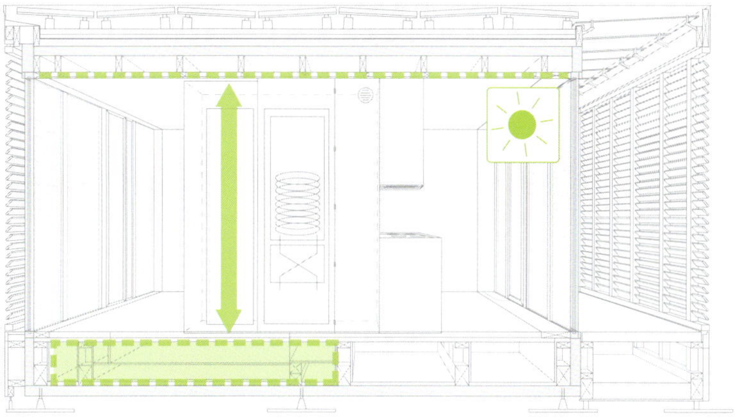

PCM cooling during day

Solar Decathlon 2007 at night

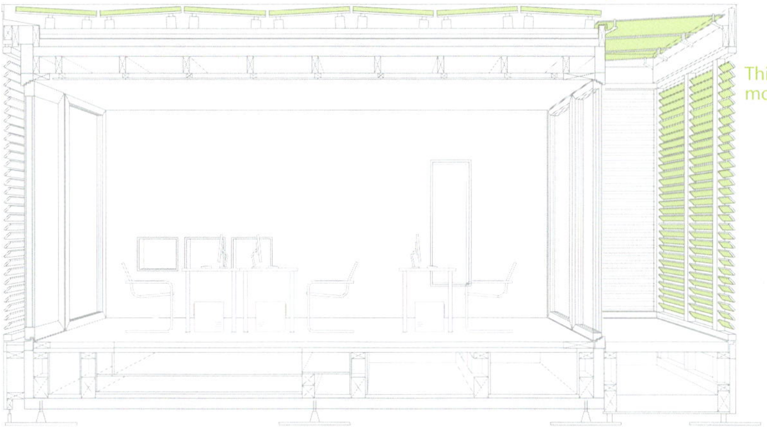

Active systems

The rules of the competition meant that only the sun could be used as an energy source. The building has a very high energy demand due to the electricity consumption of the specimen household and the operation of an electric car; consideration of solar active technologies therefore concentrated on the use of photovoltaics. The heat demand was reduced as far as possible by the building's characteristics described above and therefore plays a subordinate role for energy provision.

ELECTRICITY GENERATION BY PHOTOVOLTAICS

The designers discussed a variety of photovoltaic systems, each with specific advantages for different areas of use.

Standard modules

The electricity-generating roof consists of monocrystalline photovoltaic panels connected to one another by contacts at the rear. This type of electrical contact leads to an enlargement of the effective cell area because the module area is not partially obscured by cable tracks connecting contacts on the front. In all, these modules have a peak capacity of 8.4 kWp.

For the integration of the modules into the architecture of the building envelope, several variants were checked for their aesthetics and achievable yield. In the variant chosen, the modules were integrated without significant loss compared to an optimum mounting at an angle of 3° on the flat roof to allow drainage.

View of the roof with monocrystalline photovoltaic modules

Right: Module substructure

Glass-glass modules

The solar modules in the horizontal cantilever to the south covering the external terrace were glass-glass modules. These are shading elements but they allow a view out and light to enter.

The glass-glass modules consist of two panes with photovoltaic cells with clear space between them in the glazing cavity. The individual cells consist of monocrystalline silica and have a finely perforated structure. This reduces the yield of course, but provides a soft light with an interplay of light and shadow. The opaque parts of the modules prevent the high summer sun from entering the building, and create electricity, without fully blocking off daylight. The peak capacity is 1.0 kWp.

Thin-film modules

A further plane of horizontal slats surrounds the thermal envelope. The distance between the slats and the thermal envelope varies according to the orientation of each side of the building. The slats are installed in hinged and sliding shutters, which run on rails. The slat wall can be open or closed as required. The slats themselves are mounted on their own longitudinal axes in each shutter. Motor driven, they can be rotated by the user, or set automatically to follow the sun, for an optimum compromise between daylight, shade and energy generation.

During the design, this led to the idea of making the slats to the east, west and south activated by photovoltaics. Because the slats follow the path of the sun, they also ensure the optimum yield. Particularly in the winter months, facade activation in the vertical direction by the low sun leads a much higher energy gain. This type of active energy creation must be designed to take into account the need for daylight and views out.

The photovoltaic cells are based on thin-film cell technology. This cell type provides a particularly high yield in diffuse light. Therefore their integration into the facade is seen as particularly advantageous when viewed over the whole year. The peak capacity of all the facade modules is just short of 2.0 kWp. The orientation of the building prevents the cells producing electricity all at the same time. The east and west modules are connected to the south modules, depending on the time of day.

The photovoltaic modules are capable of producing a total of 11 kWp. More than 50 % of the thermal building envelope is photovoltaically active to achieve this, in total 99.98 m^2 of PV panels working at an average efficiency of 11.5 %. For a yield of 900 kWh/a*kWp (at Darmstadt), the theoretical output is about 10,000 kWh/a or almost 170 kWh/m^2a effective area.

View from below through the glass-glass modules in the area of the south veranda

The thin-film cell modules in the sliding shading elements on the south veranda

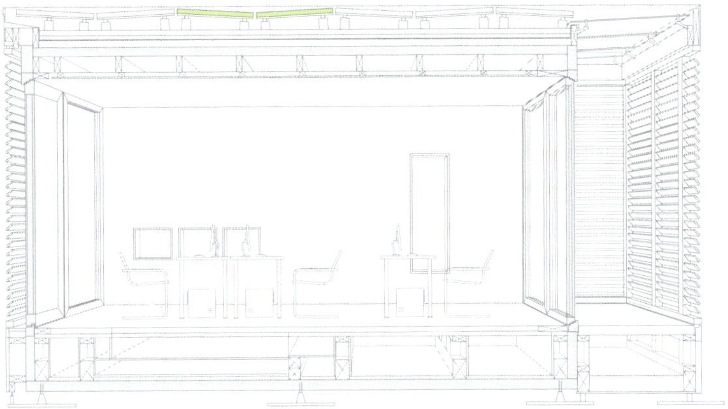

Solar thermal technology

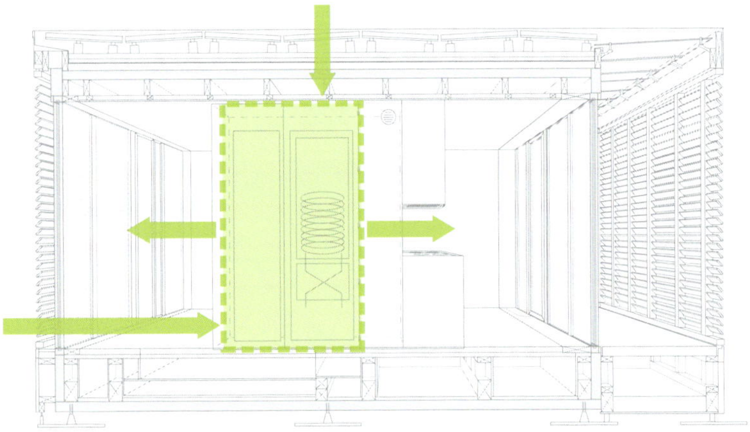

Compact unit with crossflow heat exchanger, heat pump and thermal store

Heat extraction from regenerative energy

Hot water preparation with solar-thermal collectors

Solar-thermal collectors provide potable hot water. They are flat collectors, which are made to measure and therefore can be better integrated into the modular grid of the photovoltaics to make optimum use of the roof and avoid wasting space.

In order to keep pipework lengths as short as possible, the collectors are positioned directly above the 180 l hot water tank of the compact unit. An additional option is to pass the water from the hot water tank not only through the collectors but also through the underfloor heating in the bathroom.

The two collectors have a total of 2.3 m^2. The flat collectors have an efficiency of 65 %–70 %, and the calculated yield is about 1,700 kWh/a for a Darmstadt location. The achievable proportion of the theoretical yield depends greatly on the type and amount of use of the building. In winter, the solar-thermal collectors make hardly any contribution to hot water preparation.

Ventilation, cooling and heating using a reversible heat pump

Because the design concentrates on passive components, the space conditioning requirements placed on the building technical services equipment are low. A compact heat pump unit combining a heat recovery system, heat pump and storage tank was planned as the core of the building technical services equipment. The unit, with a footprint of only 60 × 60 mm and approximate height of 2.30 m, is easily integrated into the building. The efficiency of the air-to-air pump was crucial to the final choice. The design team carried out tests to determine the efficiencies of the various heat pumps and compare them with the published figures. The most reliable device of this type with the highest efficiency was chosen.

The mechanical ventilation system feeds outdoor air through a passive counter-flow heat exchanger and uses the extract air to precondition it. The heat pump extracts further heat from the extract air and uses it to heat the potable hot water tank and further heat the supply air. The ventilation system can be used with the refrigeration circuit in reverse to cool fresh air in summer. The waste heat is used to heat the service water. This precools the supply air, which does away with the need for an air-conditioning plant.

The combi-unit controls can be programmed for weekly programs, temperature reduction at night, free or active cooling and many other user-dependent settings.

Energy-efficient domestic appliances and lighting

With the house's heating energy demand much reduced, the other energy demands, which may have appeared negligible at one time, now take on increasing importance. This concerns mainly the electricity demand for lighting and domestic appliances. The increasing demand for ambient comfort plays a role through technology. Many small domestic appliances increase the electricity demand.

This aspect was also taken into account. Therefore all the domestic appliances had to be energy efficiency class A+ or better. Appliances that use water, such as dishwashers and washing machines, were tested for their economy and efficiency. The refrigerator was given an extra layer of insulation to increase its efficiency further and improve consumption figures. Temperature measurements were taken, including inside the refrigerator, to confirm the effectiveness of these measures. The basic lighting of the building is provided by energy-saving LEDs.

Existing building

Appearance after refurbishment

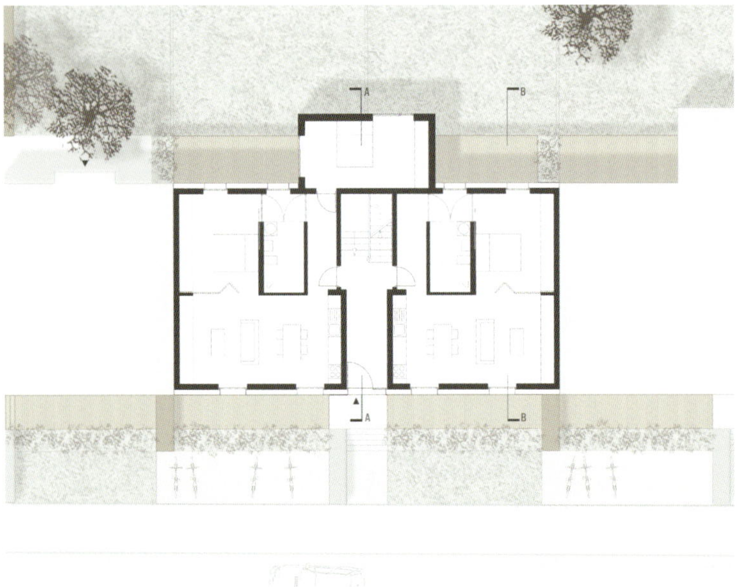

Ground floor plan

Refurbishment

Compared to a new build, a refurbishment allows much less scope for action. However, it usually makes sense from an efficiency and sustainability point of view to choose a refurbishment of the old building fabric over a replacement new building. Compared to a new build, a refurbishment focuses on other issues and poses other problems. These differences are discussed in the following example of the design for an Efficiency House Plus in an existing building. In 2012, the German Ministry of Transport, Building and Urban Development (BMVBS) and the New Ulm Housing Association (NUWOG) in the city of Ulm issued a competition for the refurbishment of a 1930s multistorey housing block to Efficiency House Plus standard. The annual energy balance, calculated on a monthly basis, must be less than zero for primary and final energy (see the chapter Energy Balance).

The solution discussed here is the entry by a team from the Faculty of Design and Energy-efficient Building at the TU Darmstadt, o5 architekten bda and ina Planungsgesellschaft mbh. The design won one of two awards in the competition. The design earned one of two awards in the competition. Shown here is purely the competition entry, to which changes came about during construction planning. The refurbishment was successfully completed in 2015.

Development of a conceptual idea

The objective of the design is the sensitive refurbishment of the existing building fabric. The basic layout of the old building should be retained as far as possible to keep the use of new materials for the new building works as low as possible. In other words: as much as possible of the embodied energy already built into the building should be retained. Key to the design process is the consideration of the whole life cycle of the building from construction and use to maintenance and demolition.

Architecturally, the objective was to create a calm, graceful appearance from the outside primarily using sustainable materials such as wood. Despite the highly limited space available in the interior, the design should offer a great deal of flexibility.

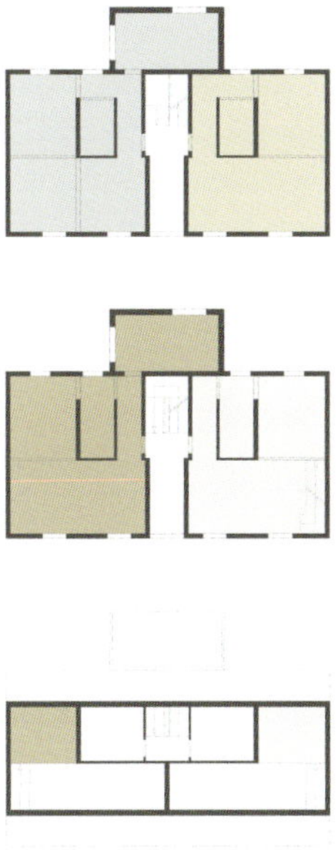

Roof storey

Upper storey

Ground floor

2-ROOM APARTMENT
3-ROOM APARTMENT
4-ROOM APARTMENT
5-ROOM APARTMENT

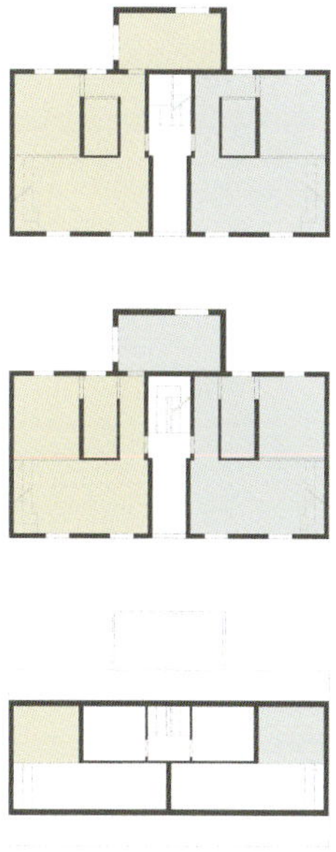

Roof storey

Upper storey

Ground floor

2-ROOM APARTMENT
3-ROOM APARTMENT
4-ROOM APARTMENT
5-ROOM APARTMENT

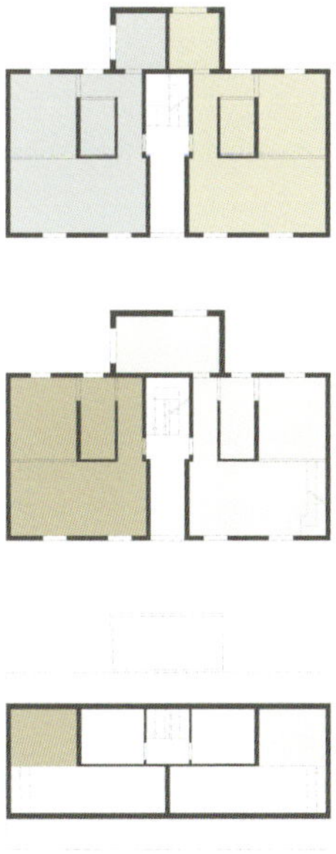

Roof storey

Upper storey

Ground floor

2-ROOM APARTMENT
3-ROOM APARTMENT
4-ROOM APARTMENT
5-ROOM APARTMENT

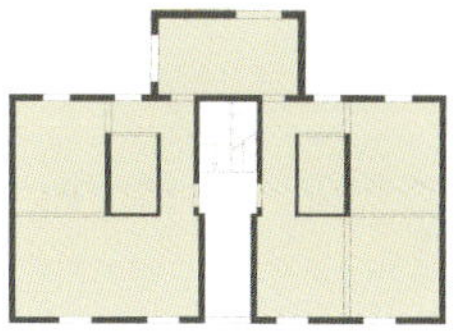

5-ROOM APARTMENT

An extension on the north of the main body allows a great deal of flexibility in designing the floor layout.

The space provided by the extension can supplement the east or the west units. It also permits the combination of both units on one floor into one large apartment or house.

The diagram shows schematically the different configurations from 1- to 5-room apartments made possible by the extension.

Passive measures

BUILDING DESIGN

The existing building layout limits the measures available to develop the building. The design is reduced to creating an extension that positively affects the energy performance of the building while increasing the flexibility of the interior. Balconies in the south offer an appropriate extension of the indoor space in summer and act as fixed solar shading from the summer sun at its highest. In the north, an extension would provide the flexibility to add space to one of the two adjoining units, enlarging one apartment per floor by one room. Alternatively the same room could link two apartments.

The plan layout is zoned around an inner core in a similar way to the new build previously discussed. The core is where all the building services and sanitary plant are installed.

BUILDING ENVELOPE DESIGN

With the core objective of retaining as much of the existing building fabric being key, the designers had to look for efficient ways of improving the thermal properties of the existing masonry walls. A very well insulated wall complying with Passivhaus requirements would have been achievable using conventional insulation materials, although here it would have created an inhomogeneous construction, which would have been problematic to dispose of at the end of the building's life cycle. A life cycle assessment completed during the preliminary design stage compared several variants and provided the basis for the final decision. Taking into account these life cycle considerations, the designers opted for a mineral insulation board installed in front of the existing wall. A thick layer of mineral external plaster is then applied to create a completely homogeneous wall. This wall does not reach the Passivhaus standards, but is only slightly over (U-value = 0.20 W/m^2K). The single disposal path, the overall solid effect in sympathy with the earlier form of construction, and the continuation of appearance have priority. The heating energy demand is 24 kWh/m^2a.

To keep the impact on the environment small and to generate further positive effects in the LCA, all extensions to the original building are designed in timber construction with very good insulation. The existing window openings now extend down to floor level to increase natural light and to allow residents to go out onto the terraces and balconies. The old lintel remains at the top of the opening to minimise interference with the existing structure. The entry of solar radiation can be regulated and shade provided by an external folding-sliding shutter system in the window reveal.

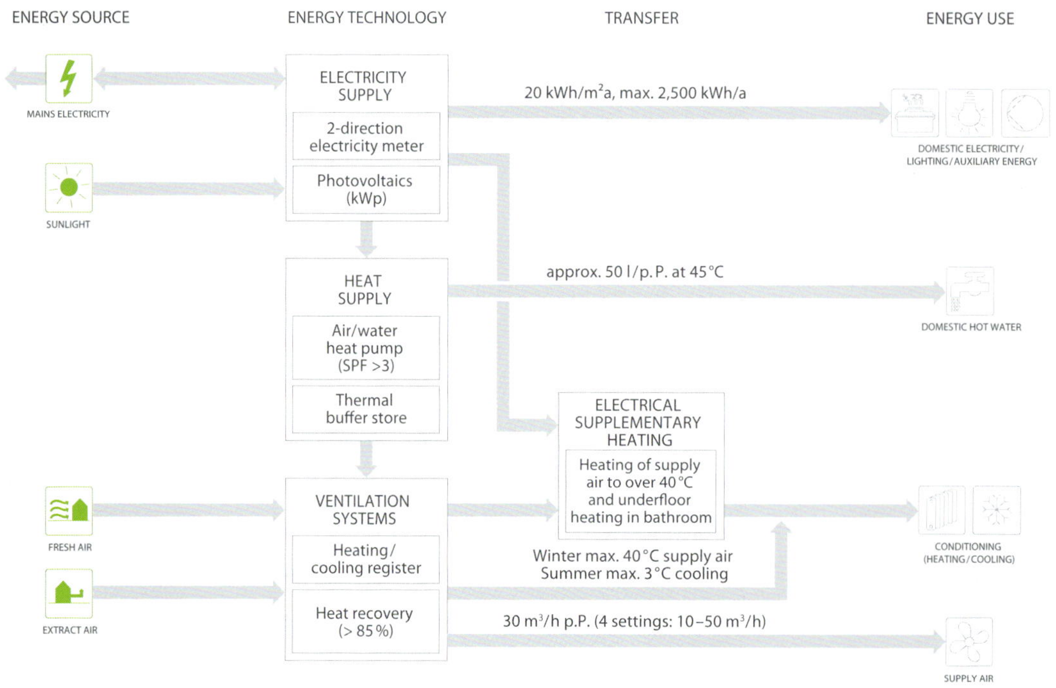

Energy supply concept of the building

Active measures

ENERGY SUPPLY

A services room in the roof floor contains the building technical services systems, which supply the houses with energy. The position of this room allows direct distribution of the pipework through a central services wall in the inner core. All this can be maintained from the roof-level storey without having to enter the rented housing units.

A wet heating system distributes heat to the housing units through a heating register in the ventilation system and panel radiators. The heated supply air enters the rooms through outlets in the services wall. The switchable room in the northern facing extension is heated by underfloor heating. With its own decentralised reversing regenerator ventilation system, the extension is decoupled from the energy supply systems for the rest of the house. This provides the flexibility described above because the space can be assigned to either of the two adjacent housing units.

An air-water heat pump supplements the building technical services system and charges a buffer tank with integrated hot water tank. This covers the hot water demand. It also increases the own-use proportion of the electricity generated on the site by the photovoltaic array mounted on the roof. The hot water tank can cover different day and night cycles and offers higher energy efficiency and economy than an electrically heated tank (storage tank). The tank can be used if necessary for heating through a heat exchanger coupled with the ventilation system described above.

The process also has a secondary effect that allows the system to provide some additional cooling: when the heat pump heats the tank, it removes heat from the air. This cooled air can be used for cooling by the ventilation system. The supply air temperature can be reduced by as much as 3°C without additional expenditure of energy.

During the design process, the energy balance for the whole system showed that the photovoltaics would perform considerably better with a module type designed to provide good yields from diffuse and zenith light, even at a latitude of 32° north. The designers therefore opted for CIGS (copper indium gallium selenide) thin-film modules. Although these are slightly less efficient than crystalline cells, they provide a higher yield in the energy balance because of their better performance with diffused radiation. A uniform appearance of the roof surface is an additional advantage. Thin-film modules perform better in the LCA because they are considerably more material efficient and cheaper to manufacture than crystalline modules.

The described examples, one a new build and one a refurbishment, show the advantages of following the design stages described above for every building project. A detailed analysis of the internal and external boundary conditions led to both concepts having an optimum building energy concept that works to best effect in any context. The slightly higher design costs are more than repaid during the operating phase of the building.

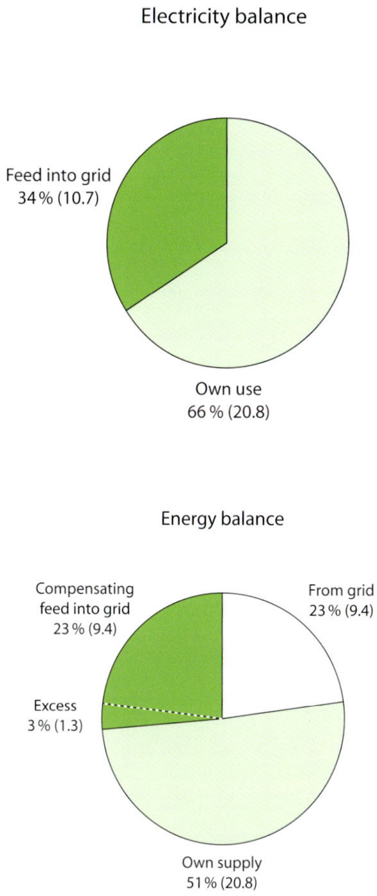

Summary of the energy balance for electricity and total energy

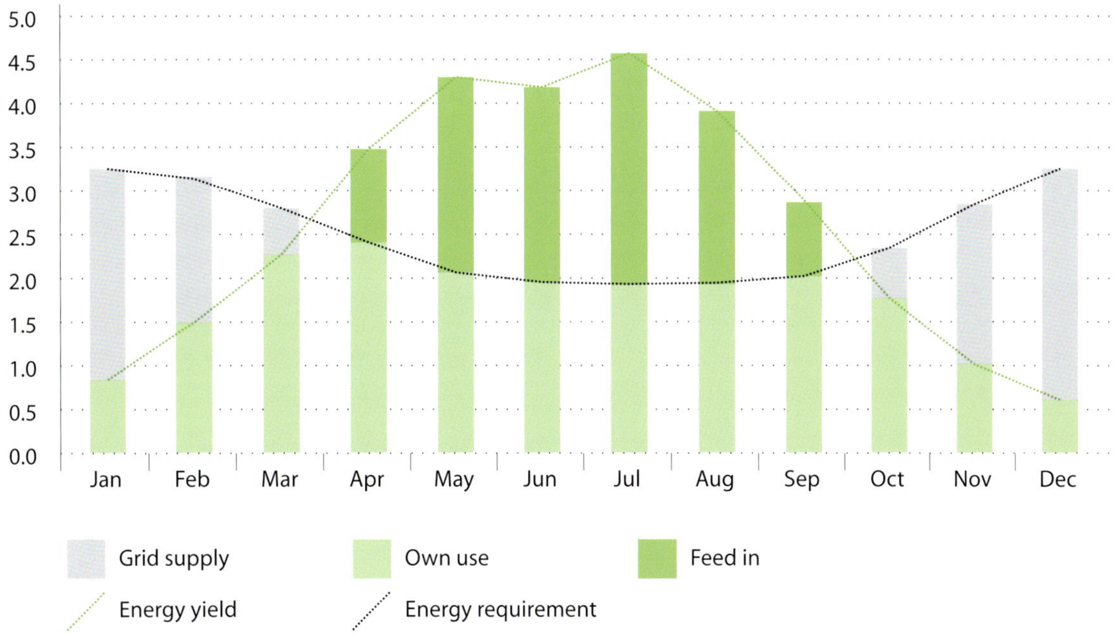

Monthly balance of demand and yield for the refurbishment concept (kWh/m²a)

Toolkit

The building envelope and the building engineering services systems provide architects and engineers with a wide range of tools for designing and constructing Aktivhaus buildings. The building envelope offers people a third skin to protect them from the effects of the external environment. This boundary between inside and outside can create pleasant, healthy, comfortable and safe surroundings, whatever the season or time of day, and provide some of the energy for operating the building. The building engineering services systems feed this energy by simple means into the energy services for the building, when it cannot be used directly.

Building types optimised for the conditions prevailing in the various climate zones have evolved over the history of construction. The developments of the 20th century towards an internationalisation of architecture have led to a general uniformity across all climate zones. Building envelopes inappropriate for their local climate have been duplicated in all the climate zones, meaning technical systems have to create the desired interior conditions. The associated cost and technical complexity are considerable; this approach has been possible only because cheap energy was freely available.

Building envelope

The Aktivhaus adopts an opposite approach that establishes a stronger link between the building and its context, and in particular the local climatic conditions. In addition, the building envelope can and should use energy from regenerative sources. Local factors, such as the climate and the locally usable regenerative energy sources, cannot therefore be ignored when developing an energy-efficient Aktivhaus. Only by considering them can designers develop an envelope that is suitable for the specific building location, ensures the building operates energy-efficiently, and provides a high level of ambient comfort for the users.

The envelope takes over the function of excluding or at least filtering outside influences that may occur to different extents, depending on the cultural and climatic surroundings. Its main function is to prevent the entry of unwanted influences such as the weather (wind, rain, snow, solar radiation, excessive heat or cold, noise, fire, air pollutants) into the building. To achieve this, the building envelope regulates the flows of heat energy and, in many climate zones, adjusts itself to changes in external ambient conditions. With rising comfort requirements, the envelope takes on complex climate-regulating functions and finds itself increasingly at the focus of building planning and design.

As the interface, the building envelope also has to fulfil conflicting requirements. For example, transparent surfaces have to master complex tasks. On the one hand, sunlight should penetrate as intensively as possible into the interior of the building so that daylight can be used and views out can improve the building users' feeling of contact with the surroundings. This gives the opportunity of using insolation, directly or indirectly, to support the heating of the building in winter in Central European latitudes. In summer, on the other hand, it should prevent overheating, for example through external shading elements.

The many other functions of the building envelope, such as load transfer, providing a surface for installations, possibly as an energy-generating surface (photovoltaics, solar thermal technology), but also the important building physics requirements and environmental conditions, require careful, thoughtful design for a building to succeed in all respects.

The technical and the architectural requirements must be in harmony. The building envelope or facade defines the appearance of the building. The characteristics of these surfaces in particular largely determine the architecture and the building's interaction with the environment.

Consequently, the tasks of the building envelope are manifold. It must protect, react, envelop, present, and create energy. It exercises great influence over the efficiency of the building, its economic performance, durability and character.

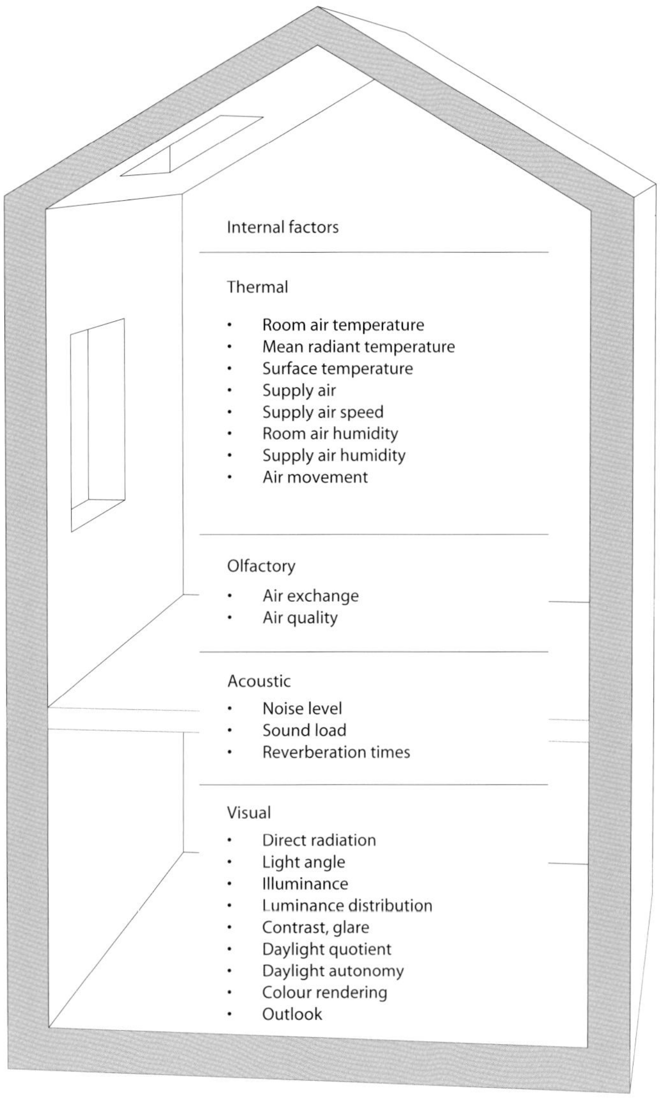

External factors

Light

- Intensity of solar radiation
- Angle of solar radiation
- Illuminance
- Horizon
- Surrounding buildings
- Vegetation

Air

- Air temperature
- Atmospheric humidity
- Air speed
- Wind direction
- Air quality
- Sound
- Precipitation

Ground

- Soil temperature
- Soil moisture content
- Soil storage capacity

Internal factors

Thermal

- Room air temperature
- Mean radiant temperature
- Surface temperature
- Supply air
- Supply air speed
- Room air humidity
- Supply air humidity
- Air movement

Olfactory

- Air exchange
- Air quality

Acoustic

- Noise level
- Sound load
- Reverberation times

Visual

- Direct radiation
- Light angle
- Illuminance
- Luminance distribution
- Contrast, glare
- Daylight quotient
- Daylight autonomy
- Colour rendering
- Outlook

Building envelope

Protection

- Moisture protection
- Wind protection
- Winter heat protection
- Summer heat protection
- Solar shading
- Glare protection
- Noise protection
- Visual protection
- Intruder protection

Services

- Lighting
- Ventilation
- Outlook
- View
- Passive thermal gain
- Active thermal gain
- Solar electricity gain

Properties

- Transparency
- Translucency
- Opacity
- Thermal conductivity
- Total solar energy transmittance
- Weight
- Sound reduction index
- Storage capacity
- Water vapour diffusion resistance

Functions of the building envelope

Receiving and retaining heat

In temperate and cold climate zones, the task of the building envelope is to ensure a pleasant climate inside the building. To achieve this in winter, the building should incorporate suitable measures to retain as much as possible of the heat within. In summer, on the other hand, it should also be able to prevent overheating. A heat balance analysis should be performed as early as possible in the design process based on the preliminary design and the environmental conditions (climate data, orientation, microclimate etc.).

In calculating the losses, the distinction should be made between transmission and ventilation losses. The gains are divided into internal loads (due to people, electrical appliances) and insolation. As much as possible of the necessary difference should covered from local regenerative sources.

The parameter H_T' is a measure of the passive thermal performance of the building envelope in W/m^2K. It describes the average heat transmission coefficient of the envelope as a heat transmitting surrounding surface.

The following principles should be observed in order to design an efficient envelope for temperate to cold zones:

- Optimisation of the geometry of the envelope (A/V ratio)
- Thermal zoning of the usable space (layout design)
- Floor area optimisation (possible reduction of the gross floor area or optimisation of the usable floor area)
- Passive use of insolation
- Optimisation of the thermal insulation of the opaque components
- Optimisation of the thermal insulation of the translucent components
- Reduction of ventilation losses (e.g. through highly efficient heat recovery systems)
- Active use of phase insolation (photovoltaics, solar thermal technology)

Insulation of a building envelope joinery workshop for design s., Deppisch Architekten, Freising (DE)

Thickness of insulation s to achieve a heat transmission coefficient of 0.15 W/m²K

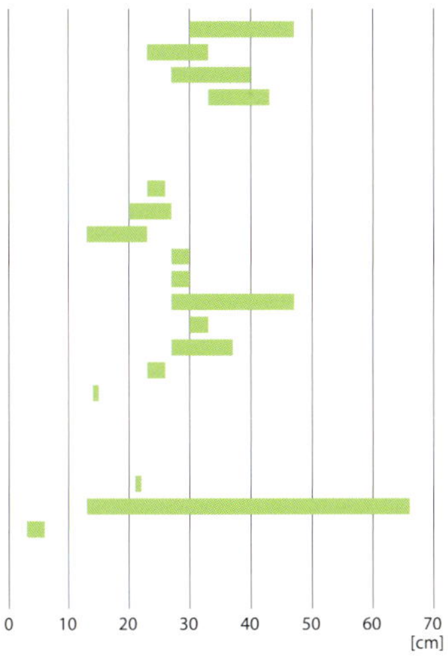

	Bulk density [kg/m³]	Weight[4] [kg/m²]	Thermal conductivity [W/mK]	Flammability class[1] [-]	Global warming potential (GWP100) [kg CO_2-equiv./kg]	Primary energy non-renewable [MJ/kg]	Product form [-]
Inorganic							
Calcium silicate	115–290	60.75	0.045–0.070	A1–A2 / up to A1	1.83	24.37	Board
Mineral wool	12–250	30.57	0.035–0.050	A1–B1 / up to A1	1.33	19.76	Board, nonwoven, wool
Foamed glass	100–150	33.33	0.040–0.060	A1 / A1	2.43	41.00	Board, loose fill
Expanded perlite (EPB)	60–300	60.00	0.050–0.065	A1–B2 / up to A1	0.51	7.07	Board, loose fill
Organic							
Rigid expanded polystyrene foam (EPS)	15–30	5.25	0.035–0.040	B1 / up to B	5.77	101.00	Board
Extruded polystyrene foam (XPS)	25–45	7.00	0.030–0.040	B1 / up to B	25.97	103.75	Board
Rigid polyurethane foam (PUR)	> 30	4.00	0.020–0.035	B1-2 / up to B	4.93	105.41	Board, in-situ foam
Cotton	20–60	10.67	0.040–0.045	B1 / up to B	0.02	31.60	Mat, felt, wool, blown-in material
Hemp fibre	20–70	12.00	0.040–0.045	B2 / up to D	0.08	18.57	Board
Wood fibre insulation boards	45–450	66.67	0.040–0.070	B2 / up to D	-1.06	35.57	Board
Coco fibre	50–140	28.50	0.045–0.050	B1–2 / up to B	–[3]	42.00	Mat, felt, wool
Expanded insulation cork board (ICB)	80–500	77.33	0.040–0.055	B1–2 / up to B	-1.08	12.70	Loose fill, board
Cellulose fibre	30–100	15.17	0.035–0.040	B1–2 / up to B	0.39	9.94	Blown-in material, board
Fumed silica	300	42.00	0.021	A1	–[3]	–[3]	Board, mat, panel
Innovative insulation materials							
IR absorber – modified EPS	15–30	4.80	0.032	B1 / up to B	–[3]	–[3]	Board
Transparent thermal insulation	–[2]	–[2]	0.02–0.1	–[2]	–[3]	–[3]	Panel
Vacuum insulation panel (VIP)	150–300	6.00	0.004–0.008	B2	–[3]	–[3]	Panel

[1] The stated flammability classes are for guidance only. They must be considered alongside the actual product data.
[2] Very dependent on the product.
[3] No information.
[4] The information relates to the lowest measured value of thermal conductivity. The bulk densities are mean values.

Insulation properties compared

Insulation

Insulation is essential in climate zones where temperatures fluctuate substantially over the year or the day. The insulation extends completely around the usable volume of the building. Penetrations of the insulation are kept to a minimum. These two measures ensure that the temperature of the interior can be maintained by keeping losses very low and protecting the interior against widely fluctuating climatic conditions. The internal wall surface temperatures fluctuate significantly less and therefore so do the room air temperatures. This provides the best conditions for higher levels of thermal comfort, disregarding internal heat loads.

The overall transmission loss (H_T') of a building envelope is the sum of all the U-values of all the components of the envelope surface of a heated building (wall, window, roof, foundation/floor) adjusted to represent the proportion of each. The result is the average H_T' in W/m²K. A low value indicates a well-insulated envelope.

External walls

The wall parts of the envelope usually make the most contact with the outdoor air. The opaque wall surfaces are very important in avoiding heat losses. The difference in area between wall surface and the roof and floor surfaces increases with the number of storeys.

The heat insulating performance of the wall surface is mainly determined by the choice of thermal insulation and the construction of the wall. The thermal insulating performance of an envelope is expressed by the U-value in W/m²K, otherwise known as thermal transmittance. It describes the flow of heat under standard test conditions between the two surfaces (internal and external) in watts per square metre and kelvin. It can be calculated as the U-value at a particular point on the envelope or as an average value, for example, of an area of wall or roof. The opaque wall surface of a Passivhaus normally has a U-value of < 0.15 W/m²K.

A 360 mm thick, solid single-skin external wall constructed from high-porosity blocks with a thermal conductivity of 0.08 W/mK can achieve a U-value of 0.2 W/m²K. The same thickness of wall constructed from a combination of sand-lime bricks and insulating porous concrete blocks can achieve a U-value of 0.12 W/m²K. A multiskin wall construction separates the loadbearing function from the insulating function into different (external, core and internal insulation) layers or, in the case of framing, combines them in the same layer. This form of construction helps to reduce member thickness in buildings with large thicknesses of insulation.

EXTERNAL INSULATION

A facade with external insulation is the current way of constructing an externally insulated wall. From the point of view of building physics, this form of construction is preferable to internal insulation. The thermal mass of the loadbearing components on the inside is able to influence indoor climate positively because, depending on the choice of material, it can regulate the moisture content and even out temperature differences throughout the day.

The choice of usable insulation materials is huge. It extends from natural products, such as cork and cellulose, to mineral wool and oil-based extruded foam or even vacuum insulation. The choice of the right insulation material depends on many variables, such as the type of construction of the facade, the statutory requirements (e.g. fire protection), individual preferences for synthetic or natural materials, and the envisaged cost frame. Durability and environmental compatibility play a special role in the choice too.

The most popular external insulation system is the external thermal composite insulation system (ETICS), which can be applied directly to an outside wall. It consists of several layers permanently bonded to one another and to the main construction. Starting with the inside layer, the construction is as follows (in principle only, it may have additional layers, depending on the system manufacturer): Adhesive on the external wall (masonry, concrete), mineral or organic thermal insulation, levelling coat, fabric tape, external finishing plaster. The complete facade construction is a sealed system. This form of insulation offers a cost-effective and highly energy-efficient solution. Retrofitting insulation to a building envelope is equally simple. Critics question the environmental compatibility of ETICS. The plaster often incorporates fungicide to prevent mould from forming on low-temperature surfaces. This may be washed out during the use phase and enter the groundwater. In addition, the organic insulation is combustible and could create a hidden smouldering fire, which would be difficult to control behind the plaster coat. This material is difficult to recycle because the many layers of different materials are glued together and not readily separated.

Alternatively, the envelope can have a gap between hydrophilic external and insulation layers. In the case of a curtain wall facade at a distance from the insulation, rear ventilation carries away the moisture. A large number of materials are available for cladding the envelope, including natural stone, cement-bound boards, wood and wood-based material or metals. Fixings have to pass through the insulation to anchor it to the loadbearing wall. These act as cold bridges, which have a negative effect on the thermal performance of the envelope, and should therefore be minimised. They are subtracted in the calculation of the thermal transmittance.

CORE INSULATION

A twin-skinned facade, sealed on both sides, can accommodate core insulation in the cavity between the external and internal skins, whether they are loadbearing or not. The insulation completely fills the cavity, which may vary in width. The insulation material itself may be extruded rigid foam, mineral fill, or mineral or cellulose fibres. The fixings between the external skin and the structural wall pass through the thermal insulation layer. These cold bridges have a detrimental effect on the envelope's insulating performance. In some forms of construction, such as twin-skinned solid walls with a cavity, the cavity may not be able to accommodate enough insulation for the wall to reach the standard of a highly insulated facade.

INTERNAL INSULATION

In the energy upgrade of an existing building, in particular one which is listed or has a culturally significant facade, no external insulation can be applied to the wall and therefore the required layer of insulation is usually placed on the inside. A number of different insulation materials, mainly some form of board, are suitable for the internal insulation. The selected thickness of insulation is usually less than 100 mm in order to avoid building physics problems, such as condensation within the insulation layer, which might otherwise occur.

Placing the insulation on the inside of the structural wall separates it from the interior. At the same time, the wall's thermal mass, which regulates temperature, and its ability to absorb and release moisture are likewise made unavailable to the interior. The result can often be the type of unpleasant climate found in overcrowded barrack rooms. Good ventilation is essential to remove moisture. The installation of a controllable ventilation system is recommended.

Because of these building physics peculiarities, internal insulation is recommended only in exceptional circumstances. The use of calcium silicate insulation can contribute to finding a good solution. The material has moisture-regulating properties. It can cope with a certain amount of condensation within the insulation layer and contributes to the regulation of the humidity within the room.

Framing combines the insulation and the loadbearing elements in a single layer. The loadbearing elements are usually studs, columns or frames at regular intervals. They are mainly metal or wood. In residential buildings, the framing is generally timber frame construction. Integrating the insulation and the studs within the frames optimises the wall thickness. The presence of the studs must be taken into account in the calculation of the U-value of a wall surface of this type. In highly insulated building envelopes, the studs are insulated additionally on the outside to minimise cold bridging.

Insulation properties compared

Energy-Plus House Luchliweg,
dadarchitekten, Bern (CH)

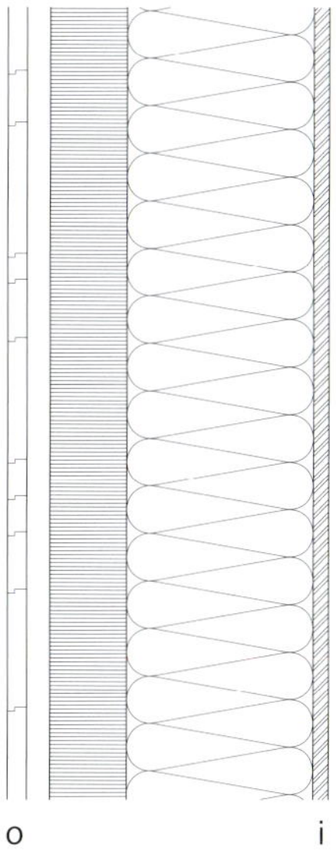

Timber stud wall
Wall construction from
outside to inside

Horizontal boards spruce 25 mm,
Tongue and groove, rough sawn, painted
Battens 30/60 mm
Wood fibre board 100 mm
Timber studs 240 mm/sheep's wool 240 mm
Three-ply board 20 mm

U-value: 0.14 W/m²K

Solar Academy,
HHS Planer + Architekten AG, Kassel (DE)

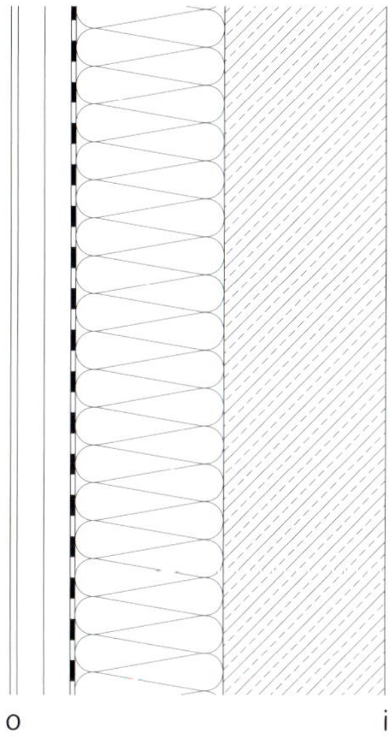

Reinforced concrete wall with external insulation

Wall construction from
outside to inside

Boards
Rear ventilation cavity 30 mm
Studs 30 mm
Insulation 50/220 mm
Reinforced concrete wall 240 mm

U-value: 0.17 W/m²K

Wall with core insulation

Wall construction from
outside to inside

Reinforced concrete wall 100 mm
Core insulation 200 mm
Reinforced concrete wall 200 mm

U-value: 0.19 W/m²K

Zero-energy house Driebergen,
Zee Architekten, Utrecht (NL)

Internal insulation

Wall construction from
outside to inside

Masonry wall 360 mm
Insulation 50/200 mm
Plaster 16 mm

U-value: 0.16 W/m²K

Solid reinforced concrete wall

Reinforced concrete wall 450 mm

U-value: 2.53 W/m²K

External doors

External doors are designed as insulated and tightly sealed components. The degree of airtightness must ensure that energy losses are as low as possible. These doors typically have a series of seals. External low-maintenance variants have automatic magnetic seals. The door profile may incorporate a central seal that creates a stationary cushion of air between the main seals to increase the thermal insulation performance further. The inner seals are normally designed as rebate seals that nestle up against the fixed part of the door profile when the door is closed, to seal the joint against draughts.

The door consists of many layers, which together achieve a U-value close to that of a wall, but with a much reduced thickness. Vacuum insulation panels (VIP) are integrated as insulation between the stiffening and cosmetic outer layers. These provide a U-value of, for example, 0.15 W/m^{2}K, similar to that of a typical wall, but are not much thicker than a triple glazing unit. This form of construction offers huge potential for reducing today's insulation thicknesses, for example, in walls. Vacuum insulation panels are prepared in the factory and not on site, therefore they require detailed design at an early stage (including all penetrations and through holes) and an installation drawing. As a result, their price structure is nothing like that of other insulation, and their use is confined to special applications, such as doors, external roller blinds and internal insulation, and not for large areas. In all these situations, the thickness of the VIP, which is a fifth or even a tenth of the thickness of a conventional insulation solution, plays a decisive role.

Roof

The roof surface forms a large proportion of the building's contact area with the outside air, especially in low-height buildings (such as a detached house). In this situation, an efficiently insulated roof can make a large contribution to minimising the heat losses through the building envelope. Roof structures are usually one of three main types: lightweight structures, rafter and purlin roofs, or solid roofs.

Flat roofs are predominantly designed as warm roofs in solid construction. Lightweight steel roof constructions with trapezoidal profile sheets or insulated sandwich panels fixed directly on to the steel members are used only in industrial buildings and large shed structures. As a consequence of their anticipated use, the thermal insulation of flat roofs must be able to resist compressive loads, and is usually placed directly on top of the roof construction along with suitable waterproofing layers. The thermal insulation is usually rigid foam or wood-based insulation boards. Foamed glass, for example, can be used for its high strength in special cases where roofs carry high imposed loads (for example from vehicles or from building technical services equipment). The installed depth of the above insulation types is more than 20 cm in order achieve a U-value of $<$0.15 W/m^{2}K.

Sloping roofs are usually rafter and purlin constructions. The only common exceptions to this are in-situ concrete and large precast member constructions. They are insulated in accordance with the flat roof principle but with an additional shear restraint to stop the roof covering from slipping. Rafter and purlin roofs can be insulated like a timber stud wall. A combination of insulation between and on top of the rafters, which, by passing over the rafters, minimises cold bridges. Insulation between the rafters would not normally achieve the desired U-value of at least 0.15 W/m^{2}K, because the calculation of the U-value of a roof of this type must take into account the rafter surfaces.

Floor slab

The floor slab of a building and the outside wall surfaces of any basement in direct contact with the ground are insulated far more effectively from outside. Perimeter insulation is used here. This perimeter insulation must be strong in compression, and moisture and rot resistant. Extruded foams or, if very heavy loads are to be applied, foamed glass are both suitable. The thickness of the insulation may be less here, in contrast to the other envelope surfaces in contact with the outdoor air, because the ground temperature fluctuates much less than the air temperature over the year. Solar radiation also has no influence here either. Having an adequate and properly installed insulation layer is wise because this part of a building can be insulated after completion only at great expense.

Energy-Plus House Luchliweg,
dadarchitekten, Bern (CH)

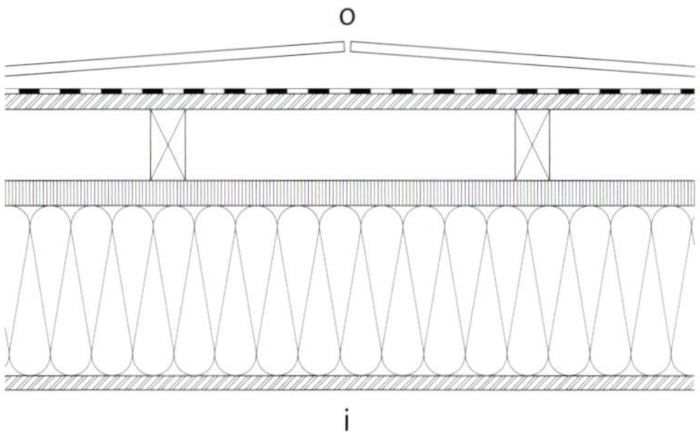

Flat roof timber (lightweight construction)
Roof construction from outside to inside

Photovoltaics
Waterproofing bitumen granulated slate surfacing
OSB board 22 mm
Battens, rear ventilation cavity 100–175 mm
Wood fibre board 35 mm
Timber ribs 240 mm/sheep's wool 240 mm
Three-ply board 20 mm

U-value: 0.17 W/m²K

Solar-Werk 01,
HHS Planer + Architekten AG, Kassel (DE)

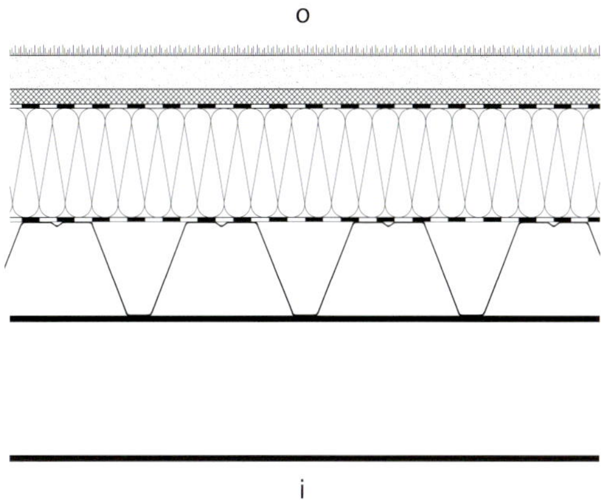

Green roof – structural steel (lightweight construction)
Roof construction from outside to inside

Sedum
Single-layer substrate 55 mm
Drainage layer 25 mm
Protective and storage nonwoven
Waterproof membrane 3 mm
Mineral wool insulation 180 mm
Metafol 250 mm
Trapezoidal profile sheet 150/280 × 1.5
Discharge pipe HEB240

U-value: 0.23 W/m²K

Solar Academy,
HHS Planer + Architekten AG, Kassel (DE)

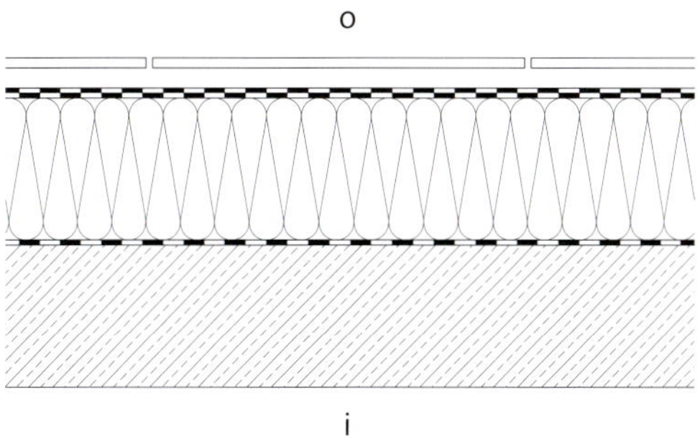

Flat roof – reinforced concrete (solid construction)
Roof construction from outside to inside

Photovoltaics
Waterproofing membrane, two-layered
Insulation 200 mm
Vapour barrier
Reinforced concrete roof slab 200 mm

U-value: 0.19 W/m²K

Various types of roof construction compared

Windows and glazing

Windows and other glazed areas in the building envelope present a special challenge to designers, manufacturers and builders. By their direction, numbers, areas and positioning, windows influence not only the energy-related performance of the building but also, and crucially, the wellbeing of the users. Analysis of the energy performance of a double-glazed window shows that it has a U-value 4–6 times worse than the wall in which it sits. On a dull winter's day, the window area could be responsible for about 50% of the energy lost through the building envelope. In contrast, windows offer high passive heat gains from solar radiation, depending on their direction. In Central Europe, the U-value of glazing is very important because of the high number of dull days.

In Germany, buildings that are to fulfil the requirements of EnEV 2014 should have triple glazing as standard. The glazing system should achieve the following properties: $U_g \leq 0.8\,W/m^2K$, $g > 50\%$. The glazing units need to have a thermal break at their edge seals, well-insulated frames, and careful design and installation. The U-value of the glass part is expressed as the U_g-value. The g-value (solar energy transmittance) describes the ability of transparent components to transfer energy. It is the sum of the direct solar transmission and the heat given off to the interior by transmission and convection. A g-value of 0.5 means that 50% of the heat from the solar radiation falling upon the transparent component reaches the room behind the glazing.

An analysis of solar gain in Central Europe shows that south-facing windows and those deviating to the east and west by up to about 30° should achieve net solar gains. Windows facing east, west or north experience a net loss over the year. This should be considered when positioning the usable areas within the building. In terms of overall energy consumption, it is advantageous for the rooms in which people spend much of their time to face south, while functional rooms should face north, assuming that the window areas and internal heat loads are limited and therefore the risk of overheating is low, as is usual in residential buildings.

Currently available glazing systems for use in energy-efficient buildings are triple glazed with a U_g-value of 0.5 to 0.7 W/m^2K and an overall energy transmission value (g-value) of 0.4 to 0.6. In most glazing systems, the U_g-value and the g-value are interdependent. Improving the U_g-value usually results in a worse g-value. This is caused by the cumulative effects of the various layers and the gas used to fill the glazing cavity.

Another parameter is the solar heat gain coefficient (SHGC), which is primarily used in the United States of America. Like the g-value, this can relate to the energy transmission of a glazed unit or the whole door or window system. In addition to the quality of the glazing, any assessment of a system should also take into account the frame, window bars and the shading devices. Lower overall transmission values can be achieved by adjusting the proportion of frame area, the quality of the window bars and shading devices. Like the g-value, the SHGC-value is expressed on a scale of 0 to 1. An SHGC-value of 0.5 means that 50% of the heat from the solar radiation striking the window reaches the room.

Developments in heat-insulating window systems are moving in the direction of quadruple glazing and vacuum glazing units. Systems with a vacuum in the glazing cavities offer a high U_g-value, as well as a reduced installation depth and weight. Their cavities are not filled with noble gases. They also have a long-term cost advantage over current glazing systems. It must be possible to maintain the vacuum over the life of the system.

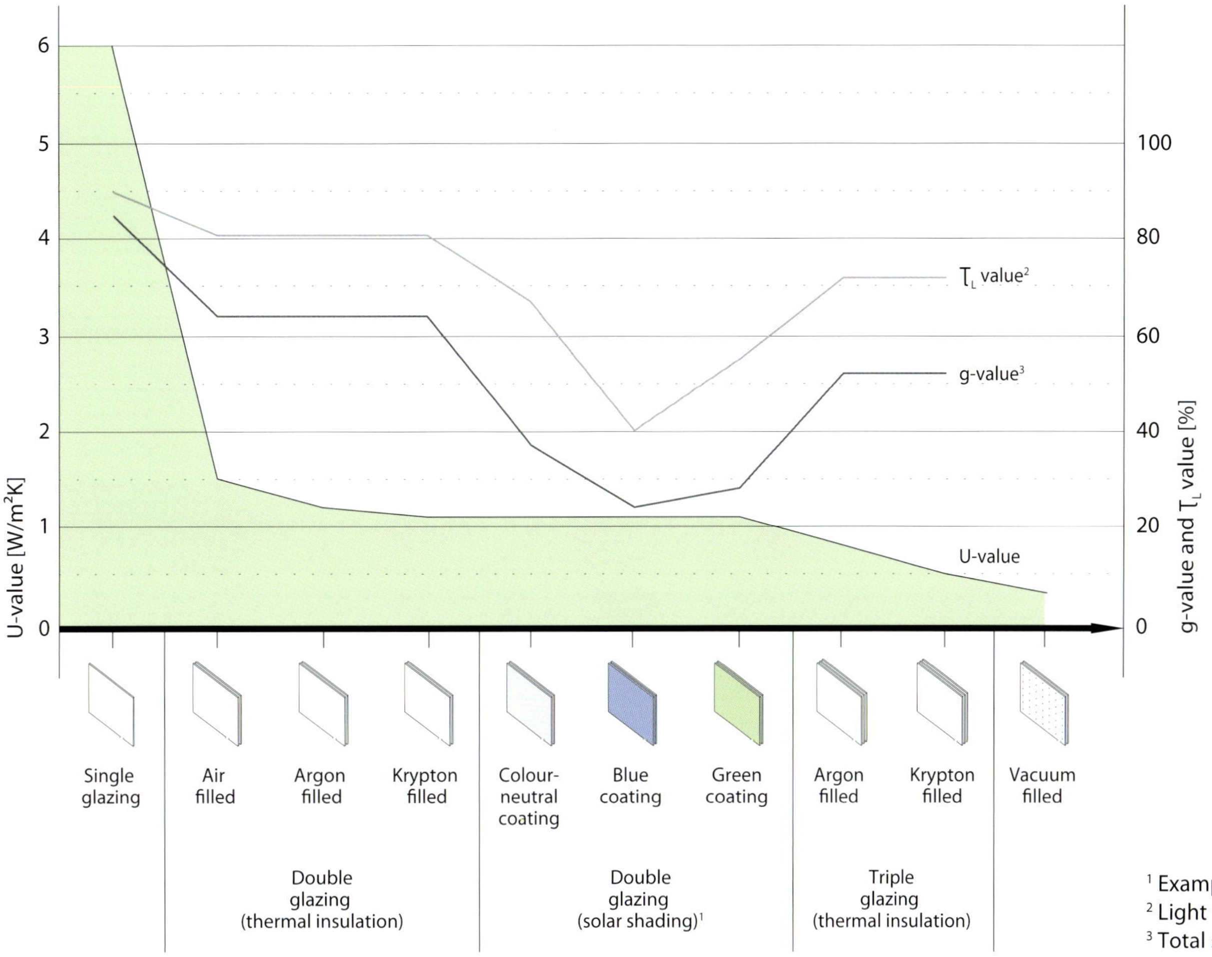

[1] Examples only
[2] Light transmittance
[3] Total solar energy transmittance

Development of glazing system standards
Building physics parameters for glazing systems

Noble gas filling

To achieve a U_g-value ≤ 0.6 W/m^{2}K requires the glazing cavities to be filled with a noble gas such as argon, krypton or xenon. Krypton and particularly xenon are considerably more expensive and energy-intensive to produce than argon, thus their energy advantage is reduced when looked at holistically.

Solar control glazing

Solar control glazing uses coatings to reduce the g-value. The coatings are usually applied to the inside of the outer pane because this can reduce the heating of the glazing cavity and hence the thermal stresses within the solar control glazing. This type of solar protection operates constantly and cannot be regulated over the year. Adaptive solar shading, on the other hand, can offer better controllability of sunlight and increase the use of natural light (see the section on solar shading p. 112).

Sound insulation

Windows represent a sound insulation weak spot compared with the opaque components around them. This can be improved by the use of sound insulating glazing with different glass thicknesses. A flexible seal should be used when installing the window.

Edge seal/spacer

The edge seal/spacer in thermal insulation glazing should incorporate a thermal break; unfortunately this is not standard practice. Aluminium spacers are still used between the glasses. The material's good thermal conductivity creates numerous cold bridges, which can lead to condensation. This must be alleviated, especially where wooden frames are used. Thermal comfort could also be adversely affected because it may not be possible to avoid surface temperatures of 13 °C or less. A better choice is an edge seal with a thermal break using stainless steel or plastic spacers. A glass edge cover of 25 to 30 mm in the frame is recommended for optimum insulation.

Frames

From the points of view of energy and building physics, window frames must be carefully designed and installed. The U_f-value of a good frame is more than twice that for a triple glazing unit and, with the area of the frame being as much as 25–40 % of the window area, its importance should not be underestimated. A great deal of energy can be lost through these components. Typical U_f-values of conventional frames are 1.5–2.0 W/m^{2}K.

A well-insulated frame is essential to achieving high thermal comfort. Insulated frames with U_f-values of 0.7–0.8 W/m^{2}K are now available at reasonable prices. Lower-quality frames with U_f-values of 0.9–1.0 W/m^{2}K should be seen as the minimum standard.

Profiles and insulating cores incorporating a thermal break are required. The insulating cores inside the frame cross section may be of various materials including expanded polystyrene (ESP), rigid polyurethane foam (PUR) and cork, depending on the product. The profiles can be of wood, wood with aluminium cladding, aluminium or plastic.

Plastic windows, which normally have foamed internal chambers and therefore are difficult to separate into different materials and dispose of at the end of their life cycles, can be criticised from an ecological point of view. The various profiles in timber-aluminium windows, on the other hand, are easily separable and this combination therefore offers a better alternative, as do purely aluminium windows. However, using timber windows normally entails higher maintenance costs.

The increasing weight of thermal insulation glazing and the higher frame insulation values are leading to greater frame face widths. However, over recent years attempts have been made to rectify the design and daylighting disadvantages, and reduce face widths. This is done by reducing the glass edge cover and increasing the depth of the frame. Embedded in an insulation layer extending over the part of the front of their frames, systems can achieve inner face widths of about 75 mm and outer face widths of 0–20 mm compared with conventional face widths of 120–140 mm. The frame depth in this case is 125 mm.

The thermal transmittance of window systems calculated from the frame and glazing combined is expressed as the U_w-value. The calculation takes into account the U_g-value of the glazing and the U_f-value of the frame and their proportions on each window.

Positioning and dimensions

Glazing allows views in and out, and the use of daylight. Building regulations (such as the regional state building regulations (LBO) in Germany) prescribe a minimum size for glazing or windows in occupied rooms. The designer should provide an adequate number of opening vents. The psychological effect of having opening windows is particularly significant in summer and the months either side – even though the natural ventilation they create is associated with high ventilation heat losses in winter. If climate concepts with night cooling and cross ventilation are developed, an adequate number of suitably sized opening vents will be essential, especially during the summer.

Double glazing with conventional glass insertion depth

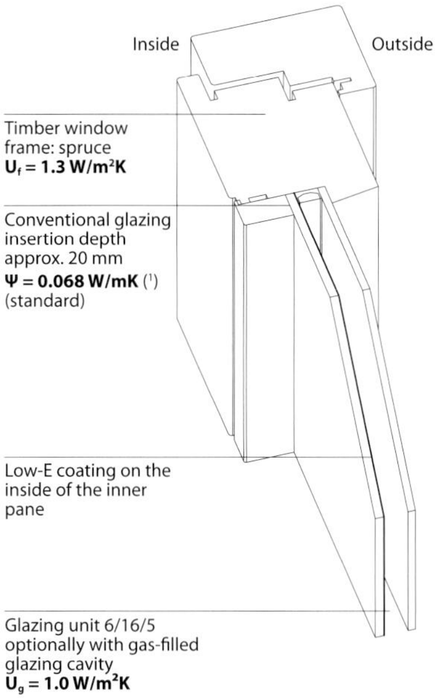

U_w = 1.29 W/m²K

Triple glazing with conventional glass insertion depth

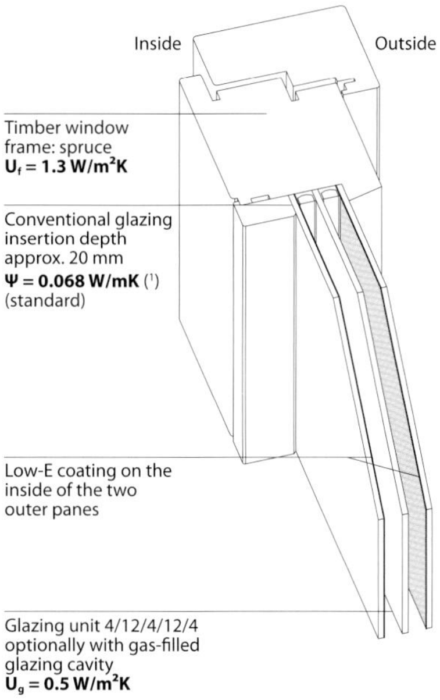

U_w = 0.98 W/m²K

Triple glazing with increased glass insertion depth

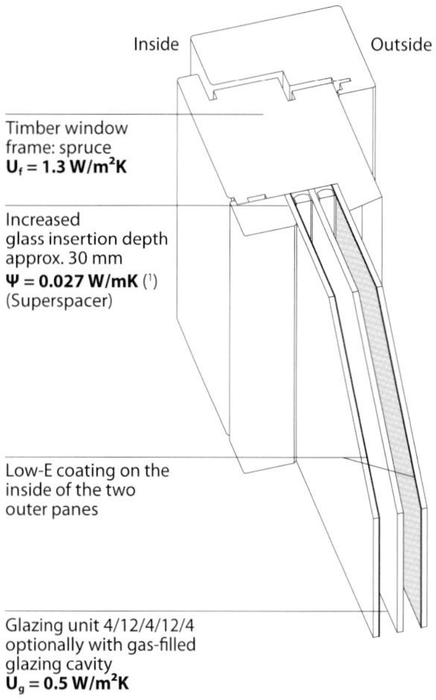

U_w = 0.87 W/m²K

¹ Linear thermal transmittance for glass edge zone

Thermal insulation glazing and frames

Timber frame profile with core insulation

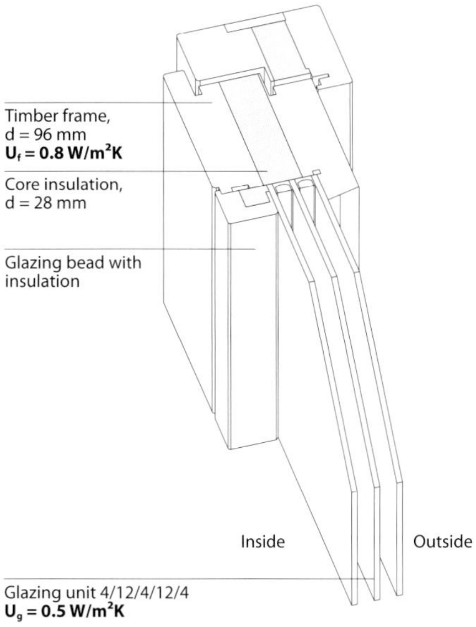

U_w = 0.68 W/m²K (¹)

Plastic frame profile, foam-filled

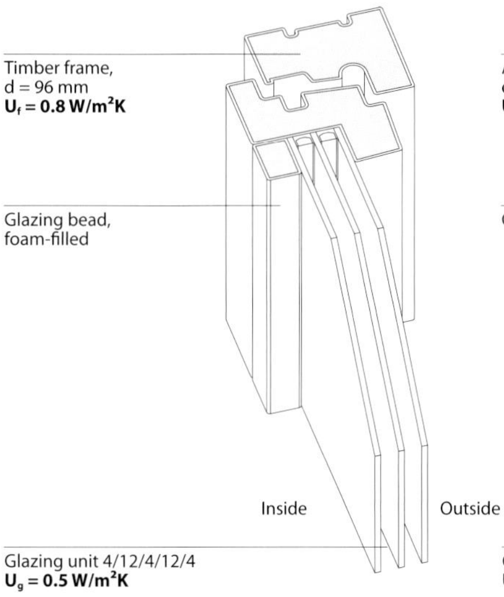

U_w = 0.68 W/m²K (¹)

Aluminium frame profile with thermal break

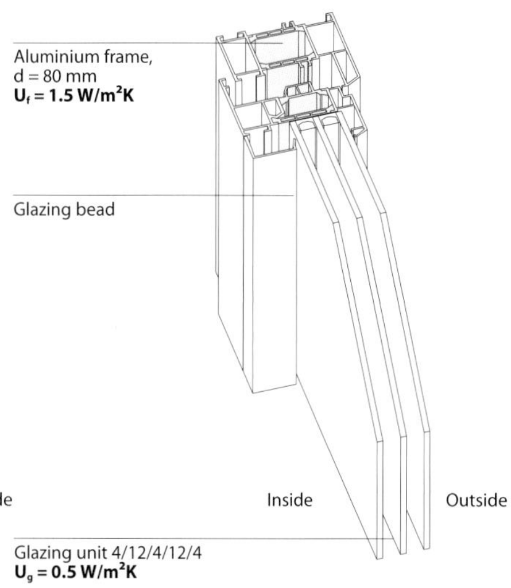

U_w = 0.94 W/m²K (¹)

¹ Values with improved thermal transmittance for glass edge zone
Source: u_w-Rechner fur Fenster (U_w calculator for windows): www.energiebedarf-senken.de; www.nachhaltiges-bauen.de

Frame insulation compared

Ventilation

The improvement in the energy qualities of building envelopes depends greatly on how well they are sealed. This means that uncontrolled ventilation through leaks and gaps, the usual situation with earlier buildings, does not happen in Aktivhaus buildings. A high level of comfort through excellent air quality and minimising ventilation heat losses, and the greatest use of natural ventilation are therefore the requirements to be met for any system providing hygienic ventilation corresponding to need. A controllable mechanical ventilation system is essential to ensure optimum air quality for the user while avoiding energy losses. It is supplemented – where possible – by natural ventilation in the spring and autumn months and cross ventilation during the night in summer (see p. 130 ff., p. 152 ff.).

Solar shading

Solar shading reduces heat loads from insolation on transparent building components. This can prevent the overheating of buildings. Before solar shading is addressed here, there needs to be some discussion of the ways building geometry and alignment, proportion of window area, and construction can strongly influence passive solar gain. In principle, in Central European latitudes unshaded south-facing windows forming more than 30 % of their facade area should have external solar protection.

The winter sun is a welcome source of passive solar gain and can penetrate deep into the building through south-facing facade openings. On the other hand, this heat is not as welcome in the summer months. With the sun high in the sky in summer, the glass in south-facing windows reflects a large proportion of the heat and therefore reduces the possible heat load from the windows. Depending on its design, a building can incorporate a projecting roof outstand, or fixed or moving shading elements. Solar shading is indispensable on the east and west sides to prevent the building from overheating. External moving solar shading is recommended here. Many different types are available, with the final choice often reflecting the region and culture. External moving solar shading, such as window shutters with moving slats, Venetian or fabric blind systems, are popular in Central Europe. There are many other varieties available to suit the particular purpose and use of the building, aesthetic design requirements and wishes. Options include the intentional use of vegetation and solar shading forming a permanent part of the structure (e.g. a roof overhang).

Translucent, light-deflecting or at least variable shading systems help to maximise daylight use while reducing heat loads and positively contributing to the overall energy performance of the building. Solar protection glazing uses coatings within the unit, known as low-E coatings to provide all-year-round solar protection. The design of the building must take into account the continuous nature of this solar protection. One effect is to reduce the possibilities of exploiting insolation inside the building. The versatility arising from the separation of shading and glazing elements is not available either. Effort is being put into developing new types of glazing systems that provide variable solar protection. No product is currently available that can offer an economic or aesthetically interesting alternative.

Controls

Automatic solar shading control linked to insolation metering can bring solar and glare protection elements automatically into the desired settings or positions according to a predetermined program. If the intensity of radiation exceeds a defined value over a preset period, solar and glare protection elements deploy or, in the case of slats, move into the shading position. If the recorded insolation is too little over a defined time period (e.g. under an overcast sky), the elements return automatically to the undeployed or an intermediate position. Solar shading slats, for example, would be brought into a position parallel to the angle of incidence of the solar radiation to maximise the use of daylight, ultimately being fully retracted again under more overcast skies.

These systems, which can be used anywhere in the world, can be optimised to take into account shade from a building or vegetation. This can avoid excessive use of artificial light in rooms affected by external shading.

Further components can be added to external solar protection systems to guard against weathering. Sensors for temperature, precipitation, wind speed and direction ensure that the systems return to their original positions in conditions where they could be exposed to excessive loads, thus protecting them from outside influences. Motor-driven windows can be programmed to close in reaction to measurements of icing, wind and rain to prevent damage caused by water penetration.

Building envelope qualities

Minimising cold bridges

In any consideration of the building physics and energy qualities of the building envelope, after achieving a good standard of insulation and seal, the focus moves to avoiding cold bridges. Good design and site supervision are essential to their elimination. The objective must be to create a conditioned building with an insulating and sealing envelope that has no penetrations or as few as possible. An envelope with no cold bridges, or very few, can be achieved by following this principle.

If the cost of good design and construction is set against the potential savings to be made by eliminating cold bridges, then it is a very economic route to building an efficient Aktivhaus.

The critical details are those at the interfaces and transitions between well-detailed standard building components – connections, corners, penetrations and edges.

Penetrations are one of the points of weakness that can be avoided simply by good design. A classic example is the balcony. The conventional construction of a concrete slab cantilevering through a penetration in the facade presents a significant weak point in a highly efficient building envelope. If, instead of this arrangement, a construction is thermally separated from or placed in front of the facade, it would almost completely avoid penetrations and cold bridges.

Cold bridges can lead not only to energy losses but also to damage to the building envelope initiated by building physics processes such as condensation. If this type of damage is not detected and rectified at an early stage, in the worst case it can lead to structural damage.

Thermography can be used to make the weak points on buildings visible in digital images. This non-destructive technology provides an excellent means of revealing cold bridges in existing and new buildings.

The aim during design and construction should be a structure free of cold bridges.

Thermographic image

Airtightness

Airtightness is another marker of the quality of the building envelope. An airtight building envelope avoids undesirable heat losses through ventilation and keeps the hot or cold air inside the building. A well-installed airtight seal also reduces sound transmission. In addition to the positive effects on energy demand and user comfort, a good airtight seal reduces the susceptibility of the building to other defects.

Air changes, movements and flows should never be unintentional. Gap ventilation is not adequate to provide a good, continuous and adequate rate of air change. Draughts from gap ventilation can detrimentally affect comfort. Uncontrolled air changes increase the chance of building defects. The moisture in the air can condense within the building envelope, leading to mould growth or structural damage.

The details to be aware of during design and construction crucial to achieving good airtightness are similar to those for avoiding cold bridges. They are likewise found in the transitions between well-detailed standard elements, typically at connections, corners and penetrations.

The airtight seal forms a layer like a continuous skin around the usable volume of the building. In most types of construction, it is normally found on the inside surface of the wall. The seal is achieved using membranes, wood-based boards with airtight joints or interior plaster. The best position for the seal is between the structural and the installation layers so that penetrations through the sealing layer, for example, by electrical installations, can be avoided. In the roof, the airtight layer is achieved using a vapour barrier film.

Mechanical ventilation and highly efficient heat recovery systems should be used to ensure hygienic ventilation of the building. This arrangement can ventilate a building without incurring large energy losses. At the same time, it could, for example, incorporate suitable CO_2 sensors that would make window ventilation superfluous. At all times when the outdoor temperature is within or near the ambient comfort window for interior rooms, it is sensible and user friendly to work with natural ventilation by opening a window. The user should have the option of opening a window, because this is an important enhancement that adds psychological and user-acceptance advantages to any mechanical ventilation system.

The familiar blower-door test (pressure test) proves the airtightness of the building after completion of the sealing layer. Fans in the sealed envelope create alternating negative and positive pressures inside the building for the purpose of the test. A factor for the airtightness of the building envelope can be calculated from a typical difference in pressure between inside and outside of 50 Pa. A value of $n50 \leq 0.6$ 1/h should be the target.

In the event of the test indicating leaks in the building envelope, the weak points can be located quickly by creating smoke inside the building. After rectification of the weak points, the blower-door test can be repeated if necessary.

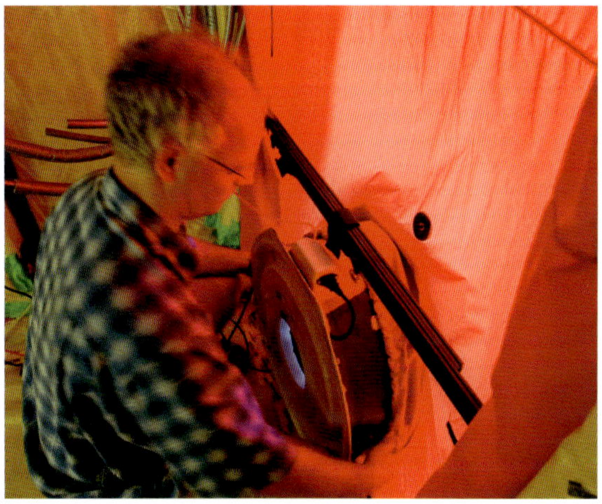

A blower-door device built into a door opening

Storage mass

Storage mass in a building helps to even out temperature fluctuations over several hours or days, in winter or in summer, and suppresses temperature peaks. The solar radiation entering a room through the glazing in summer is stored in the solid components surrounding the space. A phase-shifted discharge of the components then takes place during the night when they are cooled by natural cross ventilation. The stored cool of the night is ready again the following day to help to compensate for the heat of the incoming solar radiation and to avoid overheating the Aktivhaus.

During winter and the months either side, the heat stored during the day can contribute to tempering the Aktivhaus at night and preventing the well-insulated building envelope from cooling.

To maintain a good room climate, we recommend using materials with a high thermal storage capacity. Particularly suitable materials include loam, natural stone, brick, mastic asphalt, ordinary concrete and many other heavy materials. Some lighter materials, such as wood, also have good storage properties. The designer should bear in mind that only the first centimetre of the material is effective for short-term temperature compensation. The deeper layers of the material play no part in short-term storage. An alternative is phase change materials (PCM). A PCM is usually a special salt or paraffin that undergoes a change of phase over a suitable effective range of temperatures. A phase change from solid to liquid takes up a lot of heat. The same quantity of heat is given out again and the PCM solidifies when the surroundings cool. One advantage of this property is that the PCM changes phase over a narrow range of temperatures. Although a PCM has comparatively little mass, it can emit or absorb a lot of energy. Using phase change, the PCM can be set for a temperature range conducive to ambient comfort and can help to stabilise temperatures. The PCM is contained in small capsules or placed in cavities between two layers, or integrated into materials such as plasterboard.

The components used to store heat must not be separated from the interior space by suspended ceilings, double floors or wall cladding, since this may be detrimental to the thermal storage capacity.

How great the storage mass of a building should be to ensure as high a level of comfort while minimising the use of mechanical equipment can be calculated from a

The thermal storage capacity of loam and its humidity-regulating properties can contribute to increased ambient comfort inside a house.

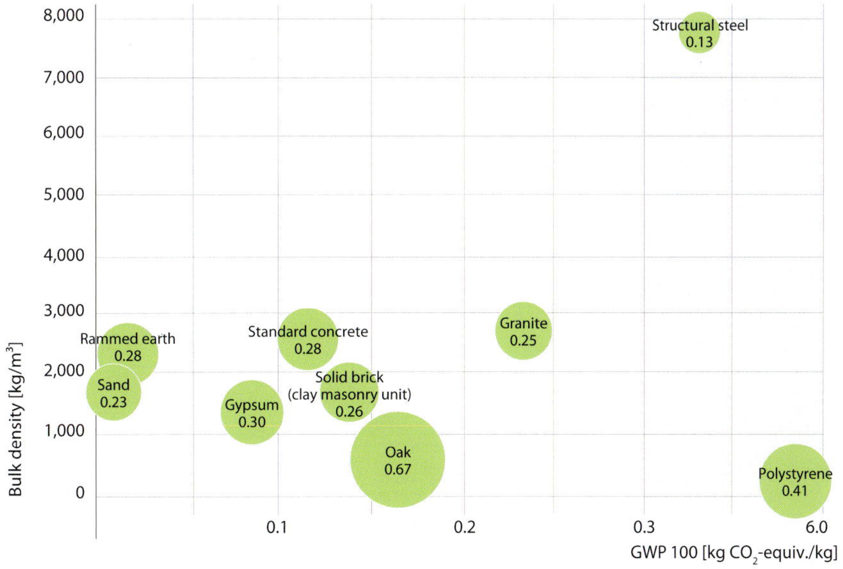

Building material	Heat storage capacity [Wh/m^3K]	Specific heat [Wh/kgK]	Bulk density [kg/m^3]	Global warming potential (GWP100) [kg CO_2-equiv./kg]	Primary energy non-renewable [MJ/kg]
Water (at 20 °C)	1,157	1.16	998	-[1]	-[1]
Structural steel	1,015	0.13	7,850	1.820	29.850
Standard concrete	690	0.28	2,500	0.120	0.820
Granite	660–710	0.25	2,600–2,800	0.230	3.700
Ice (water at 0 °C)	523	0.57	918	-[1]	-[1]
Rammed earth	470–610	0.28	1,700–2,200	0.004	0.080
Sand	410	0.23	1,800	0.001	0.023
Solid brick (clay masonry unit)	360	0.26	1,200–2,000	0.142	2.610
Paraffin[3]	357	0.42	849	-[1]	-[1]
Wood	350–465	0.58	600–800	-[1]	-[1]
Solid wood, planed (oak)	450	0.67	670	0.165	23.700
Gypsum	290	0.30	850–1,600	0.085[2]	1.830[2]
Polystyrene (PS)	12	0.41	15–30	5.770	92.500

[1] No information
[2] Base value for gypsum plaster
[3] Product RUBITHERM® GR 50 (1-3)

Storage capability of various materials

consideration of the effects of the environment on the building and the relevant parameters of the building (such as glazing fraction, orientation and type of use).

The ability of materials to store heat is described as their specific heat c. The value is expressed in watt hours per kilogram and kelvin (Wh/kg K). Typical values of specific heat for a selection of materials:

- Solid brick 0.26 Wh/kg K
- Lime-cement plaster, concrete, screed 0.31 Wh/kg K
- Steel 0.14 Wh/kg K
- Copper 0.11 Wh/kg K
- Water (at 20 °C) 1.16 Wh/kg K

It should be noted that the volumes per kilogram of these materials vary greatly. For example, polystyrene at 0.35 Wh/kg K has a much higher heat storage capacity than concrete, but in order to store the same quantity of heat as a given volume of concrete, the volume of polystyrene would need to be many times larger. For this reason, the storage capacity of materials is also expressed as $s = Wh/m^3K$.

Another aspect to be taken into account is the storage or regulation of moisture in the air. Materials such as loam, gypsum or wood are very suitable for storing excess moisture from the room air and then releasing it again when the air is drier. If a building has free or mechanical night cooling, the normally moister night air is brought into the building and the moisture stored in these materials. Next day this moisture is given off as soon as the moisture level in the room air drops. This regulates humidity throughout the day, which not only benefits comfort but also prevents moisture-related damage such as mould.

Generating energy

In times of rising energy prices, increased worries about the security of energy supplies, concern over environmental damage from high CO_2 emissions and the politically backed transition to renewable energy sources, facades and roofs are becoming increasingly important as energy creating surfaces. The possible ways of integrating heat- or electricity-generating technology into building envelopes are many and varied.

By reducing consumption, optimising the building skin and the remaining required technical components, it is possible, not only in residential building construction, to create buildings that cover their own energy consumption with regenerative energy created on or around the building.

If the approach taken is one of reducing the energy consumption of a building, thermally optimising the building envelope and finally creating the remaining required energy and heat from local regenerative sources, the following technologies for generating energy are available (see p. 122 ff.).

Photovoltaics

Photovoltaics create electricity from daylight on the building. Insolation can contribute to autonomous or even grid-independent operation of a building.

Photovoltaics find overwhelming use as adaptive elements on the building envelope, usually on the roof. However, the actual potential of photovoltaics and solar thermal technology for generating energy lies in integrating it into the building envelope. An integrated approach allows these modules to provide privacy and views out through areas of glass as well as protection against the weather and sun. Taking this idea further, complete building envelopes favourably oriented can act as energy-efficient and visually distinctive power plants. Technical and aesthetic integration presents a great challenge.

Solar thermal technology

Solar thermal technology provides the means to create energy for heating, cooling and the provision of domestic hot water. Flat or tube collectors capture insolation. They then give up their heat through a transfer medium to a hot water storage tank. The various consumers can draw off the stored heat.

Thermal solar collectors use the whole spectrum of sunlight and convert insolation into heat at an efficiency of 60%–80%.

Geothermal technology

Geothermal technology can exploit near-surface (down to 400 m) or deeper geothermal heat at depths of 400–4,000 m. Most near-surface geothermal systems use the heat stored in the top layers of the earth for heating and cooling. They use collectors, ground loops, energy piles and hot spring systems to extract heat from or give up heat to the ground through a circulating liquid.

Heat pumps make this heat, which is usually taken from low-temperature sources, available for use. Buildings can be cooled by similar systems without using heat pumps.

Heat pumps

Through the use of the energy needed to drive them, heat pumps raise thermal energy from a low to a higher temperature level to produce useful heat for heating a building. The reverse of this principle allows heat pumps to be used for cooling (refrigerator principle). The possible media include outdoor air, near-surface geothermal, ground, surface and waste water.

Comparison of different photovoltaic cells

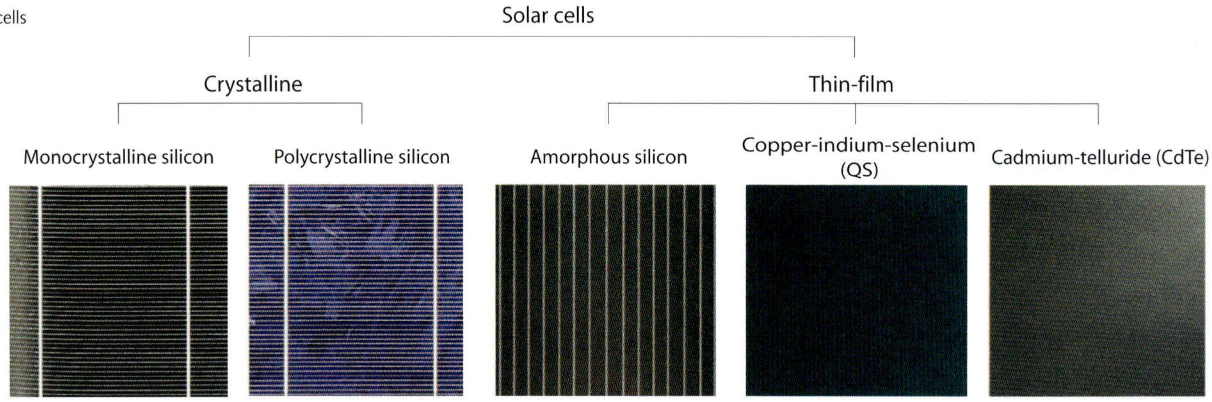

	Monocrystalline silicon	Polycrystalline silicon	Amorphous silicon	Copper-indium-selenium (QS)	Cadmium-telluride (CdTe)
Efficiency	15–24%	13–18%	5–7%	8–13%	8–13%
Efficiency laboratory cells	up to 33%	up to 19%	up to 13%	up to 21%	up to 20%

Lighting

Natural lighting/daylight

The optimum use of daylight is a key aspect of building design and must be considered in the preliminary planning process. External influences such as daylight factors and the path of the sun are analysed and conclusions drawn about the shading from surrounding buildings and trees. The detailed design examines the best use of light from natural sources to cover the lighting demand.

Building and room depths, room heights, positioning, the size and character of the building openings, and the forms and finishes of surfaces have a decisive influence on daylight qualities.

Daylight entering rooms through windows and the views out into the open contribute greatly to the building user's feeling of wellbeing. Skylights can create an even level of lighting inside a building. To do this, the windows should not be more than the height of the room apart.

The effective use of daylight demands a flexible means of control by shading and glare protection systems. In residential and workplace situations, the first concern is to create good visual conditions. The systems also help to regulate the entry of solar heat and therefore allow the user to react to the changing conditions over the year. Other passive and active measures improve internal daylight conditions, views in and out, and the mood of a room. These include light-deflecting blinds and glass, slatted blinds integrated into the glazing units, angled window reveals, lintel heights, skylights and roof windows, and light pipes, which conduct daylight to places not adjacent to an external facade. Material and colour choice for shading and glass tint must be made with the anticipated colour impression in the interior and the desired colour perception in mind. The efficient use of daylight in buildings is essential to promote and maintain people's physical and psychological health, and reduce energy consumption by minimising the use of artificial light.

Building openings must be appropriately sized to avoid overheating in summer, yet keep heat inside in winter. They generally have poorer U-values than the surrounding walls or roof surfaces, and solar radiation brings heat into building through them. The energy performance of the building envelope is therefore closely linked to its use of daylight. In addition to the form of the building, the design must aim for the right balance between the use of daylight, the entry of solar heat, losses and, above all, user comfort.

Daylight consists of proportions of direct and diffuse light. These vary greatly, depending on location and season. In northern Europe, the ratio of direct to diffuse light is about 1:1 in June, while in winter the ratio is

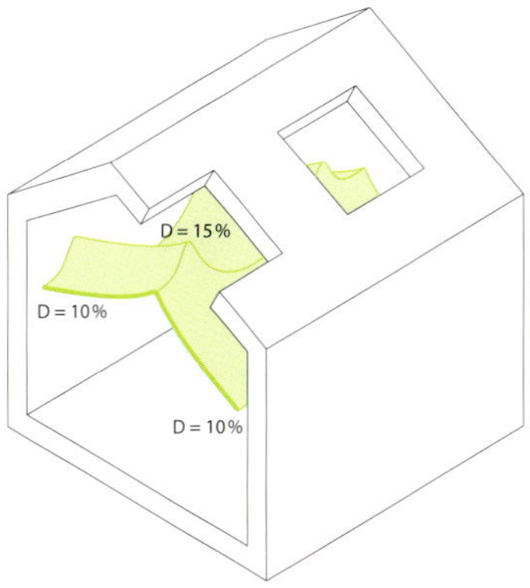

Daylight factor in a room with roof skylights

Natural lighting with skylights, VELUX Sunlighthouse (AT)

Daylight factor in a room showing dependence on room depth and facade opening

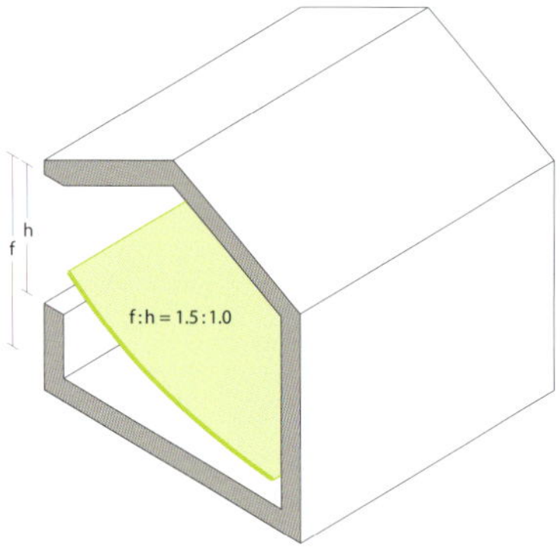

Light entering through a photovoltaic facade, Solar Academy, HHS Planer + Architekten AG, Kassel (DE)

1:2–1:4, i.e. the direct light striking an object is only 25–50% of the diffuse light. However, in southern Europe the summer ratio is about 2:1; this drops to 1:1 in winter. The diffuse proportion of the light is suitable for providing uniform levels of light in rooms. Direct light, on the other hand, can be specifically deflected and used inside deep rooms. The daylight factor is a measure of the usable daylight in a building. This compares the illuminance of a horizontal surface in the open air with that of the same surface inside a building. Good visual comfort in a room requires illuminance to be as uniform as possible. The ratio of the brightest and the darkest surfaces in the room should not be greater than 1:6; where there are roof skylights, not greater than 1:2.

Artificial lighting

Recent years have seen many new developments in the field of lamps, in particular light-emitting diodes (LED) and organic light-emitting diodes (OLED). LED lamps are electronic semiconductor building components. When an electrical current passes through them they produce light, converting energy very efficiently from one form into another. Advantages include a great variety of available colour temperatures, long life (low maintenance costs), high resistance to on/off cycling (approx. 1 million cycles), the ability to be dimmed without loss of colour or efficiency, low-temperature compatibility and low energy consumption (lower by a factor of 10 or more compared with incandescent lamps). LED retrofit light units can be used to replace conventional light bulbs and fittings.

LED lights, with their compactness and minimised light source, offer completely new directions for the development of lights.

In comparison, fluorescent lamps offer an efficiency almost as high as LEDs (factor of 4 to 10). In the optimisation of energy consumption, attention moves to the ballasts normally used with these highly efficient lamps. Compared to conventional ballasts, electronic ballasts can increase efficiency by a factor of 2. In spite of their somewhat higher purchase price, electronic ballasts are preferred because the lower amount of heat they give off reduces the heat load on the building. Using a cut-off electronic ballast, which has lower losses and runs cooler because the electrode heating is switched off after the lamp starts, can achieve a further 20% saving.

Lighting controls present further significant savings potential. Among the controls available are motion/presence, acoustic and infrared sensors, and time switches. Their use in stairwells, basements, toilets and the like can result in considerable energy savings. The sensors can also take on other functions, for example, ventilation control. Artificial light required over longer periods and mainly in outside areas is controlled by twilight sensors, if necessary in conjunction with presence detectors or time switches. Combination with presence detectors allows dimming when nobody is present. Daylight sensors are best sited in areas where people use daylight and/or (supplementary) artificial light. Deployed in this way, sensors can, for example, automatically adjust the amount of artificial light in offices to suit the current lighting conditions and supplement the available natural light. Depending on room depth, it can be worthwhile to install several sensors, either at intervals into the depth of the room or near workstations, if the lighting units do not already have integrated sensors. In combination with a motion or presence detector, the artificial lighting can react autonomously to the occupation of the room, dim the lights when everyone leaves the office, and switch them off completely after a specific period of time. The user should always be able to intervene and adjust the light intensity to suit individual requirements, above all because a user deprived of the ability to influence a preset system will have difficulties accepting it, even if calculations show it is the best one possible.

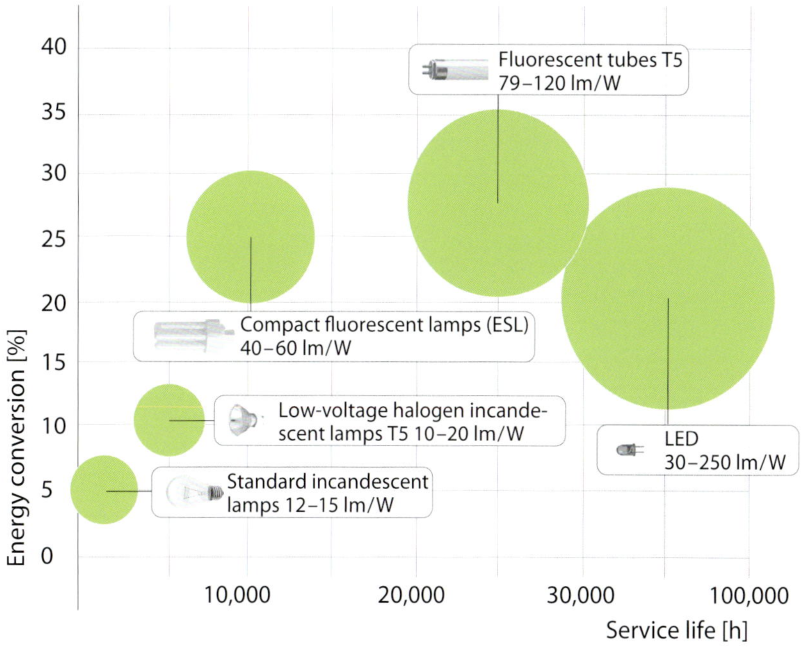

Efficiency of various light sources

Qualities and details

The interface between the interior and exterior determines the energy, technical and architectural quality of the building. The form of the building envelope is closely linked to the design of the building. Solid construction, often preferred for residential buildings, offers inexpensive solutions albeit with some limitations on the positioning of openings. With external insulation, also subject to constraints in monolithic construction, the thermal mass of the structure can be used to improve comfort and smooth out temperature peaks.

The loadbearing structure and the building envelope are separate layers in the vast majority of office and industrial buildings. This often tempts the designer into using large areas of glass in office buildings. However, excessive glazed areas can jeopardise comfort and energy management.

The integration of effective insulation and storage, well-designed windows and ventilation openings, and effective, controllable shading into the building envelope, plays a key role in attaining an efficient and comfortable building. The elements of architecture such as building form and materials, mass and transparency, texture and colour are also the elements of energy-efficient building.

An energy-efficient building envelope can succeed in providing the required indoor conditions throughout the year almost completely by means of passive measures. Only a small contribution from active measures and the associated energy supply technology is necessary to maintain the desired requirements. In an Aktivhaus building this should be covered from regenerative sources, if possible generated and used locally. The technical means to do this are already available today for a Central European climate and can be used economically, taking life cycle costs into account.

Building services

In an Aktivhaus, after ensuring the best possible use of all passive measures and the optimisation of the envelope, the focus falls on energy gain from regenerative sources. Regenerative sources in this context include solar radiation, geothermal, hydropower, wind and renewable raw materials. The use of regenerative energy sources rather than fossil energy carriers can move the world in the direction of CO_2-neutrality. Furthermore, regenerative energy carriers such as the sun, wind and geothermal heat are free and, in terms of their security of supply and price stability, they are a better choice than fossil energy carriers. Regenerative energy sources such as biogas and wood are often locally harvested and therefore their use incurs no high transport costs.

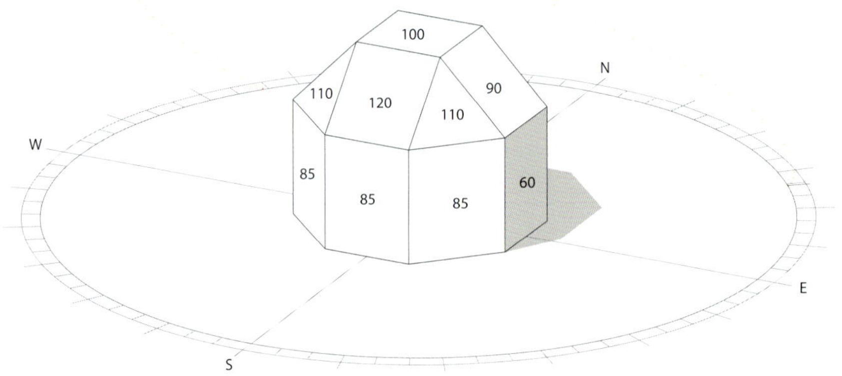

Yield from a photovoltaic system and its dependence on orientation

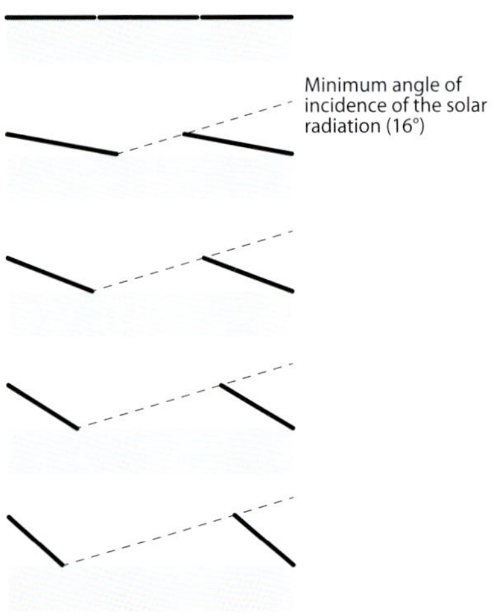

Inclination of module surface [°]	Effective solar surface [%]	Specific insolation [%]	Effective insolation [%]
0	100	100	100
10	75	106	80
20	61	111	68
30	53	113	60
40	48	113	54

Influence of the arrangement of roof photovoltaic systems on the effective insolation area

Collecting and converting renewable energies

Solar radiation

Solar radiation has a long tradition as an energy source going back many generations. Over time, its use has developed to the extent that today solar radiation is not only used passively, but is converted, stored and used in many different energy forms. Photovoltaic, solar thermal and air collectors are three of the technologies employed.

Photovoltaics

Insolation can be changed by photovoltaics (PV) into electrical energy.

The created energy can be used or consumed directly in the building, fed into the public grid or stored in batteries.

Solar modules consist of individual solar cells. Light striking the cells causes them to emit electrons, which creates a direct current. This phenomenon has been known since the end of the 19th century and is called the photovoltaic effect. Several solar modules are connected together to produce higher outputs and make more efficient use of the available area. The direct current is converted to 230 V 50 Hz alternating current by an inverter. In this form, the electrical energy created from sunlight can operate any household device or, if it is not required for household use at that moment, fed into the public grid.

The designer of the PV system should ensure that minimal full or partial shade falls on any module, as otherwise this would have a negative effect on the output. Integration into a vertical building facade is perfectly possible, though the usable insolation or the expected yield may be less than that of an optimally oriented roof installation. Photovoltaic modules are available as standard products, but can also be manufactured as specials to suit the project. Likewise, they can be integrated directly into building products to create, for example, photovoltaic roof tiles or roof windows.

By integrating photovoltaic elements into the building envelope, mainly the facade, in addition to generating electrical energy, they can also provide privacy, protection against the weather and sun, and take over further functions normally fulfilled by the outer layer of the facade. The use of integrated PV systems can save at least one building component, which can make a worthwhile difference to the overall building cost.

The direct integration of photovoltaics, for example, into the facade layer can be done in a variety of ways. The choice is generally between installation as transparent glass-glass modules or as an opaque surface.

Glass-glass photovoltaic modules can be inserted as single or insulated glazing units into a mullion and transom system or into window frames, or be used as overhead glazing. These modules are usually all manufactured specifically for the project. At the same time, this allows parameters such as the type of photovoltaics, disposition of the cells, details of the glass specification, the amount of available daylight, the shading effect, precise dimensions, shape and other particulars to be specified. Alternatively, the facade design can be based on standard sizes, which should reduce costs.

Setting the modules in the best position and alignment with respect to the sun for the particular site will produce the maximum yield per square metre of PV. Deviation from the optimum setting is certain to reduce the efficiency per unit area. On the other hand, it grants greater design freedom and the ability to maximise use of the available building envelope surfaces for energy creation. Tracking is another option for increasing the efficiency of the panels. A comparison of various orientation angles and placement densities of an on-roof PV system shows, however, that taking advantage of an arrangement at the optimum angle results in a sub-optimum use of many parts of the envelope. From the point of view of economy and energy yield, the latest recommendation (in view of falling PV module prices) for European locations is to integrate the modules into the building envelope in order to make maximum use of the available building envelope surfaces.

The yield is optimised if the modules are rear ventilated to transport away the generated heat. The optimum operating temperature is 25 °C. The output drops approximately 0.4 % per °C. Dirt on the PV surface also reduces output. A slight slope of 3°–5° is enough for rain to self-clean the units.

The efficiencies of the different PV technologies vary greatly. Prototypes are already achieving an efficiency of 33 %. A further improvement of the efficiency while module prices fall is expected. The choice of variant is often decided on the basis of module price and the predicted yield. In facades, the appearance of the units also comes into play. For units used as design elements, the manufacturers have a large choice of visually attractive photovoltaic cells and modules available as standard or they can manufacture them to suit individual or project requirements.

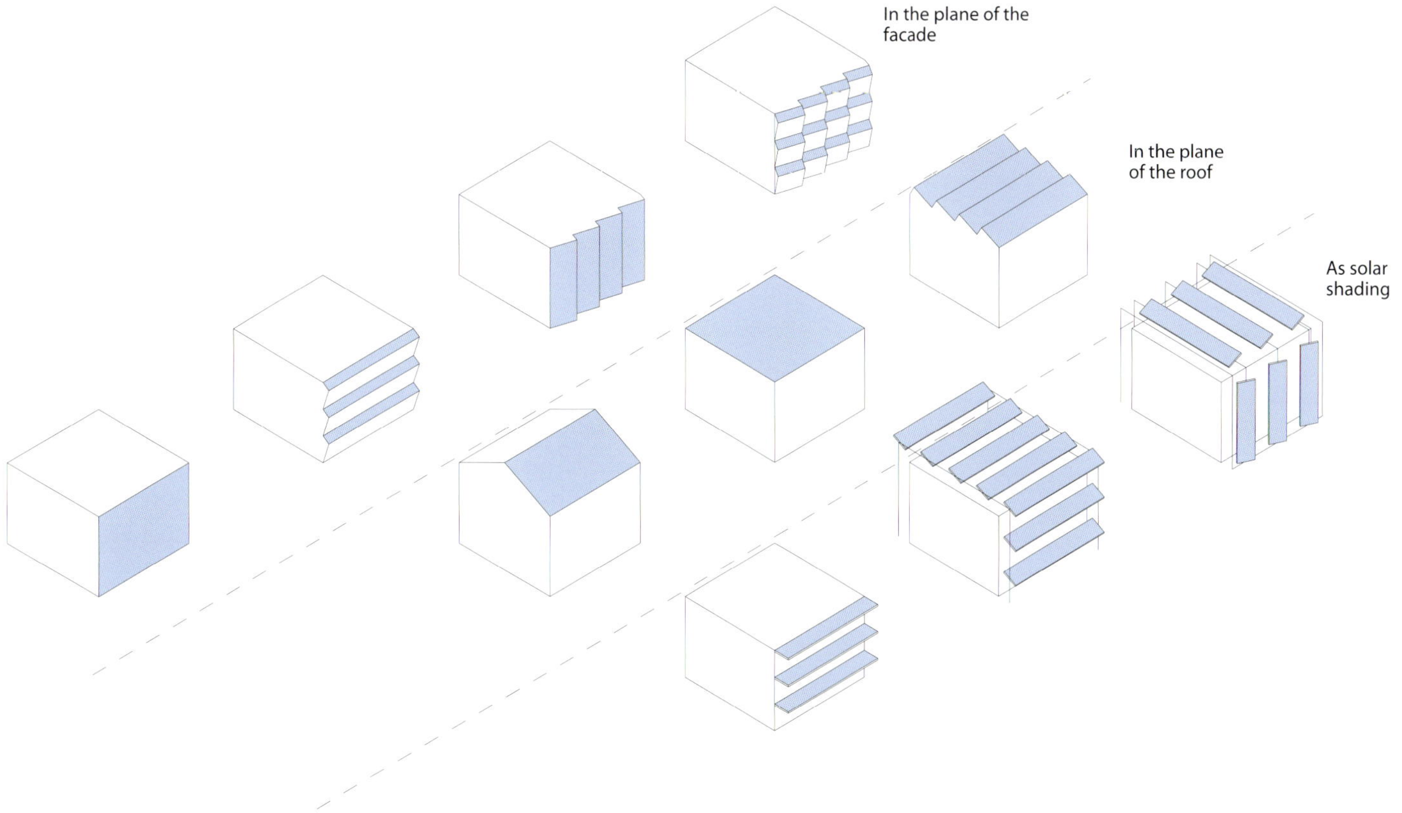

Selection of possible modes of integration and arrangements of photovoltaics on the building

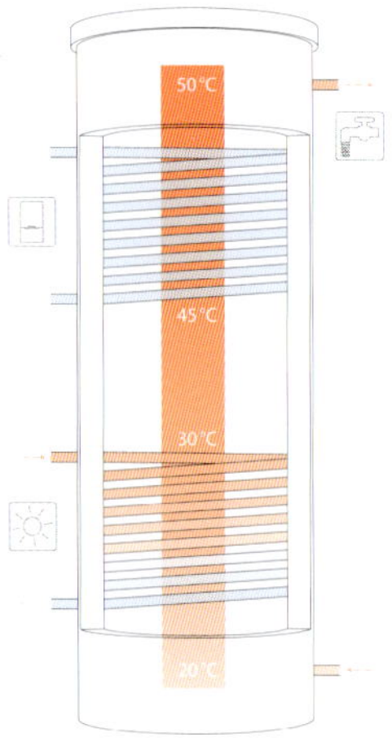

Solar thermal technology providing only domestic hot water

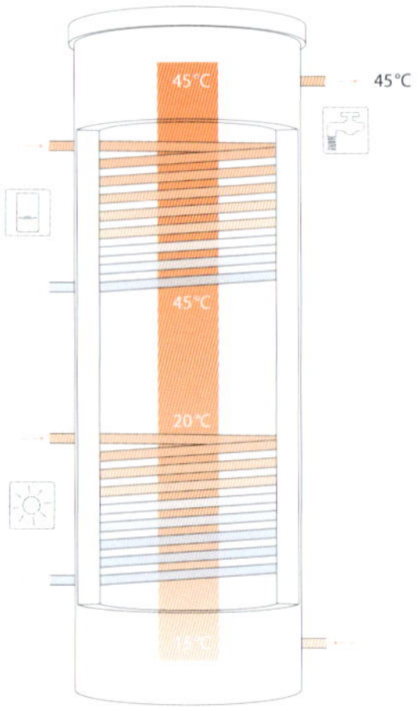

Domestic hot water also providing space heating support

Solar thermal technology

In addition to supplying electricity, solar radiation can also be used for heating domestic hot water and for providing support to space heating systems through the use of solar thermal technology. A carrier medium is heated by solar radiation falling on a particularly good heat-absorbing surface and transfers the energy to a hot water storage tank, which acts as a buffer. The storage tank is normally a bivalent type. The carrier medium is fed into the lower part of the tank and transfers its heat to the cold water. The heated water rises and can be drawn off from the upper part of the tank as domestic hot water. If the solar hot water supply is inadequate or falls below a specified value as a result of being drawn off, the system tops up from another energy source, such as a gas condensing boiler, wood pellet boiler or an electric immersion heater.

To be able to use the solar hot water for space heating, the storage tank requires a further heat exchanger, through which the heating circuit's carrier medium circulates. Monitoring of the storage tank temperature ensures the desired room temperature can be reached. If the temperature of the solar hot water is too low for this, it can be additionally heated by a boiler.

Hot water preparation presents a huge savings potential. In the Central European context, the poor overlap of the high space heating demand in winter and the availability of insolation in summer instigates vigorous debates about the sense and economic viability of combining domestic hot water preparation with space heating support. The fuel savings for systems with simultaneous hot water preparation and space heating support are up to 35%. In the case of a system providing support for hot water preparation alone, even a small area of solar thermal collectors can cover a much higher proportion of the energy required, resulting in fuel savings of up to 60%.

Vertical collector surfaces can be advantageous to increasing the gain in winter because, with a high heat demand and a shallow angle of solar incidence, maximum use can be made of the facade layer to generate heat. In summer, on the other hand, the angle of solar incidence to the collector is so small that only a little heat is generated, which correlates better with the lower summer demand for heat. This provides a good reason for integration of solar thermal technology into the facade. A particular advantage of facade integration is that the insulation on the back of the flat panel collector can also function as thermal insulation.

Solar thermal systems are advisable and very cost efficient for continuously servicing high demands for hot water, for example, in multistorey residential buildings, hotels or swimming pools.

Under these conditions of use, it is worthwhile designing a solar thermal system to cover the minimum demand or to work with a larger buffer tank. This offers a means of dealing with the phase shift between the period of insolation and the use of the regeneratively heated hot water.

Use of heat created from solar energy for cooling by means of an adsorption-type refrigeration system can be of interest for office, industrial and commercial buildings.

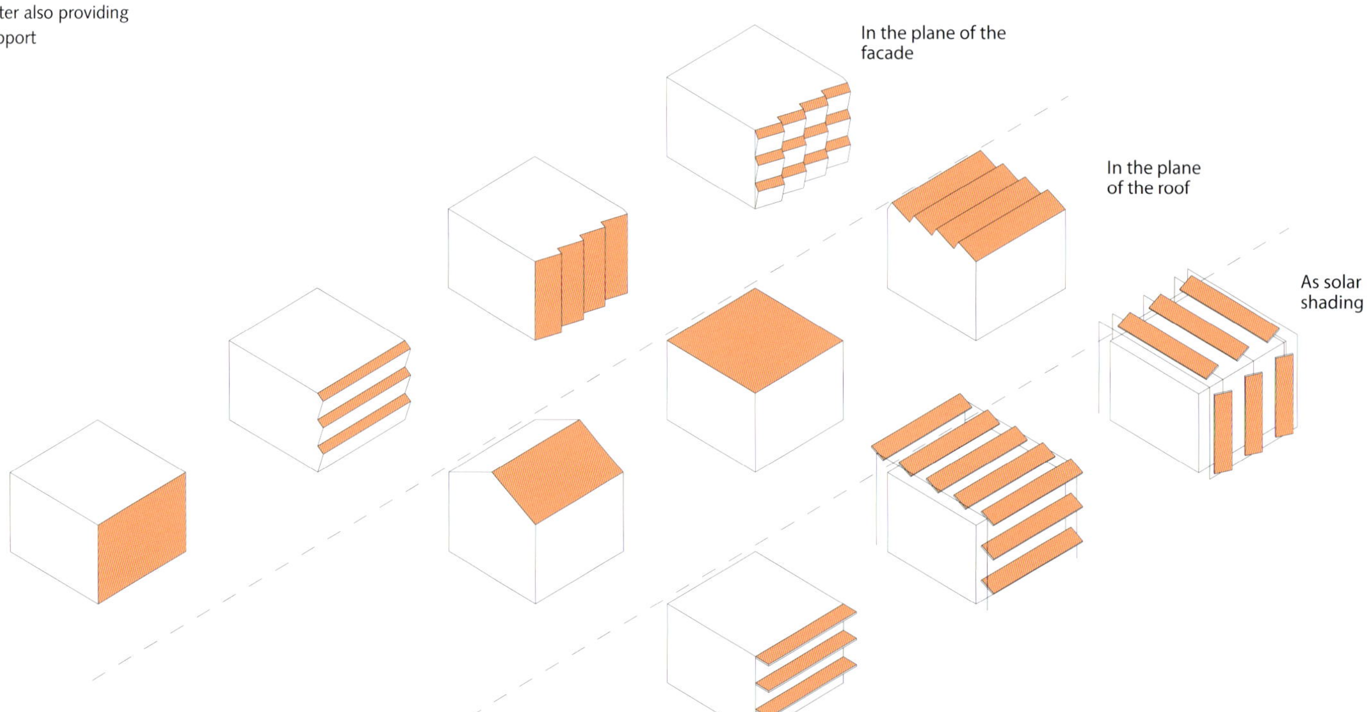

Selection of possible modes of integration and arrangements of photovoltaics on the building

Integration of solar thermal technology –
House Satteins (AT), Unterrainer

Collectors of various types and with different efficiencies are available. Collectors commonly used for open-air swimming pools are of particularly simple construction, consisting of black rubber mats with the water flowing directly through them.

Flat-plate collectors have a highly heat absorbent metal plate, through which normally a water-propylene glycol mixture (60:40 ratio) flows. They are covered with a glass plate and insulated at the back to optimise heat creation in the same way as a greenhouse and so as not to lose the heat too quickly.

Hybrid collectors combine flat-plate collectors with a cover of photovoltaics. Combination of electrical with thermal energy creation is still in the early stages; the heat transferred away by the solar thermal technology can contribute to the cooling of the photovoltaic modules.

Vacuum-tube collectors have the highest efficiency. Their parallel tubes orientate themselves by rotating axially to the optimum inclination to the sun. For surfaces with a near-southerly alignment, they can be integrated in the horizontal and vertical directions into the architecture.

Solar thermal collectors and their efficiencies

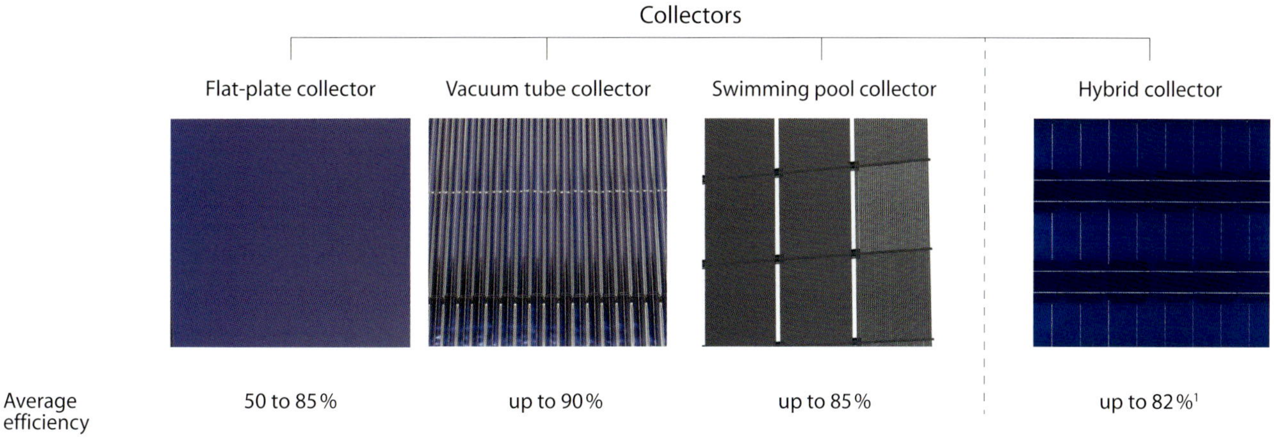

[1] Examples from manufacturers' information

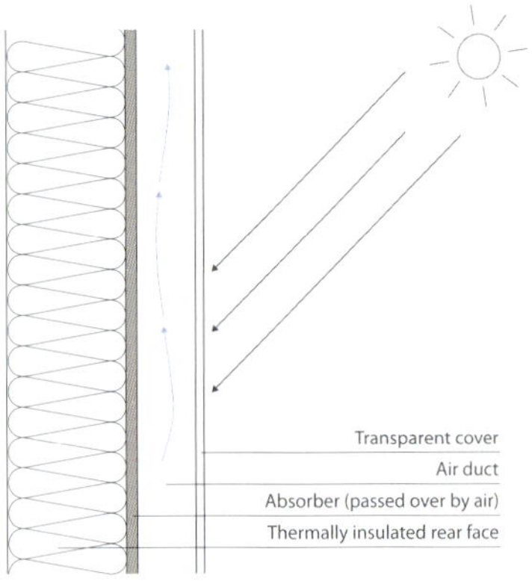

Air collectors

Air collectors use insolation to heat air and can help to temper or precondition room air. They are similar to solar thermal collectors in construction and function, but air has much less storage capacity.

They are based on simple technology. The collector consists of a dark absorption surface at a distance from transparent cladding fixed in place at the front. Outside air is let into the collector at one end. The air flows through the collector and is heated by the solar radiation striking the absorption surface. This heated air flows directly into the building by the stack effect or is introduced mechanically. Ideally, and for better control, the hot air flows into the ventilation system, where it is heated further or ducted to the rooms requiring heat. Air collectors have efficiencies of 55 % to 70 %.

Air collectors find application as prepreparation stages in mechanical ventilation systems or air-air heat pumps. They are of most use in winter and the months either side. In summer, the ventilation system must have an alternative outdoor air inlet to avoid overheating.

Air collectors are not widely used in residential and office buildings. In agriculture, on the other hand, they have been employed successfully for years for drying hay, grain and biomass. Storage of the created heat for use at night, for example, in a pebble-bed is worth considering. A further use could be to convert the heat of very hot air by means of an easily boiled medium (water, alcohol etc.) in a steam engine into mechanical or electrical energy. The waste heat from this process can be used for applications with lower heat demands. This multiple use of the created heat is known as cascading.

The use of very hot air for creating cold energy can also be done with an absorption type refrigerating plant (see p. 140).

Air collector facade – Gründerzentrum, Hamm, HHS Planer + Architekten AG, Kassel (DE)

Biomass

Generating heat from wood

Heating with wood offers an almost CO_2-neutral way of creating hot water for providing space heating and domestic hot water using mostly locally available resources. The CO_2 released by burning the wood is equal to the CO_2 it stored during the growing process. This view does not take into account the energy used in harvesting, processing and transporting the wood. The procurement, storage and feeding of the fuel must be planned for the site to ensure the procurement and heat provision chain runs smoothly for the operator. Keeping a store of enough fuel for a heating season or arranging deliveries of the required amounts under contract is recommended. Wood can be bought as logs, chips or pellets. Automatic boiler feeding and cleaning are possible in systems using wood pellets or chips. Logs can be used in natural draught boilers and wood gasification boilers. They are manually fed. A large buffer store is necessary for both systems.

Pellets offer the best alternative for satisfying the low heating demand of an Aktivhaus. Pellets are uneconomic for larger buildings or for providing district heat because their low energy content per unit volume would require considerable storage space. The use of chips or logs would be preferred here despite the higher effort required from the user during operation. Boilers are available in all sizes to suit the heat demand.

The stimulation of the local economy by the purchase of fuel, short transport routes and the use of a regenerative raw material are positive characteristics. Security of supply of wood is assured, for example, in Germany because low demand over many years means the potential of sustainable forestry management has not been exhausted.

However, the emissions of the fine dust associated with the burning of wood are viewed critically because of its effect on health.

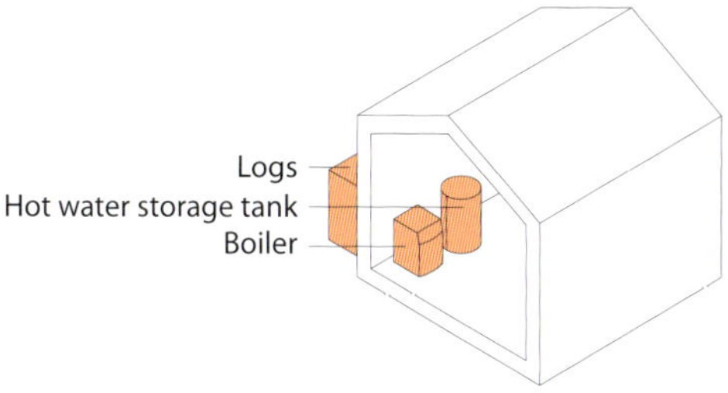

Schematic for logs as fuel

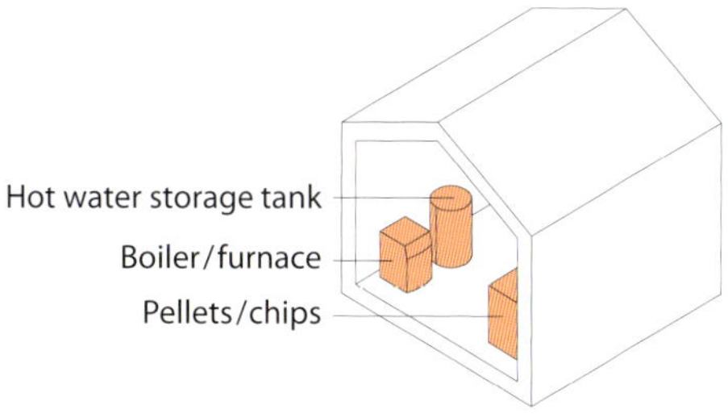

Schematic without an automatic fuel feed

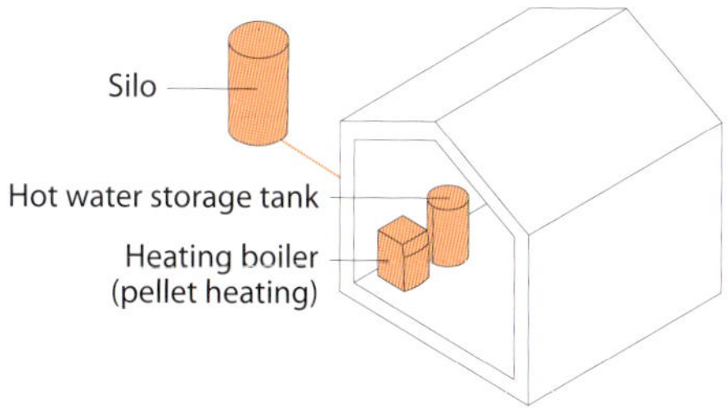

Schematic for wood as heating fuel fed from a silo

Energy carriers based on wood: logs, chips, pellets

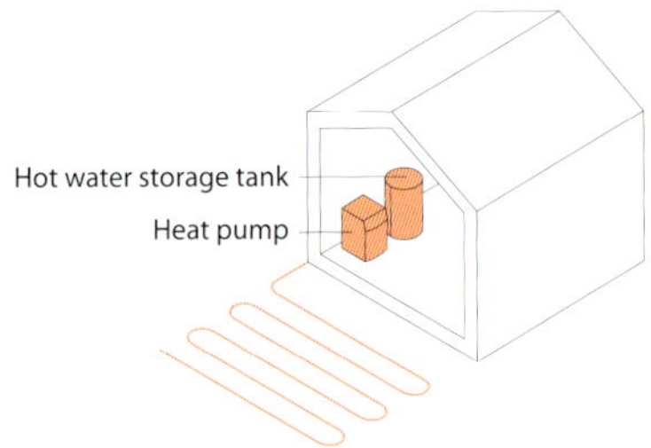

Schematic for near-surface geothermal energy

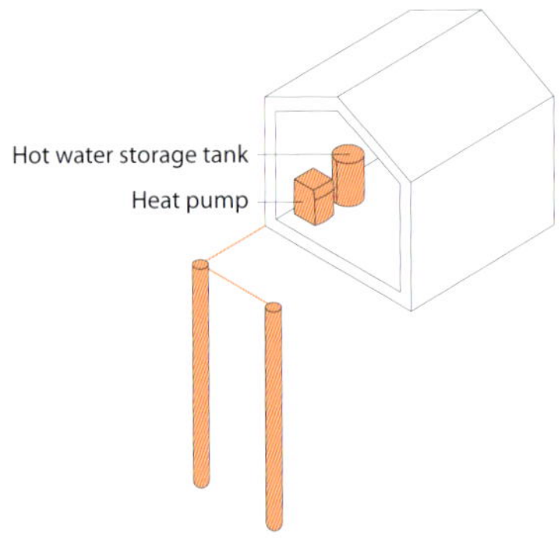

Schematic for deep geothermal energy

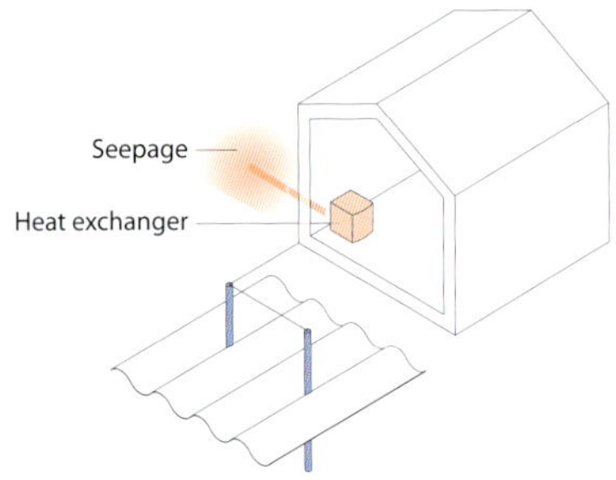

Schematic for groundwater as an energy source

Water, groundwater, ground

Generating electrical energy from water involves the conversion of kinetic energy by turbines and is hardly ever performed directly in or near Aktivhaus buildings. The thermal use of water, in particular groundwater, flowing or collected rainwater, is feasible much more often, given the right circumstances.

Direct cooling

There are two methods available for generating heating and cooling energy from flowing water, groundwater or rainwater. The water can be used directly for preconditioning and cooling at its natural temperature or brought to the required temperature level by a heat pump. The first method does not require any heating or refrigeration plant. Cooling comes to the fore when designing Aktivhaus buildings, particularly for uses with high internal heat loads such as offices. In this context, the direct use of flowing water, groundwater or rainwater to cover the cooling demand load suggests itself because normally the primary energy input and the operating costs can be greatly reduced; only pump energy is required.

The water is extracted from the ground through suction wells and used for cooling through a heat exchanger. The ideal water temperature for cooling is approximately 14 °C. The warmed water can be used for toilet flushing and irrigation of gardens, but most of it is transferred into an adjacent retention basin or stormwater overflow system. During times of low outdoor temperatures, groundwater can be used to precondition the incoming fresh air before it is heated by the air-conditioning system and hence contributes to reducing the use of other energy carriers.

Heat pumps

A heat pump uses the compression cycle with running or groundwater as a medium to provide energy for heating or cooling and condition an Aktivhaus according to the user's wishes or to fulfil specified requirements (see p. 133 ff).

Whether running water or groundwater is used as the medium depends on the location and normally requires some sort of official approval.

Heating or cooling with a heat pump is very economically efficient. Examples of buildings where heat pumps are used include server rooms, commercial kitchens and photographic studios. The cooling demand of these rooms is high, and in winter, late autumn and early spring, buildings used as offices have a significant heating demand. Combined heating and cooling units provide cooling while transferring the heat produced by their heat pumps directly into the heating grid. CO_2-based heat pumps ($COP < 4.5$) operate with higher feed temperatures on the heating side and with high system efficiencies, which makes them suitable for use in existing buildings.

Near-surface geothermal energy

The ground contains stored heat, which can be used by exploiting geothermal technology to provide heating and cooling energy in an Aktivhaus.

Boreholes to allow the use of geothermal energy (geothermy) are usually 50–100 m deep. An almost constant temperature at least 10 °C prevails at these depths. An alternative arrangement to make use of this relatively constant temperature, if circumstances on the site allow, is to bury a ground collector array consisting of a continuous series of loops of pipe placed horizontally at a depth of approximately 1.5–3 m. The geothermal heat is transferred through a brine medium to a heat exchanger. The temperature at this point is raised to a useful level by a water-based heat pump.

In some parts of Germany, for example, the price of energy produced in this way is already competitive compared to that of conventional technologies.

The potential of geothermal systems exceeds the actual demand for energy by a sizeable factor and it is inexhaustible on the human scale. A small heat pump is as suitable for single buildings as larger systems are for supplying districts.

Deep geothermal energy

The use of deep geothermal energy is of particular interest at locations where geothermal activity takes place relatively close to the surface. Germany, Austria and Switzerland have power plants in which turbines generate electricity from deep geothermal energy. For example, in Unterhaching, Germany, 40 MW of electricity are produced from a borehole some 3,500 m deep providing water at 120 °C. A number of projects with boreholes up to 5,000 m deep with a geothermal output as high as 80 MW are in design or being built.

Wind

The generation of electrical energy from wind for use in an Aktivhaus building can be done at the site or in the immediate neighbourhood.

Wind turbines

A wind turbine uses the wind to create electrical energy. The currently available wind turbines have three profiled rotor blades on the windward side of a horizontally rotating nacelle, which is attached to a gondola. In windy conditions, the aerodynamic shape of the rotor blades causes a dynamic pressure to build up on them so that they rotate. A generator converts this rotation into electrical energy. In most cases, the electricity is fed into the public grid.

Large wind turbines produce electricity at an economic rate today without the need for subsidy. A production cost of 6 ct/kWh, which corresponds to a final consumer price of 29 ct/kWh (green electricity, Q1/2016), is achievable taking into account all other costs. Increases in efficiency over the next four years should lower prices by 1.5–2 ct/kWh. This would be the equivalent of the achievable carbon electricity price on the electricity market.

Small-scale wind turbines with an output of up to 5 kW can make a contribution to the use of regenerative energy sources. Such systems can be operated even in densely populated urban areas. Use directly on site should always be the priority, because the investment cost per kW is relatively high and the corresponding feed-in tariff for smaller turbines is usually below cost. For installations in urban areas, the wind conditions depend greatly on the surroundings and building geometry. Field trials are under way at the moment to obtain valuable knowledge about wind conditions and the possibilities of using these systems in urban areas for use in future designs.

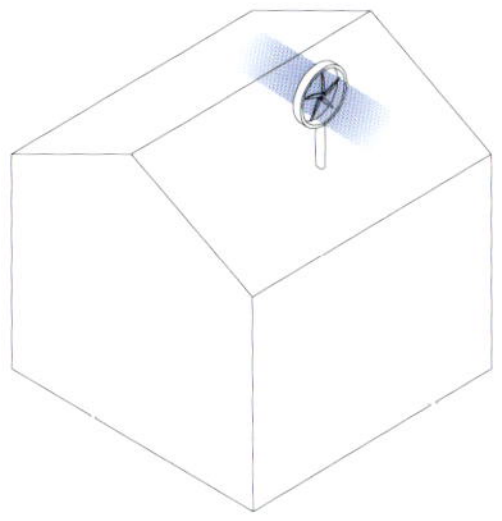

Use of a wind turbine attached to the building for generating electricity

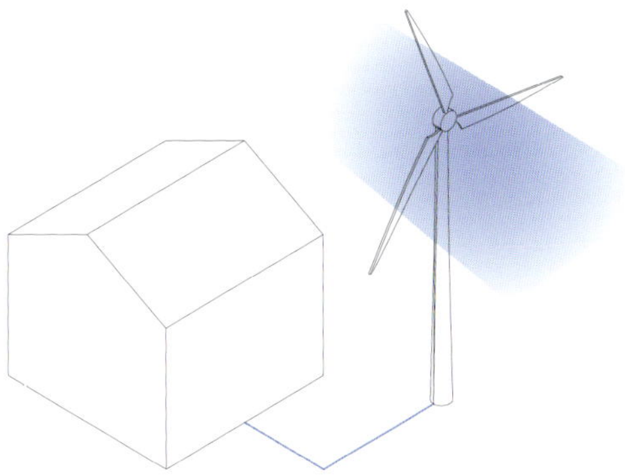

Use of a free-standing wind turbine for generating electricity

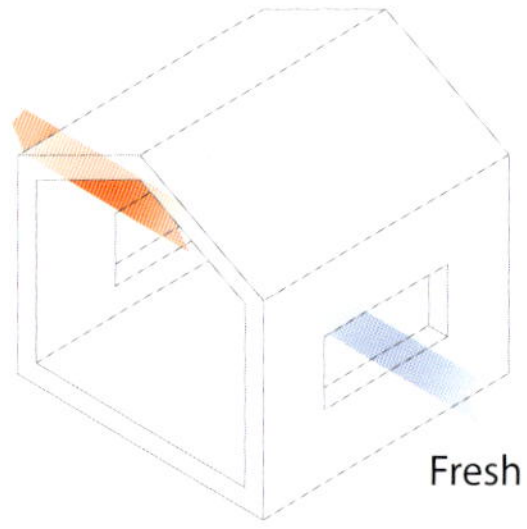

Natural window ventilation

Outside air

Natural ventilation

The energy qualities of building skins are very firmly based on attaining a superior seal of the building envelope. This means that uncontrolled ventilation through leaks and gap ventilation, a common property of earlier buildings, no longer happens in Aktivhaus buildings. A high level of comfort by providing excellent air quality, minimising ventilation heat losses and the greatest use of free ventilation are therefore the requirements to be met for hygienic ventilation corresponding to need. A controllable mechanical ventilation system is essential to ensure optimum air quality while avoiding energy losses. Such a system is supplemented by free ventilation whenever possible.

Window ventilation

Free window ventilation without the use of a ventilation system can be provided in an energy-efficient new building only by regular purge ventilation. This must take place several times daily at regular intervals (also during the night) for 5–10 minutes. This purge ventilation may be manual or automatic.

The manual type of natural ventilation can maintain the desired standards of comfort and hygiene only if the user exercises high discipline. If occupied rooms are not regularly purge ventilated, air quality and air hygiene fall, which can lead to building physics and hygiene problems, and even endanger health. With continuous ventilation, for example, through tilting windows, air and surface temperatures in the room fall drastically. Not only thermal comfort suffers as a result, in cold weather, a lot of energy is spent providing heat to counteract this effect, and the energy efficiency drops.

One alternative is mechanical window ventilation. In this system, an electrical motor opens the window to allow controlled natural ventilation; at its simplest, a time switch is set to regulate opening and closing to suit the ventilation requirements of each room automatically, taking into account the actual temperature, precipitation and wind conditions. Users normally have the option here of intervening directly, in other words they can open and close the windows themselves. The design should have an adequate number of opening vents.

The psychological effect of having the use of opening windows is particularly important in summer and the months to either side. The high thermal comfort of an Aktivhaus allows the user to dispense with window ventilation and eliminate the associated high heat losses during periods of low outdoor temperatures.

Free window ventilation can, however, be used in summer and the months to either side at any time of day, and can replace mechanical ventilation. Viewed over the whole year, allowing users to control the ventilation and react to any degradation of the air quality or change in weather can be seen as a worthwhile additional feature.

Cross ventilation, created by opening windows or vents ideally on directly opposite walls of a building, can be used to maintain a continuous flow of air through the building interior. A sufficient number of opening vents of adequate size should be provided. In summer, overnight cross ventilation can very effectively cool or discharge a building in which heat has built up over the day. The building then stores the night-time coolness for the following day. The high air throughput provided by cross ventilation also helps to make the high day-time temperatures feel much more bearable.

Mechanical ventilation

An automatically regulated ventilation system with heat recovery is worthwhile installing in an Aktivhaus to avoid the high energy losses from window ventilation. An efficient heat recovery system can achieve heat loss savings in winter of 75 % to 90 % and cooling energy loss savings of up to 60 % with actively cooled buildings in summer. The electrical energy for the fans recovers 8–15 times the equivalent amount of heat energy.

The mechanical ventilation system helps to ensure a constant inflow of air to achieve a minimum of 0.3 air changes per hour and a rate of air change suitable for the demand at all times. High comfort ensures any moisture is conducted away, preventing formation of damp along with the danger of mould. The control of the ventilation system can be continuous by a programmable time controller, made occupation-dependent by a switch or a presence detector, or even be completely demand controlled. A demand-controlled system with sensors measuring CO_2, mixed gas or VOC reacts to the air pollution levels in the room and can provide the optimum rate of air change to suit the circumstances at any time. This saves considerable ventilation heat losses and lowers operating costs. In spite of automatic regulation, the user should be able to intervene at any time to increase acceptance and wellbeing, especially in the regulation of temperature (+/- 5K) and air volume flow rate. The ventilation system should have a bypass so that the waste heat recovery feature does not have to be used in summer.

In a residential Aktivhaus building, the ventilation system can be integrated centrally or locally distributed in the building. The supply air flows directly into main rooms, such as bedrooms, living rooms, children's rooms and working rooms, through outlets suitably positioned to avoid creating draughts. Extract air is drawn out of rooms with high pollution loads or emissions, such as the kitchen, bathroom and WC, and ducted to the ventilation system. Here it flows through a heat exchanger, which removes heat from the air and transfers it into the incoming fresh air. This form of ventilation is called cascade ventilation. The air blown into one room is used again a number of times in several rooms. There is always a clear direction of air flow.

The fresh air is drawn in through an opening in the facade or roof, the exhaust air is blown out of the building, taking care to avoid any short circuiting of the air flows. An earth tube exploits the stable temperature of the ground to precondition fresh air passed through it before it is heated or cooled in the ventilation system. It is usually preheated in winter and cooled in summer. A second air intake directly from the outside air is recommended for the intervening months. The air passes through a filter before it enters the ventilation system.

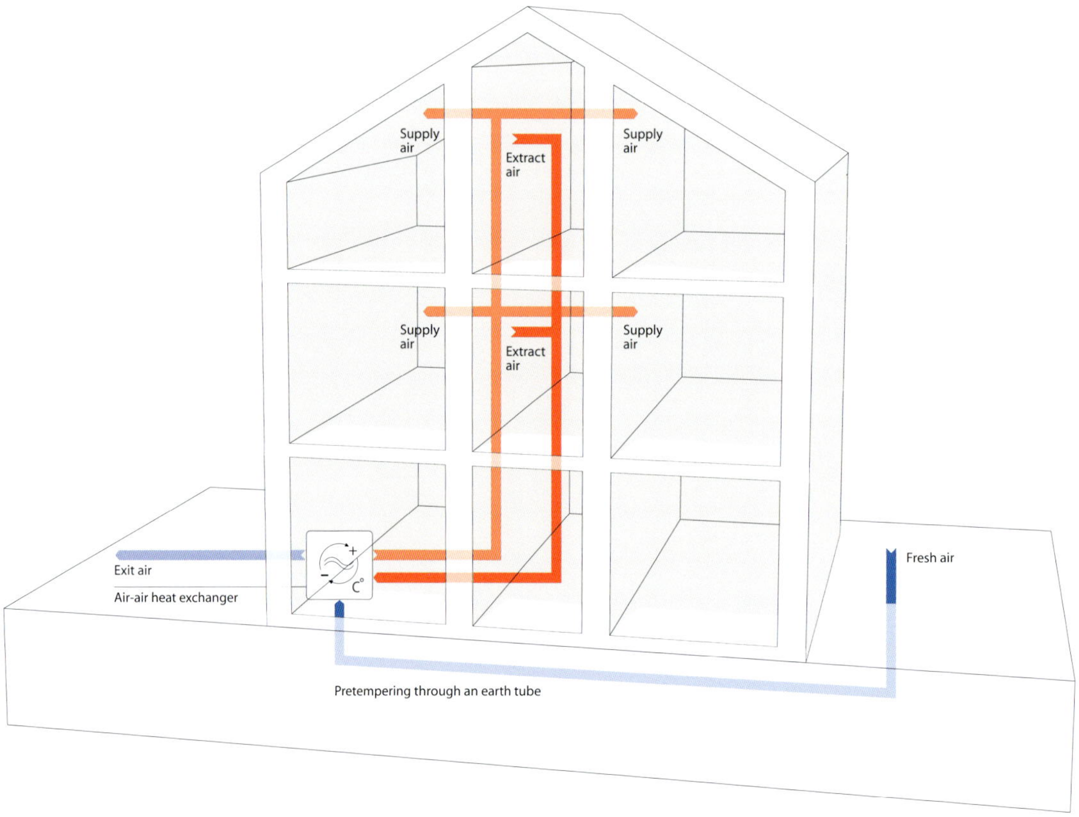

The working principle of a centralised comfort ventilation system with heat recovery and air intake through an earth tube

The filter removes even the smallest particles, such as dust or pollen, from the outside air and therefore offers a comfort gain, and not just for people who suffer from allergies. Regular replacement or cleaning of the filter is essential for maintaining the level of air quality and the proper functioning of the ventilation system.

Heat recovery

The use of a ventilation system with heat recovery reduces the heat demand and can therefore contribute to a substantial reduction of the size and scope of the heating system. With a well-designed building and a high standard of thermal insulation of the building envelope, the heat demand can be so small that only the supply air needs to be heated. This can be done centrally in the main ventilation plant or locally by a heating register at each outlet. Purely extract air systems without heat recovery are not considered here because they lead to high ventilation heat losses.

Conventional crossflow heat exchangers transfer the heat from the extract air, normally through highly conductive metal surfaces, to the supply air, without the two air flows being in direct contact with one another. This can make the air very dry, particularly in winter. There is a new development available to mitigate this effect: a high-quality paper-based material covers the exchange surfaces, which then are capable of exchanging not only heat but also moisture. The relatively dry outside air is preconditioned by the moister indoor air and the result is a more pleasant indoor climate. The use of humidifiers in air-conditioning systems should be avoided if at all possible because of the high energy demand and the danger of microbial growth.

Decentralised systems with heat recovery are also available for retrofitting a mechanical ventilation system. This option involves less building work. They are inserted directly into small-diameter holes bored into the building's external walls. Air changes take place first in one direction, then in the other. Air is transported from inside to outside and the energy transferred to the storage medium directly in the air flow. Then the air flow reverses direction and feeds outside air into the interior. The storage medium gives up its heat to the supply air. This alternating direction of flow achieves up to 90 % heat recovery.

Heat pump

A heat pump extracts the heat contained in environmental energy sources, such as the air, the ground and ground water, sometimes even waste heat and waste water, using heat exchangers. This heat is brought to a suitable level for heating or cooling purposes by another pumped circuit. Before it can be used for heating or cooling, a buffer tank ensures the final system can operate

Working principle of a heat exchanger

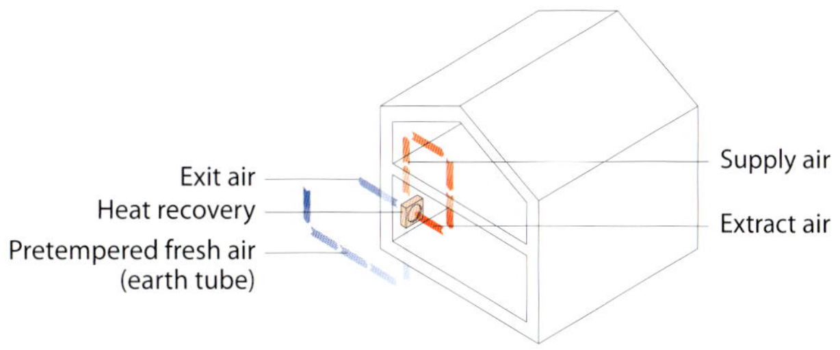

Working principle of a ventilation system with heat recovery

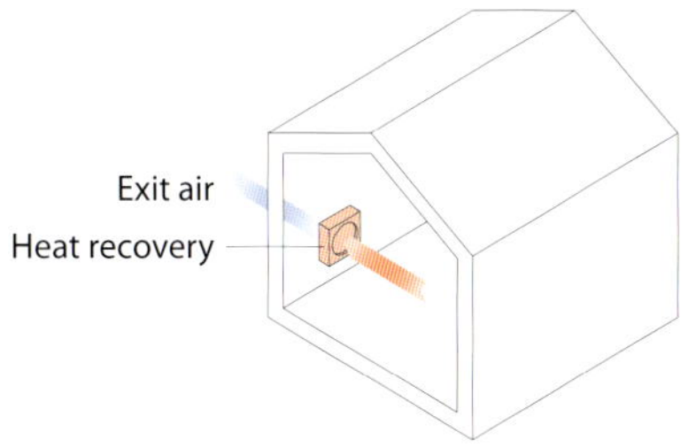

Schematic for ventilation with heat recovery

all the time, respond continuously to the changes in demand and achieve high efficiencies.

The operating principle of a heat pump is similar to that of a refrigerator. The refrigerant vapour is compressed to a temperature necessary for use as domestic hot water and space heating. Using a low temperature level for tempering a building increases the efficiency of the whole system. However, this in turn requires larger heating surfaces. Using thermally active building components appears particularly worthwhile with these systems. This involves whole building components and therefore large surface areas are available to transfer the heat or cold to the indoor space. The solid components incorporate pipes to carry a heating or cooling medium to allow them to be used as heating or cooling surfaces. Underfloor heating is recommended for use with heat pumps.

For cooling with a heat pump, the distinction must be drawn between passive and active cooling. With passive cooling, the heat in the building is extracted by a heat exchanger and transferred into the heating circuit and given up to brine or water. With active cooling, the heat pump functions in reverse so that it behaves like a refrigerator and produces active cold to cool an Aktivhaus and the equivalent heat is given up to the environment.

Using air as the medium, the temperature of the adjacent outdoor air is transferred through a heat exchanger to the water circuit of the heat pump. Even in the low air temperature conditions of winter, air source heat pumps achieve good efficiencies. Installation is simple and the initial investment costs are lower than with other heating and cooling systems. Systems that use geothermal technology, groundwater or watercourses as a medium perform even better and achieve the same high efficiencies all year round because the source medium remains at a constant temperature.

The type of compression process taking place inside the heat pump influences the efficiency of the whole system too. The coefficient of performance (COP) gives information about the ratio of heat output to the power consumed. The seasonal coefficient of performance (SCOP) describes the average COP under defined conditions over a year.

Heat pumps can be used to good effect on a wide range of projects, from detached houses up to large district heating systems. It is particularly advantageous to provide the operating electricity for the heat pump from regenerative energy sources, with the energy being created either in the immediate area or directly on the Aktivhaus.

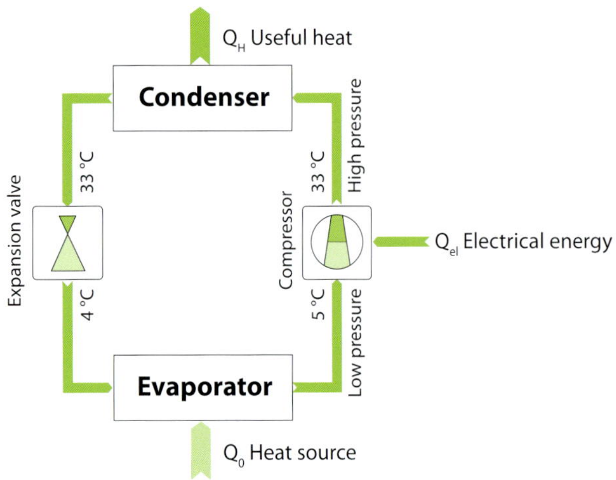

Working principle of a heat pump

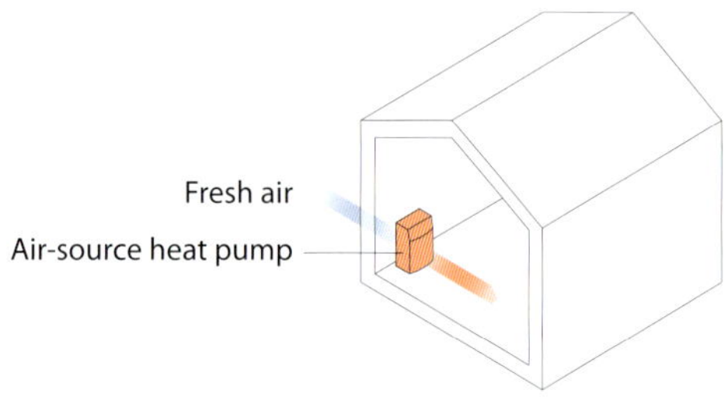

Schematic for ventilation with heat recovery

Waste heat

Heat recovery

An Aktivhaus must have a ventilation system. This system recovers heat from the extract air and contributes to reducing the heating and cooling energy requirement by minimising ventilation heat losses. The heat recovery coefficient is a measure of the efficiency of a heat recovery system. It takes the temperature differences between the supply and the extract air, and between the extract air and the outdoor air, and expresses them in the form of a ratio. Typical values for various heat recovery systems are:

Crossflow heat exchanger	50 % – 70 %
Rotary heat exchanger	50 % – 80 %
Cross counterflow heat exchanger	70 % – 90 %

The choice often falls on the crossflow heat exchanger because it offers a good compromise between efficiency and maintenance costs. In comparison, a rotation heat exchanger has higher costs for investment and maintenance (= life cycle costs). Its advantage is that moisture recovery by condensation is possible. However, this has higher maintenance costs to ensure hygiene because of the possibility of microbial growth. Highly efficient cross counterflow heat exchangers are used mainly in small decentralised ventilation systems. These take a flow of air out of the building. The warm air flow is passed through a ceramic heat exchanger, which becomes thermally charged. After a period of time, the air flow is reversed and the incoming fresh air is preconditioned by the heat stored in the thermally charged ceramic heat exchanger. A ventilation system should have a bypass that can open automatically, depending on the indoor and outdoor temperatures, to allow the air to enter or exit the building without going through the heat exchanger. This can be beneficial in summer, when heat recovery is helpful.

Waste water heat recovery

The recovery of thermal energy from waste water, for example, from industrial concerns and households, has been an untapped resource until now. Sources can include waste water from a single building, industrial waste water or the public sewer system. The technology used for heat recovery may vary as appropriate. For small demands, such as for a house, systems are designed to be integrated directly into the property's sewer connection pipe. The outer wall of the waste water pipe is enclosed by the fresh water supply pipe. Without the involvement of any further technology, the warmer waste water heats the incoming water and preconditions it ready for hot water preparation.

The whole heat demand of an Aktivhaus building could be covered by the use of the waste heat from a continuously flowing public sewer. The temperature of the waste water can be up to 40 °C. A heat exchanger removes some of this heat and a heat pump brings the temperature of the medium to a suitable level. Different types of heat exchangers can be used for this purpose. To keep maintenance costs to a really low level, they can be integrated directly into the outside wall of the pipe carrying the waste water.

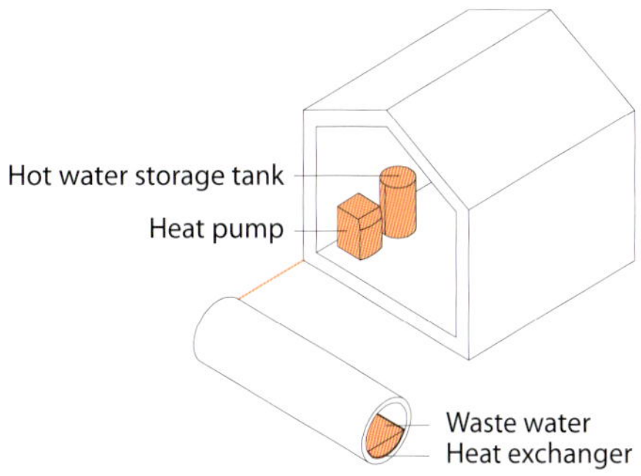

Schematic for heat recovery from waste water

Generation of electrical energy, heat and cold

Combined heat and power

A cogeneration unit uses combined heat and power (CHP) to convert fuel into electrical energy and heat (process and non-potable water). The use of both these sources of energy raises the efficiency of a cogeneration unit compared with a plain combustion furnace. To make most effective use of a cogeneration unit, there must be coincident demands for heating and electrical energy.

In a combined heat and power unit there is normally an internal combustion engine, although electrical generators, steam engines or wood gasifiers are sometimes appropriate to use. Another alternative is combining a gas burner with a Stirling engine to produce electrical energy. CHP can be operated sustainably with regenerative solid fuels or biogas.

The waste heat produced during combustion is used to heat water for heating and other processes. With this double exploitation of the energy carrier, the energy losses are only about 10 %, depending on the type of plant and use. Much higher losses occur in conventional systems where electricity and heat are produced separately. Compared to CHP, these separate systems use more fuel to produce the same amount of heat and electrical energy.

Electricity created but not used directly on site can be stored in batteries for use later or fed into the public grid. This possibility of feeding energy into the grid is of benefit to the energy supply companies, who can compensate for fluctuations in the electricity grid and cover peak loads with centrally controlled decentralised CHP units. CHP units are primarily used in buildings with a large energy demand, such as apartment blocks or offices, and operated by a local energy supplier. The advantage of the high efficiency of these units can be passed on to the customers through a lower purchase price.

CHP for own-consumption of energy is used mainly for constant, large consumers such as industrial plants, hospitals and apartment blocks. The coupling of neighbouring consumers, such as houses in a housing development, offers an interesting option for its use. The base load of heat and electricity supply can then be provided by CHP. The base heating load of the heating requirement is taken as the reference to guarantee the highest possible number of hours in continuous operation. The average load can be covered by further CHP modules with correspondingly fewer operating hours per year. This modular design is supplemented for peak loads by one or more separate boilers, which can react quickly to demand because they are, for example, biogas-fired condensing boilers.

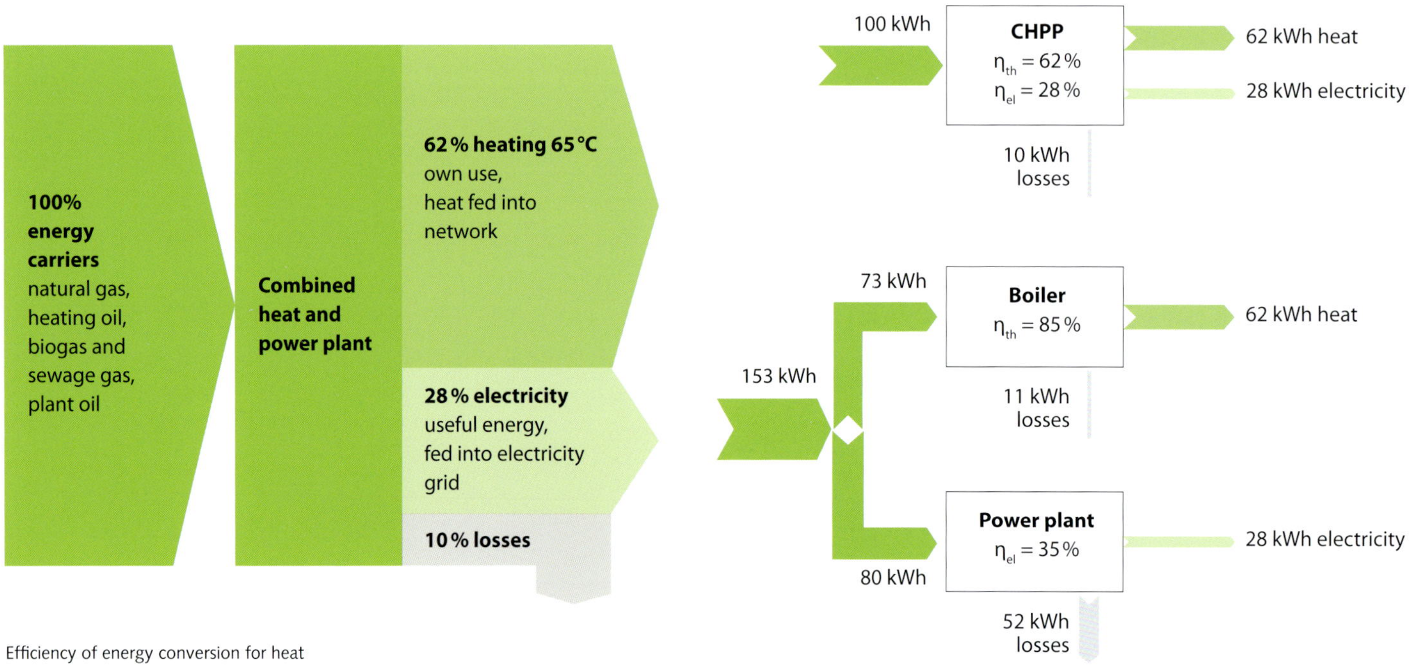

Efficiency of energy conversion for heat and power cogeneration

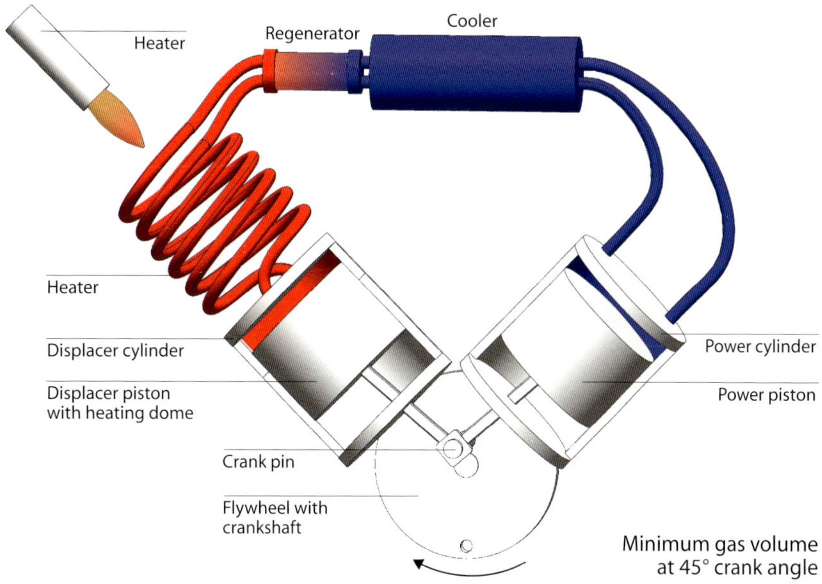

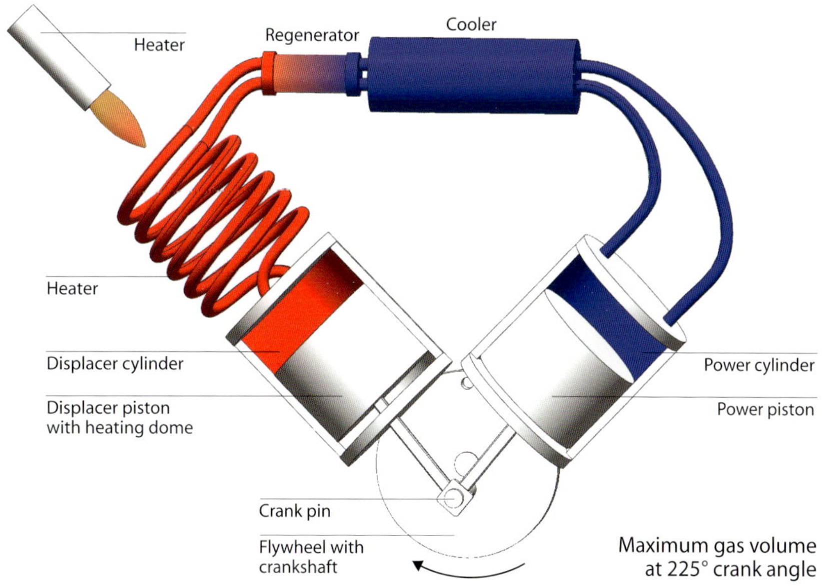

Operating principle of a Stirling engine

Smaller CHP units for residential properties achieve only low efficiencies because the demand is hardly ever constant. These smaller CHP systems usually combine gas condensing boilers with a Stirling motor in one unit. A heat store acting as a buffer makes the waste heat and the heat not directly used available for use later. Larger buffer stores enable the CHP units to run for longer and increase the efficiency of the overall system. More modular CHP units are becoming available. These can react at 50–100 % of their maximum output to a changing demand and increase the operating time and efficiency of the whole system. Combination with a larger buffer store and a condensing boiler for regulation is worth consideration.

Fuel cells

Almost all previously known methods of creating electrical energy have involved burning a fuel, in other words producing heat to create motion, which is then converted into electricity by a generator. This method of creating

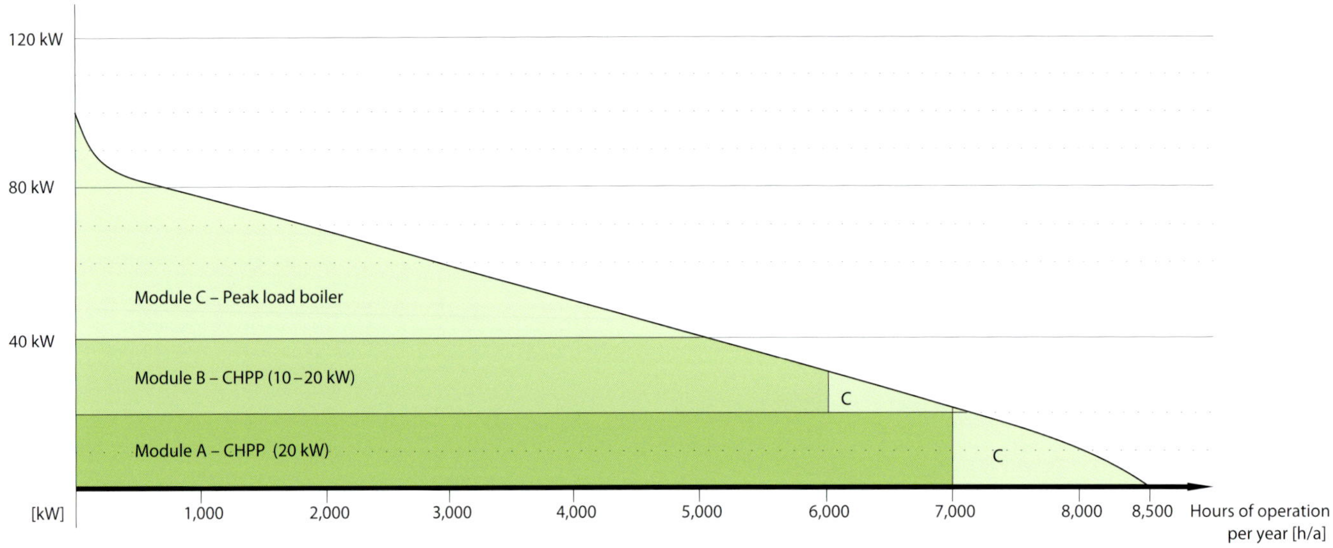

The modular CHP concept increases availability and reduces operating hours in the uneconomic part-load range. Conflicting objectives, depending on the system: having a few large CHP modules results in low operating hours of the individual modules. Having many small modules increases the operating hours and the capital investment costs. The degree of efficiency of CHP units increases typically with their size.

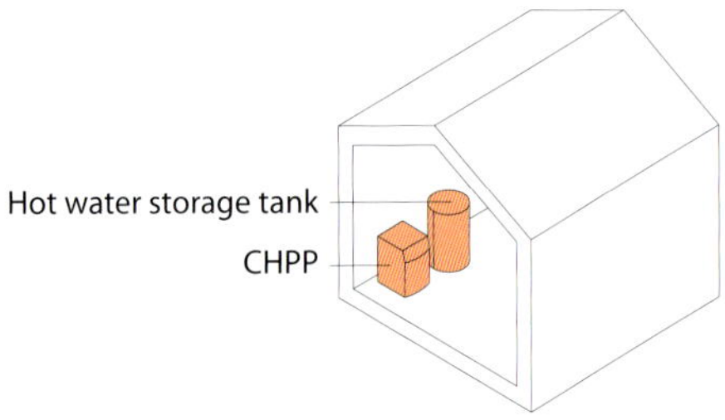

Schematic for CHPP

electricity is associated with a low degree of efficiency because of high thermal losses.

In contrast to this, a fuel cell can produce electricity and heat continuously through a controlled chemical reaction of oxygen with hydrogen (which can be removed from e.g. natural or biogas). The direct combination of hydrogen and oxygen in the fuel cell is avoided because of their high reactivity. Hydrogen is split into positively charged protons and negatively charged electrons at the anode with the help of a catalyst. The protons travel through a membrane to the cathode. The electrons move along an electrical circuit to the cathode. Protons and electrons react at the cathode with the oxygen supplied to form water. Natural gas can also be used as the fuel after first being changed into hydrogen rich gas by a reformer. Fuel cells operate at temperatures of 60–1,000 °C, depending on the technology. The heat generated by the fuel cell during operation can be used to heat process or heating system water.

The fuel cell process is considerably more efficient than the CHP process. With a fuel cell, the proportion of electrical energy obtained is higher than that of heat, which makes fuel cells more appropriate for applications where the demand for heat is less than the demand for electrical energy. However, when considering efficiency, it is important not to ignore the energy expended in producing the fuel – e.g. hydrogen – otherwise the overall performance cannot be meaningfully compared with other technologies.

As with combined heat and power, when a fuel cell is used to supply a residential building with heat and electrical energy, the focus is on the supply of thermal energy. The electrical energy generated at the same time as the thermal energy is used on site, stored or fed into the public electricity grid. The thermal output is transferred to a store to allow a continuous supply of heat and increase the running times of the fuel cell. A fuel cell combined heat and power (FC CHP) system designed to satisfy the basic demand for thermal output results in very long operating periods at optimum efficiency under constant load. An additional (conventional) heating system can cover the peak thermal loads.

The electrical output obtained must be converted by an inverter from direct current to alternating current. A fuel cell has a primary energy demand up to 25 % lower and CO_2 emissions up to 50 % lower than a CHP system. Supplying energy to larger buildings from a fuel cell as a substitute for existing old heating technology will become much more within reach with the anticipated falling prices of this technology. The development of this idea to provide small stationary supply systems with outputs of 1–5 kW of electrical energy is within sight.

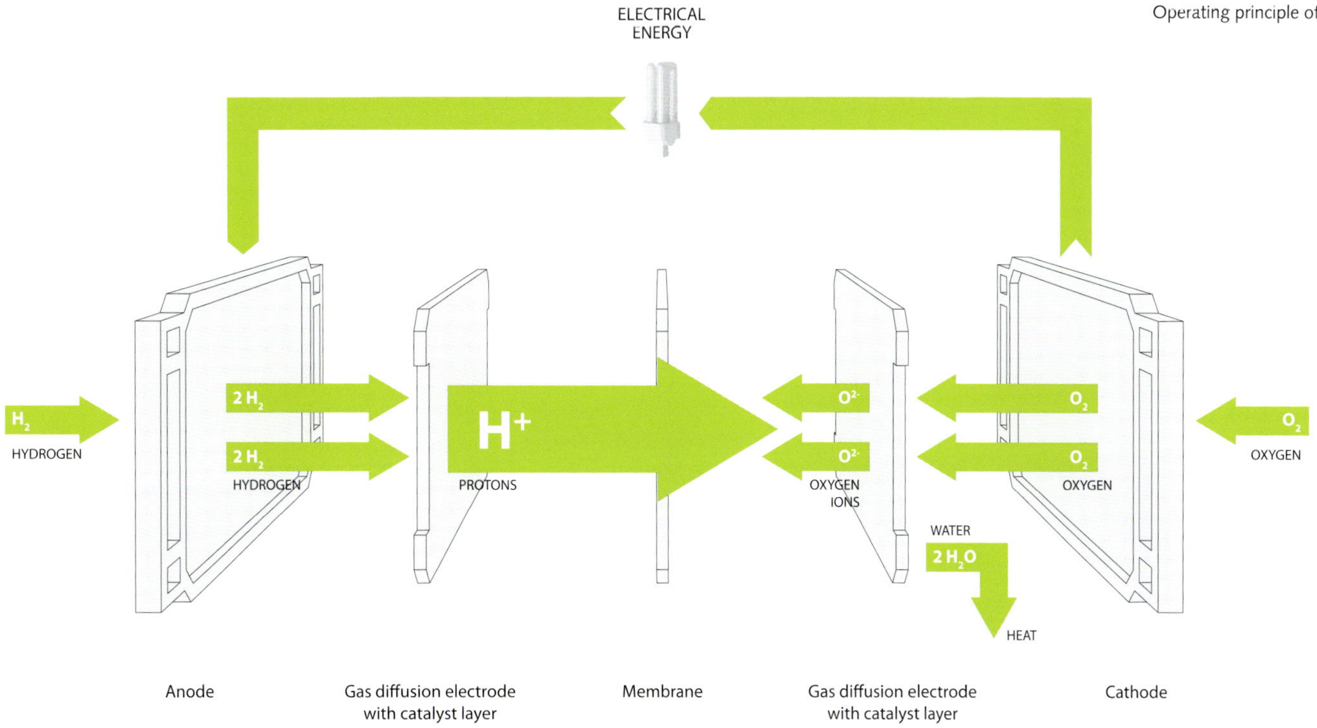

Operating principle of a fuel cell

Absorption cooling

A high comfort level is achieved when the ambient temperature is about 22 °C at approximately 50 % relative humidity. When outdoor temperatures are low, the comfort level in an Aktivhaus can be achieved without consuming a great deal of energy. In summer, on the other hand, there will be some days on which this comfort level cannot be achieved without active cooling, especially in hot and humid or hot climate zones. At a temperature of 27 °C, people's production capacity drops to 70 %, at 33 °C it is only 50 %. Recommendations in Central Europe give a maximum temperature of 26 °C to ensure people feel well and their ability to perform their tasks is not unnecessarily limited. If the high temperatures of the interior rooms cannot be brought down to a bearable level by passive or active means using minimal energy, then some sort of air-conditioning system should be considered. The difference between indoor and outdoor temperatures should not exceed 6 K. For wellbeing, the indoor temperature should be adjusted to suit. Incidences of colds increase if the temperature difference is greater (see p. 63 ff).

The energy consumption of air-conditioning systems on hot days in centres of population is a problem, despite their high efficiency. The use of such systems is sensible only where outdoor temperatures are permanently high and/or sensitive facilities such as hospitals, research centres or laboratories need to be protected. The reduction of cooling demand should always have priority over its efficient fulfilment.

Measures to do this include improving insulation, reducing the heat entering the building through glazed areas and internal heat sources, and increasing the through flow of air and cross ventilation. Life cycle costs (LCC) have to be taken into account. The initial investment in an air-conditioning system is usually less than 20 % of its life cycle costs. An efficient system and an optimised concept designed precisely to meet the users' needs and optimised for the available energy services can reduce LCC significantly.

Absorption refrigeration machines/ combined heat, power and cooling

In a combined heat, power and cooling (CHPC), an absorption refrigeration machine (ARM) is linked in series with the CHP process. This can convert the heat generated in the CHP process into cooling energy and thus take over the cooling provision.

An ARM is a binary system driven by a temperature-responsive solution of a refrigerant. The refrigerant is absorbed by a second substance at a low temperature in the cycle and then separated out of it at a higher temperature (desorbed). This process exploits the temperature dependent solubility characteristics of two substances and can work only with substances that always remain soluble under the prevailing temperatures. Substances such as lithium bromide, which absorbs water, or water, which absorbs ammonia, are often used. This process for creating cold energy is known also as thermal compression.

The cold energy can be stored in a buffer to be made available later. Working alongside an adequately sized store, an absorption refrigeration machine can run for the long continuous periods necessary for it to operate at high efficiency. Compared to a compression refrigeration machine, this method of creating cold energy uses less primary energy.

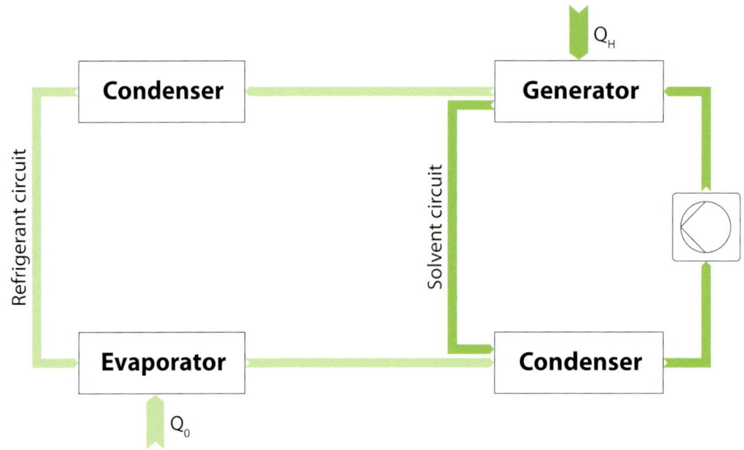

Principle of an absorption refrigeration machine

An ARM can provide cooling very cost effectively if a free or low-cost source of heat is available. Possible solutions include use in combination with industrial waste heat, superfluous district heat in summer, deep geothermal heat or solar thermal energy as a heat source with a connected ARM. The heat source should be capable of providing a temperature of 80–130°C.

The scope of application of ARMs lies mainly in buildings that require cooling at particular times to operate, such as industrial plants, laboratories, computer centres and hotels.

Air-conditioning systems can be classified as centralised or decentralised. Decentralised units can provide the necessary conditioning, ventilation and filtering, while taking up very little space. They are mainly used to condition individual rooms or small sections of buildings and are often installed under hallways or within a facade. The same unit draws in the outside air and expels the exhaust air. Most units cannot fulfil the highest requirements for hygiene, system regulation, humidity and draught control. One of the reasons for this is poor regulation. The development of decentralised air-conditioning units began with the window units often seen in Asia and America. The vertically sliding windows typically found in these parts of the world make installing or retrofitting one of these systems easy. The associated disadvantages and effects of wind pressure, increasing numbers of leaks in the building envelope, and further unfavourable conditions have long been tolerated.

Split units can compensate for some of these disadvantages. They are divided into an external unit, which contains the condenser or compressor and one or more (multisplit unit) interior units, which contain the evaporators. The technology used can be compared with principle of the heat pump (see section on heat pumps, p. 63). The principle of recirculating air cooling, which depends on fresh air being taken in from outside, means that it can cool only the air present in the room (recirculating air). This type of system is not usually combined with a ventilation system with heat recovery. The cold is lost through the necessary window ventilation and is replaced by hot outdoor air, which then has to be cooled again. In the case of conventional air-conditioning systems, the compressor either runs at maximum output or is switched off. This stop-start-maximum operating mode is not good for the durability of the unit's mechanical components. The abrupt loads placed on the electricity supply grid are seen as undesirable. Inverter technology can alleviate these bad effects by introducing demand-based control.

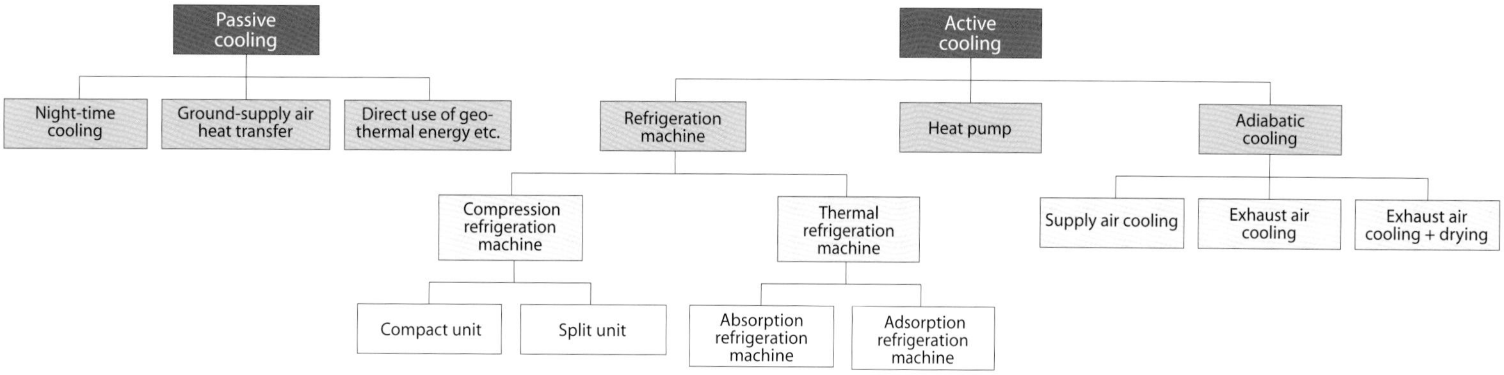

Overview of different types of air-conditioning systems

Central air-conditioning systems offer a more efficient and convenient solution.

Central air-conditioning systems provide conditioning, air handling, filtering, humidifying and dehumidifying in a centralised unit or several components forming one unit. The supply air is conditioned centrally and conducted to each room. The extract air is collected. The energy it contains can be used to precondition the fresh air through an efficient heat recovery system. As with comfort ventilation, a bypass of the heat recovery system should be provided to avoid room temperature building up. This effect is caused by internal heat loads (people, electrical consumers, lighting) and by passive solar gains. In many cases, these can lead to the extract air temperature being higher than the required internal temperature and therefore it makes sense in terms of energy to conduct the extract air directly outside without it going through the heat recovery system.

Central air-conditioning systems differ in the way they transfer cooling energy into rooms. Air/water systems condition the blown-in fresh air. At the same time, a wet transfer system, such as chilled ceilings or building component activation etc., covers the cooling demand. This system can deliver high amounts of heat and cold energy. Ease of regulation and adjustment of the output to suit the demand of individual rooms and parts of buildings provide great flexibility. Air-only systems cool the building using just the supply air. The saving in initial cost must be balanced against less flexibility, poor regulation and reduced capacity. By careful choice of the air volume flow rates, the design should seek to avoid creating draughts. Applications for air-only systems tend to be mainly larger assembly rooms such as trade fair halls and theatres, which require surfaces to be accessible and an open interior design. A panel cooling system would create obstructions. The supply and extract air flows, usually located very close to the air-conditioning plant, are often taken or expelled through the roof of the building. Installing an earth tube as an additional air intake is advisable to exploit naturally occurring differential temperatures. In summer, the fresh air can be brought in through the earth tube and cooled by the lower temperature of the ground without expending additional energy. In winter, this fresh air is prewarmed by the ground, which is at a higher temperature than the outdoor air.

The most popular method of providing cooling is an electrically driven **compression refrigeration machine**. This uses the physical effect of the vapour-compression cycle. It can be compared with principle of the heat pump (see section on heat pumps p. 133). The refrigerants begin to boil under pressure at very low temperatures. The boiling point is below the desired cooling temperature. As it boils, this medium absorbs heat and becomes a gas. The refrigerant passes from the compressor to the condenser. The pressure is increased to a level at which the gas condenses. The condensation temperature is higher than the ambient temperature and that of the cooling water. The principle is found in many applications including refrigerators and is technically mature.

Diagram showing the principle of the vapour-compression cycle

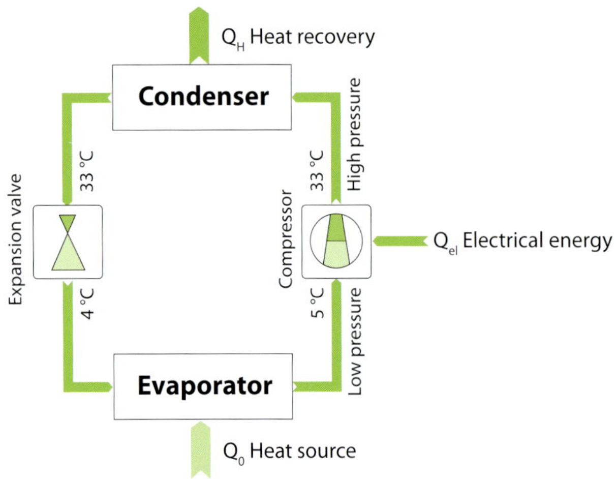

Magnet refrigeration (magnetocaloric effect)

Recent years have seen attention turn to raising the efficiency of vapour-compression cycle refrigeration following an innovation in the field of material science based on a discovery made in the 19th century and the new materials arising from the innovation. At the heart of this advance is the magnetocaloric effect.* Certain materials heat up when placed in a magnetic field. They cool again to the ambient temperature when taken out of the magnetic field. Up to now, the technology has required powerful electromagnets that consumed more energy than they created through the heating effect. The use of permanent magnets has meant this discovery can be taken up again, this time on an economical basis. Rotors with small blades made from magnetocaloric material have been developed to the prototype stage. Suitable for a refrigerator, they are no larger than 60 × 15 × 15 mm. The rotor turns at about one revolution per minute; it heats up in the magnetic field and cools down out of it. The material is the rare earth metal gadolinium or a similar substitute. It is thought that this technology could reduce the electricity consumption of cooling processes by up to 50 %.

* Emil Wartburg discovered the magnetocaloric effect in 1881.

Adiabatic cooling is a very efficient means of producing cold energy. It is based on the principle of evaporative cooling, which requires no electrical energy. Water evaporating in a closed system removes heat from the system and reduces its temperature. When this principle is used in an air-conditioning plant, spray nozzles create a fine water mist to moisten and therefore cool the air. Air with a low level of saturation can cool by 2.5 K per gram of added water. The outdoor air can be cooled by up to 10 K. There is a limit to which buildings can be cooled using this method. However, a recommended cooling of up to 6 °C below outdoor temperature is achievable in many situations. The operating costs are about a third less than vapour-compression systems because natural evaporation produces the cooling effect and electricity is saved because there is no compression cycle. There are three possible ways adiabatic cooling can be used in an air-conditioning plant:

With adiabatic cooling of the supply air, slightly saturated outdoor air is conducted through a humidifier, which cools it. Comfort needs to be maintained with respect to room temperature and humidity because this form of cooling increases the relative humidity of the incoming air. The amount of cooling is limited in practice to keep the humidity in the comfortable range of about 30–65 %. The moisture in the air needs to be less to ensure comfort. At a room temperature of 26 °C, the air should not contain more than 55 % moisture. The system requires a supply of very dry outdoor air. In Central European climate zones, this is mostly not the case. In very dry parts of the world, which would be ideal for this technology, water is in short supply and hence its use for air conditioning is not seen as a priority.

The second way is in the adiabatic cooling of extract air in conjunction with heat recovery. The extract air is moistened and therefore cooled before it goes into the heat recovery system, where a plate heat exchanger transfers the cold from the extract air to the supply air. The potential cooling capacity is considerably greater because no account needs to be taken of the temperature of the introduced water. The extract air can be completely saturated, and therefore its relative humidity can be up to 95 %. The corresponding cooling performance can be read from the h-x diagram using the values of extract air temperature and moisture content. This type of system can cool the supply air by up to 6 K.

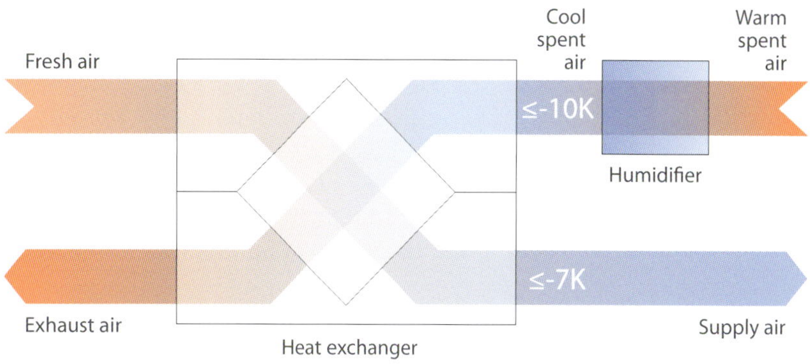

Working principle of adiabatic cooling of the extract air with subsequent heat recovery

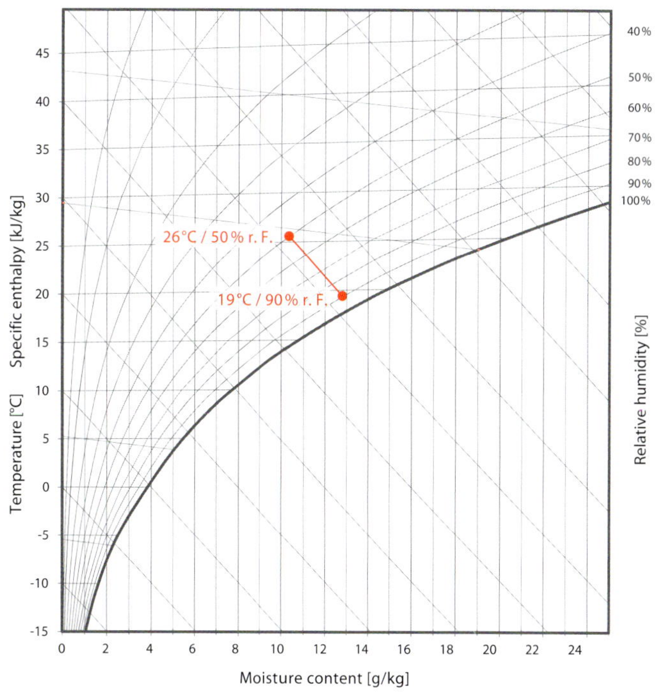

Adiabatic exhaust air cooling
Mollier (h-x) diagram for moist air
Pressure = 1,000 bar, altitude: 110 m asl

Mollier h-x diagram representing the changes in state of moist air. The potential of adiabatic extract air cooling is clearly shown here. The extract air has a temperature of 26 °C at a relative humidity of 50 %. This is raised to 90 % relative humidity by the addition of moisture. This causes the temperature to reduce to 19 °C. The cooled air can be used to cool the supply air through a heat exchanger.

Example of an evaporation humidifier for adiabatic cooling (Copyright: WOLF Anlagen-Technik GmbH & Co. KG)

By using two humidifiers in series, each with a heat exchanger, a cooling effect of up to 10 K can be achieved with a corresponding increase in cost. The supply air flows through the second and then immediately through the first heat exchanger (see diagram).

A further method of achieving a high cooling output adiabatically is by using a run-around coil system (RAC). This produces evaporative cooling separately from the supply air because the cold is transferred to a register in the supply air flow by a carrier medium (for example, a water-glycol mixture). It can be used in combination with or be retrofitted into a conventional air-conditioning and ventilation system. These heat exchanger systems must be leak-proof to ensure that the water used for humidification cannot come into contact with the supply air and detrimentally affect hygiene.

The amount of adiabatic cooling depends generally on the relative humidity of the supply or extract air taken and the amount of humidification necessary to achieve a reduction in temperature. Depending on the climate zone and the cooling requirement, the initial humidity could already be so high that the desired cooling cannot be achieved. The problem can be alleviated by actively drying the air before humidifying for adiabatic cooling. This method of air-conditioning is called adiabatic cooling with extract air and prior drying. The principle of drying and then humidifying is called desiccant evaporative cooling (DEC). Drying can be performed by a conventional mechanically driven refrigeration process or by hygroscopic materials such as zeolith, salts or silicates. The air flow is passed over a sorption material. The water vapour contained in the air is removed and bound to the material. The air, now with a lower relative humidity, is

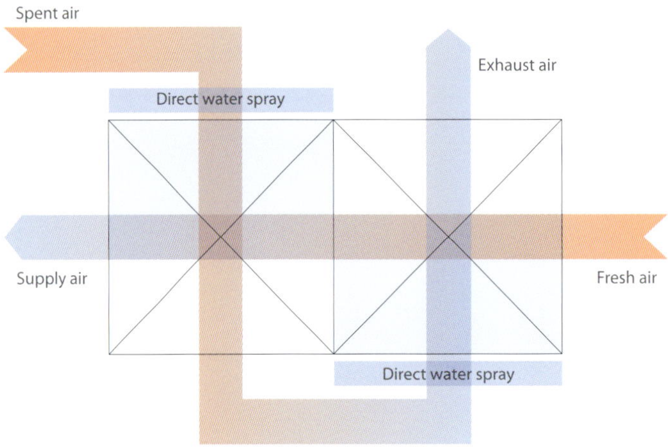

Double plate heat exchanger

heated. The ability of the drying medium to absorb moisture is not unlimited. When this material becomes saturated, it must be regenerated to allow it to take up more water vapour. This process of desorption is done by heating. The temperatures necessary for driving out the water depend on the sorption material and can be anything between about 45 °C and several hundred °C. Sorption materials used for drying air are generally those with a lower regeneration temperature in order to ensure the overall system works efficiently. The heating for regeneration can be by waste heat from a CHP plant, superfluous heat from a district heating grid, or solar thermal energy.

This strong heating of the now dry air means the air must be cooled, for example with a heat exchanger using outdoor or exhaust air, before it is humidified. Then adiabatic cooling can be performed by humidifying. The cooling potential is high because of the very low relative humidity of the air. However, this drying leads to higher technology costs and energy input. The operating costs are still lower than conventional vapour-compression systems. The water used for this evaporative cooling does not have to fulfil high quality requirements in order to be sprayed through jets. Using collected rainwater for this is preferable to using non-regenerative resources, or those that regenerate only over the very long term, such as groundwater. When humidifying in this way, the requirements on the water are more onerous because of its direct contact with the supply air. Adiabatic cooling of a 1990s office building with 1,000 m^2 usable floor area in Central Europe would consume around 1 m^3 water per day. Evaporative cooling systems are cheaper to operate than vapour-compression cooling systems because they consume considerably less electricity and the type of water used is rainwater or drinking water, which are available at little or no cost. A comparative analysis of the energy balance for both technologies – adiabatic cooling and vapour-compression – shows that only adiabatic cooling produces actual overall cooling. Vapour compression uses electrical energy to move heat from one place to another. More heat than cold is produced in the overall energy balance because of the use of electrical energy. This heat goes outside the building but remains in the immediate neighbourhood. This creates heat islands and the heat finds its way back into the building, with the effect that it must be transported outside again. This self-potentiating effect heats the surroundings of buildings, districts and cities. With adiabatic cooling, on the other hand, the heat is absorbed as latent heat by the water and the water vapour and transported away. Condensation takes place in the atmosphere when clouds form. The heat then comes back to earth or out into space as long-wave atmospheric radiation. The subsequent precipitation soaks away and replenishes the groundwater.

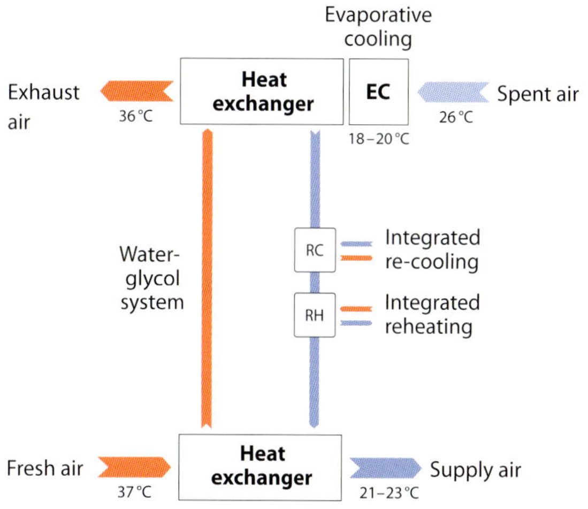

Working principle of adiabatic cooling with a run-around coil system (RAC).

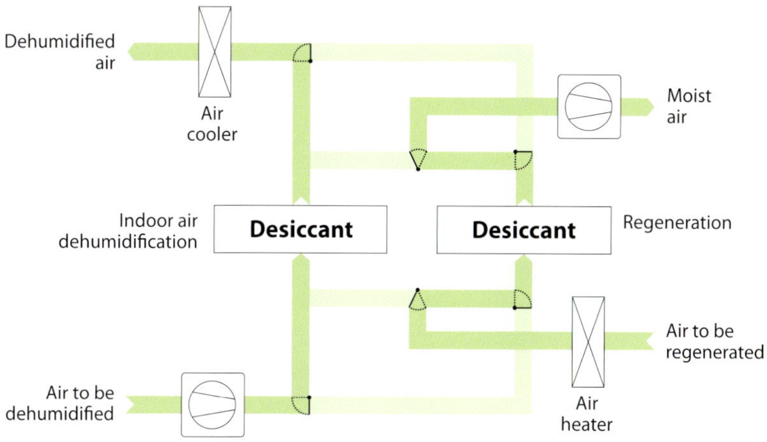

The principle of air dehumidification by adsorption

Storage and distribution

Storage systems may store energy in the long or short term. Short-term storage acts as a buffer to bridge peak loads, fluctuations in demand and periods of bad weather for up to a few days. At the same time, short-term storage helps to reduce the size of the system required to cope with highly fluctuating demands by allowing it to produce and store energy when demand is low and call upon it later when required. Another option is for the store to be charged at night when cheaper energy prices may be available or when more favourable outdoor temperatures prevail.

Long-term storage includes seasonal heat stores, which may use, for example, hot water tanks or borehole thermal energy stores to retain energy over a period of several months. In terms of physics, there are three types of heat storage available here: sensible heat storage (e.g. a water tank), latent heat storage (no perceptible change in temperature), and thermochemical or sorption heat storage.

Energy concept of a building with PCM (summer day): The phase change material (PCM) used in the ceiling and the walls smoothens out fluctuations in the outside air temperature that prevail between day and night, thus achieving a pleasant, constant interior temperature. The heat produced in the building during the day is stored in the PCM, hence cooling the room. The thermally loaded PCM releases energy at night due to the lower outdoor temperature. The colder night temperatures thus discharge the PCM, enabling it to store heat again the next day. In this process, the PCM passes through the described phase change.

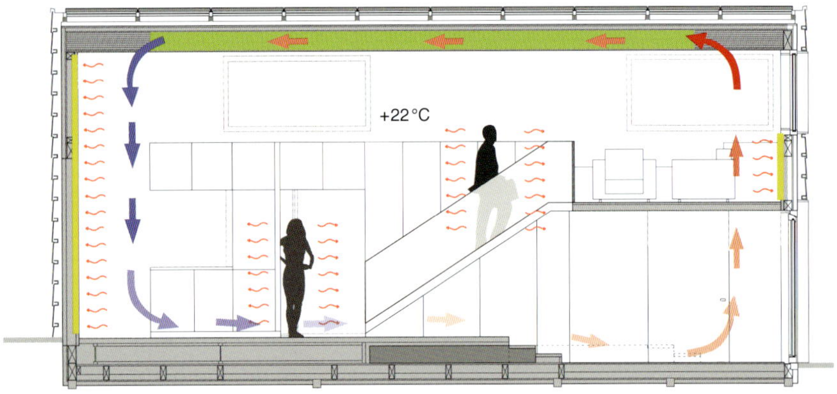

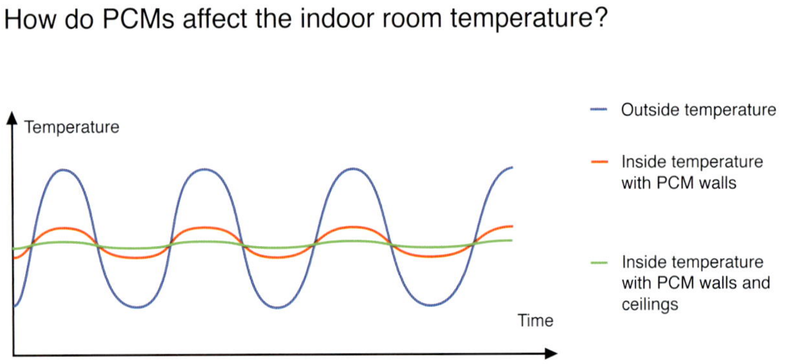

In buildings of "lightweight construction" without conventional thermal storage mass, this lack can be compensated with phase change material (PCM). As a result, outdoor temperature fluctuations can be mitigated inside and an almost constant indoor temperature can be achieved.

Heat

A number of different storage technologies can store heat and cold. Short-term storage mainly relies on water tanks and thermochemical storage systems. Long-term storage is often hot water heat storage (highly insulated tanks) or gravel-water storage, which takes place in insulated trenches in the ground. Borehole thermal energy stores with boreholes up to 100 m deep are also used. These store the heat locally in the soil and rock to be recovered on demand. Aquifer heat stores are another form of long-term storage. They use deep geothermal boreholes that penetrate a standing aquiferous layer. The water and the surrounding soil are used as the heat store. Thermochemical and latent heat stores are suitable for long-term storage. In general, for long-term storage it can be assumed that some of the stored energy will be lost and only part of it will be available for time-shifted use. The losses depend on the selected technology and the period of storage.

Conventional heat storage

Materials with a high heat storage capacity are particularly suitable as storage media because of their compactness. Water is often used as the medium because of its good storage capacity, easy availability, transportability (within the building services systems) and low cost. This type of storage consists typically of an insulated tank that stores the heated water for later use. A distinction must be drawn between service water storage with low quality requirements and drinking water storage, which contains drinking water intended for consumption and is therefore subject to high hygienic requirements.

Thermochemical storage

Thermochemical stores cycle through a chemical reaction in storing heat or cold. In the right circumstances, the reaction is infinitely repeatable. One example of this is the sorption process in which a storage medium is charged by heating and dehydrates. The reverse of the process is achieved by the application of water vapour to the storage medium: heat is released.

Latent heat storage

Latent heat stores undergo a phase change from solid to liquid in order to store heat. No or very little increase in temperature of the storage medium is noticed. The phase change material (PCM) must first be thermally discharged in order to be able to store heat again. A phase change from liquid to solid must take place. The use of PCMs, for example, in a composite material extends their scope of use in increasing the storage mass in a building and in improving room climate.

Cold

All the systems for heat storage mentioned above can also be used for cold storage. If water is the medium, it should be noted that its storage potential for cold is much less than for heat. From an assumed room temperature of 20 °C water can be heated to about 80 °C without it vaporising. Cold, on the other hand, can be taken down only to freezing point and therefore stored at a much smaller temperature difference. An ice-water mixture such as that found in an ice storage system has a large storage capacity for cold because of the phase change from solid to liquid, which corresponds to a temperature increase of up to 77 °C.

Moisture

Building materials based on animal, organic or porous mineral substances can contribute to regulating moisture in the room air because of their hygroscopicity. The ability to take up moisture and release it again quickly when necessary contributes to increasing comfort. A moisture-regulating effect also reduces the occurrence of moisture-related problems such as mould. As well as the energy effects due to thermal mass, comfort gains are also highly valued. In general, the relative humidity should be maintained within the comfort limits of 30–70 % whatever the air temperature. A relative humidity below 30 % causes dry eyes and mucosal irritation, while values above 70 % actively promote the growth of fungi, colonies of house dust mites, and property damage. The upper comfort temperature limit falls with high air moisture levels. Mechanical humidification or dehumidification of the supply air may be used if necessary, but it requires a great deal of energy. On energy grounds, therefore, using moisture-regulating building materials to accomplish this is to be encouraged. In temperate climate zones, the temperature-dependent capacity of the air to take up water vapour means that there is no need for active dehumidification in most buildings. Bringing cool outdoor air into the building is sufficient. The absolute capacity of the outdoor air to take up water vapour increases with its temperature. Warming cold outdoor air gives it a high capacity for taking up water vapour and is an effective means of dehumidification. However, additional mechanical dehumidification may be required in climate zones with high outdoor temperatures and/or a persistently high air moisture content.

Mechanical condensation drying can be used to dehumidify the air. In this process, the air is led through a cooled evaporator. This cools the air below the dewpoint. The water vapour contained in the air condenses as water on the surface of the evaporator and is cooled. The now cooled stream of air is cooler than required for the supply air. It is conditioned in an air heater, the condenser, before it is blown into the room at a comfortable temperature and moisture content. The principle is similar to a conventional air-conditioning system, which requires dehumidification as a result of mechanical cooling of the air.

Dehumidification can also be produced through **absorption** of the water vapour by hygroscopic substances. The flow of air at a high relative humidity is passed over a salt solution (for example, lithium chloride, calcium chloride). This solution absorbs the water vapour, which results in a dilution of the solution. When the hygroscopic substance is saturated, it can be regenerated by heating and conducting away the water vapour.

With dehumidification by **adsorption**, water vapour is trapped on the surface of, for example, activated carbon, silica gel or zeolith. The internal surface area of these materials can amount to more than 1,000 m² per gram. A high adsorption capacity can be obtained using very little volume or weight of material. Similar to absorption, regeneration is by heating and conducting away the water vapour. The amount of energy used in these methods of active dehumidification is correspondingly high.

Air can be humidified mechanically by the ventilation system using spray jets or steam. Use of mechanical humidification in residential buildings in Central Europe is increasing. Reasons for this include comfort ventilation systems that continuously feed conditioned outdoor air into the building, even in winter. The heated air can have a relative humidity below the comfort limit. Local ultrasonic air humidifiers can be used to increase air humidity. These take up very little space and can create a continuous aerosol mist while consuming comparatively little energy. Direct integration into a ventilation system in residential buildings is uncommon because of the higher capital and maintenance costs. An innovative development in the transfer of atmospheric moisture is the use of a new high-grade paper material for the heat-transfer surfaces of the heat exchanger, which exchanges not only heat but also the moisture from exhaust and supply air, retaining a large proportion of the room humidity at very little cost.

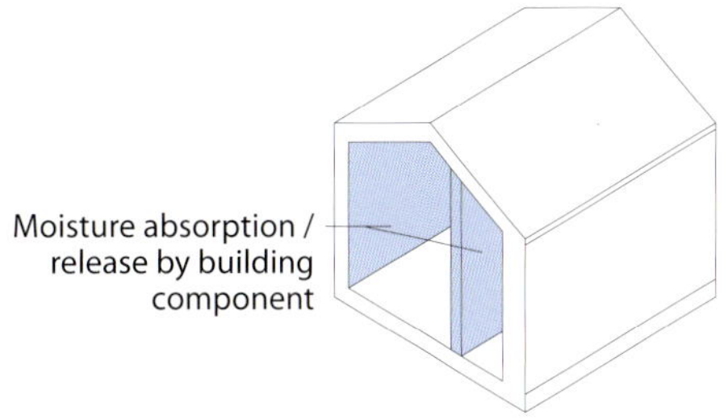

Use of building materials to regulate moisture

Battery storage system Solar Academy, Niestetal (DE)

Electricity

The politically driven decentralisation of energy supplies, the shift towards creating a higher proportion of energy from renewable sources, and the time difference between generation and demand, have focused attention on electrical energy storage systems. They can help absorb fluctuations in the energy supply, such as load peaks. Changes in feed-in tariffs are making own-use of self-generated electricity more attractive. Various technologies for storing electricity are in development or are ready for the market. Depending on the application and the required capacity, these include batteries, compressed air, methanation and pumped storage hydroelectricity. On the scale of an Aktivhaus, the most commonly available solution in practice is batteries.

The storage of electrical energy compared to thermal energy is considerably more complex and expensive. With the anticipated reduction in the prices of batteries and the increasing use of electric cars, in future, for example, battery storage systems will be more economic as a buffer store for electrical energy.

Batteries

Battery storage systems in buildings are still unusual in countries with high energy security and generally available supplies of electricity. This situation has meant there are currently very few off-the-shelf products available. Further development breakthroughs in efficiency, ranges of products, availability and prices are expected.

In addition to larger battery systems, there are back-up solutions that can be integrated directly into a solar inverter. In this way, cover for power outages can be provided without an additional battery system or emergency generator.

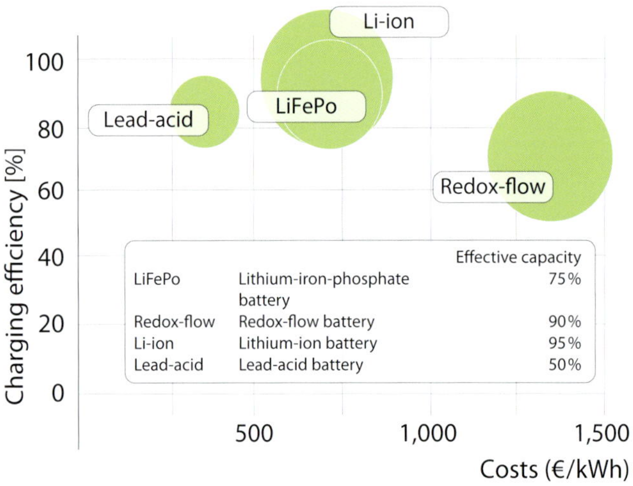

Usable capacity of
different battery storage systems

Heat and cold transfer

There is a choice of methods for transferring heat and cold into a room. The method most usually employed in a building is to use radiators. A further development is activated building components, such as surface heating systems. Any ventilation system worth recommending for an Aktivhaus should be able to transfer heat and cold. Radiant ceiling are used mainly in commercial premises. Other special solutions such as gas radiators, steam radiators, skirting board radiators, chilled beams and gravity cooling are discussed no further here.

Convectors

Convectors are radiators that work using a low volume of water as the heat-transfer medium. They have thin lamellae and convection plates, past which the air flows and is warmed. The most compact form has lamellae that run continuously around a pipe carrying hot water. They can be installed in bays off a main room, within the floor construction or in suitable voids within the building structure. They may have blowers to increase the amount of heat transferred. The use of a ventilation system in conjunction with convectors must be carefully designed because the air flows resulting from both systems may detrimentally affect each other. Convectors can also be used for cooling.

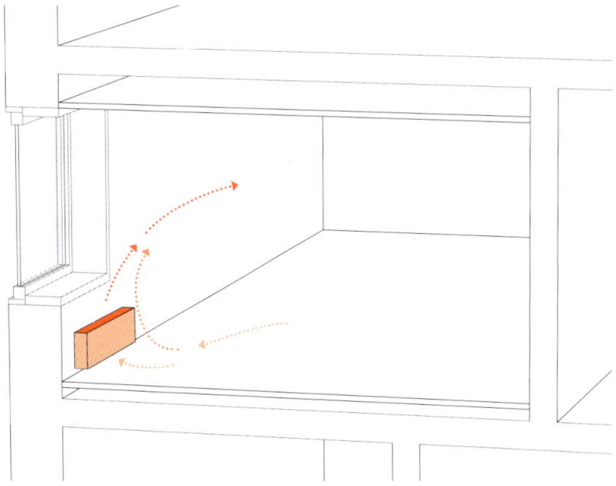

Principle of heat transfer – convector

Radiators

Radiators are larger than convectors because they have a lower area of heat-transfer surface per unit volume. Compared with convectors, radiators have much less air flowing around them but give off more of their heat by radiation. There are various different designs of radiators. Systems that offer good radiation and heat the air well often combine panels with convector plates. Radiators with several parallel panels may have the panels controlled separately. One panel is enough in normal conditions. The advantage of this arrangement is reduced reaction times.

The radiators should be positioned so that the temperature differences within a room are minimised. Placing them near glazing or on outside walls generally ensures the best levels of comfort. Their low water content leads to fast reaction times and better regulation.

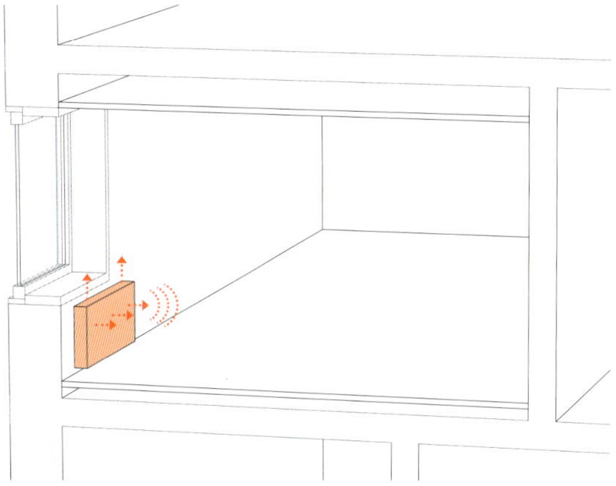

Principle of heat transfer – radiator

Off-peak storage heaters

Off-peak storage heaters use electricity at night when it is cheap to heat a thermal store, normally magnesite bricks, to an internal temperature of over 600 °C. The heat is then given off continuously over the day. Poor regulation and the use of non-CO_2-neutral electricity makes it difficult to recommend these systems.

It is possible this form of thermal store could be linked with an automatic yield and demand forecasting system in an Aktivhaus in order to make better use of electricity generated on site.

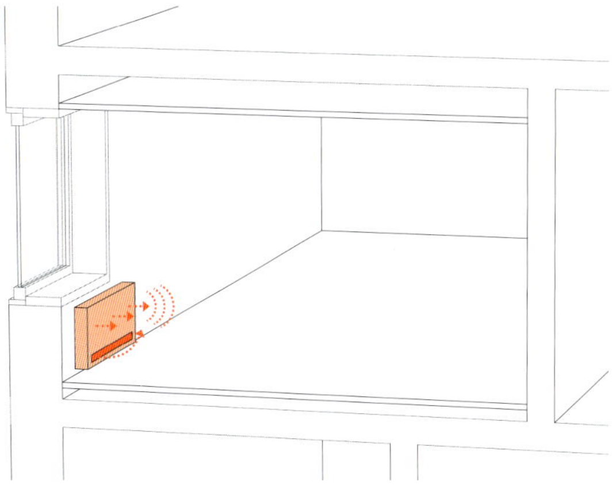

Principle of heat transfer – storage heater

Thermally active building components

Thermally active building components are components with a large thermal storage capacity through which a medium flows in pipes to make them thermally active. A flow temperature of approximately 23 °C is used all year round, with this value allowed to vary by +/-2–6 °C depending on the time of year and the requirements. To ensure good regulation of different parts of a building, various sections of it are included in separate conditioning circuits. These systems are very sluggish but offer the advantage of thermal storage capacity. Using a ceiling sail to condition the air in addition can increase the system's

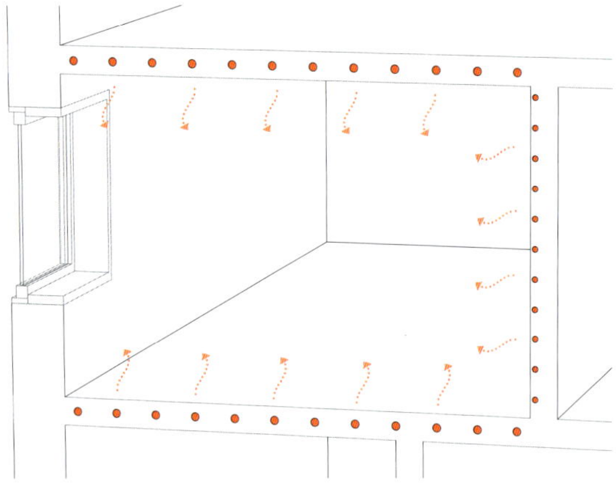

Principle of heat transfer – thermally active building component

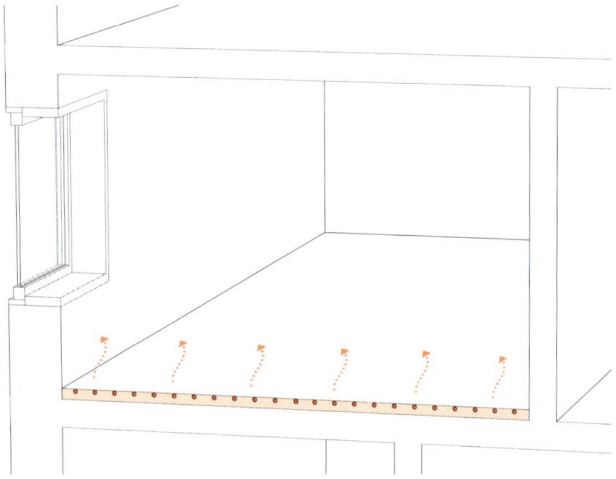

Principle of heat transfer – underfloor heating

speed of reaction. Thermally activated surfaces should not be clad or their performance could be adversely affected. People perceive heat radiation as comfortable, particularly that given off by thermally activated components, but temperatures deviating from the specified temperature range can lead to discomfort. Draughts, which can also be perceived as uncomfortable, are practically eliminated with this form of heat transfer. The choice of materials and the mass of the components influence the thermal storage capacity, which has a positive effect on maintaining an even surface temperature, and can also contribute to moisture regulation (including for example the use of loam or a covering coat of loam plaster).

Underfloor heating

A heated medium circulates through the floor increase its temperature to provide underfloor heating in an Aktivhaus. The flow temperatures need to be between 28–40 °C, depending on the quality of the building envelope. The underfloor heating system is divided into several heating circuits to allow the heating of different spaces to be independently regulated. Underfloor heating systems can be wet or dry. In wet systems, flexible plastic pipes for carrying the water are cast into the floor screed. In dry systems, prefabricated mats or plates incorporating the pipes are laid on the floor. The required construction depth is very small and the mats have a thermally insulating effect. Dry systems are installed very quickly and shorten overall construction times. The user's choice of floor covering can be laid on top. The time taken to transfer heat from the carrier medium water into the screed and eventually into the floor covering is shorter than with wet systems. Compared to thermally active building components, the reaction time of underfloor heating is shorter but is more sluggish than that of radiators. One advantage of integrating a heating system into the enclosing surfaces of a room is the freedom of the interior design, which is not limited by the positioning of equipment in the room. On grounds of hygiene as well, thermally active building components such as underfloor heating offer advantages because the reduced air circulation avoids stirring up dust in the room. Carpets or other insulating floor coverings should not be used if good heat transfer is to be achieved.

Heated/chilled ceilings

Heated/chilled ceilings can be used to heat and cool rooms. They can be used in places where high flexibility is required and little space can be provided for installing systems in the floor and walls. The elements, which are made from metal plates and have water pipes running through them, are suspended below the ceiling. They have a layer of insulation on top to ensure the heat is radiated as desired into the room. Their use is recommended where the cooling demand is greater than the heating demand because the layering of the air that occurs when they are used for heating is seen as unfavourable. The system can be subdivided into small areas, allowing the user to control it down to room level. Maintenance is low cost and can be done in sections without detrimentally affecting the whole system. In their conventional form, these systems are not easily modifiable because pipes have to be plastered into the ceiling and the heat- or cold-transferring elements are permanently fixed into or onto the ceiling. However, systems with easily accessible pipework and modifiable ceiling elements are perfectly possible.

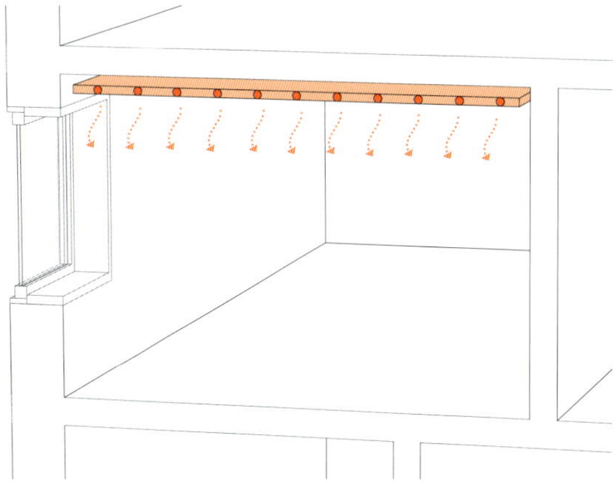

Principle of heat transfer – heated/cooled ceiling

Ceiling sails

Ceiling sails consist of metal plates with water-carrying pipes attached and can be used equally well for heating or cooling. They hang without any insulation on their top sides below the ceiling and the surrounding air flows freely around them. They also raise or lower the temperature of the ceiling surface, allowing it also to contribute to the conditioning of the room. Compared to heating/cooling sails, the more efficient transfer of energy to the room air allows the use of smaller elements, which leads to cost savings. They normally operate with higher flow temperatures. This requires them to be positioned at a greater distance from people in order to avoid imposing restrictions on their operation in order to preserve comfort. Regulating the sails appropriately should avoid condensation when used for cooling. In the case of a high cooling load or space for only a small area of ceiling sail, these systems can be operated with very low temperatures ($T_{fl} < 5\,°C$). This requires condensation collecting troughs under the ceiling sails to collect and drain away the condensate.

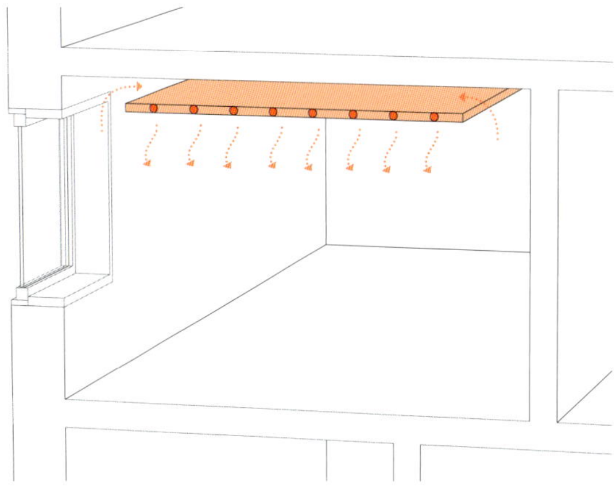

Principle of heat transfer – ceiling sail

Electric direct heating

Electric direct heating is installed close to the surface in walls and floors. Heat transfer takes place without any noticeable delay. As a compact additional heating system occupying very little area, it can be used very effectively in bathrooms. On the other hand, using it as the only form of heating, for example in a residential property, must be viewed critically because of the electricity required at peak periods. However, its exclusive use for this purpose would not be a problem if the electricity used were to be generated on the building from renewable sources.

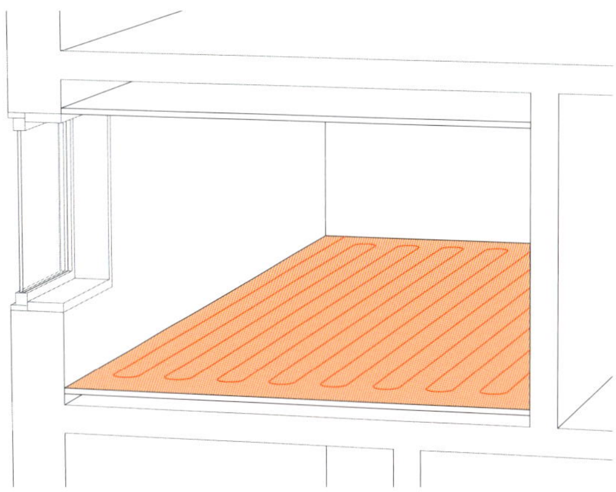

Principle of heat transfer – electrically heated floor

Ventilation systems are used in an Aktivhaus to replace stale and polluted air with fresh air. They can be used in conjunction with a highly efficient heat recovery system to help to reduce ventilation heat losses to a minimum. All the mechanical ventilation systems described here can be used in an Aktivhaus.

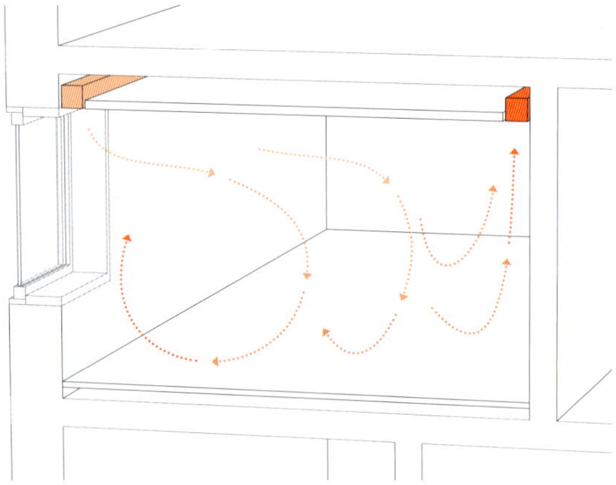

Principle of heat transfer – mixed ventilation

Mixed air flow ventilation

In mixed air flow ventilation systems, air is introduced into the room through inlets in the wall or ceiling. The increased speed of the blown-in air creates turbulence in the room air. A wide horizontal distribution is obtained by suitable positioning and appropriate choice of outlets (e.g. long-range nozzles). Use in high rooms is not recommended because it can lead to unfavourable layering and uneven conditioning of the room. Air is extracted from the room near the floor or through overflow openings into neighbouring rooms.

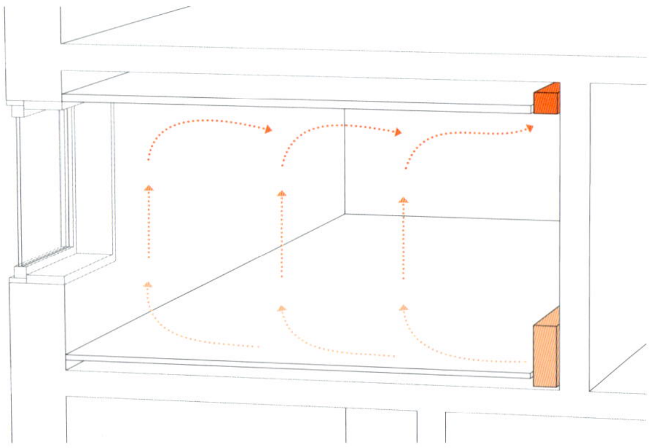

Principle of heat transfer – displacement ventilation

Displacement ventilation

Displacement ventilation uses very low-volume flows and introduces the air into the room through slot diffusers near floor level, floor grillages or a double floor. A lake of cold air at a temperature of 2–4 °C forms on the floor. This fresh air rises at warm surfaces and is therefore able to capture pollutants directly from the source and transport them away. The sources of heat anticipated for a room need to be taken into account in the design because they directly influence the thermal air movements and the required volume flow. These systems cannot be used for heating. The cooling load is limited by the lowest temperature permitted on comfort grounds.

Night ventilation

The heat entering a building over the day can be compensated for by lower nocturnal temperatures. Natural cross ventilation of the rooms at night is necessary. The thermal storage mass of the building must be freely accessible by the air flow and not covered or clad. This free cooling requires high cross ventilation air flows and protection against burglars and the weather. The chimney effect and openings in the opposite facades provide strong cross ventilation air flows.

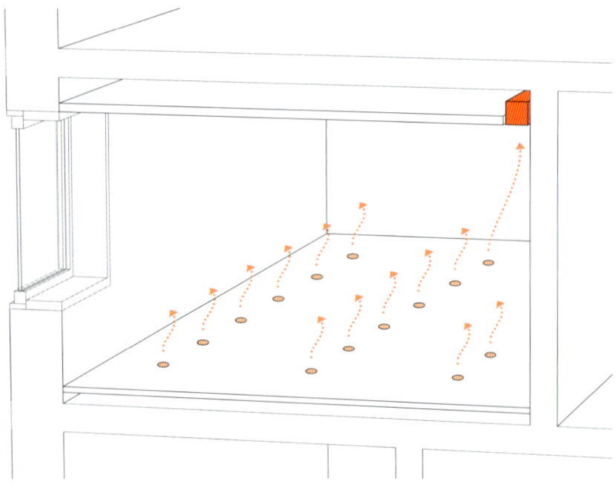

Principle of heat transfer – displacement ventilation through double floor

Night cooling can be automated using sensors and electromechanical openings. These ventilation arrangements can also be achieved mechanically using a supply and exhaust air system or an exhaust air system with a high air throughput.

Comfort ventilation

Comfort ventilation systems are mainly found in housing. A compact ventilation unit is installed in a central position in the house. It supplies the living rooms and bedrooms with fresh air and extracts the stale air, which is introduced through overflow openings into the kitchen and bathroom. The fresh air introduced into the building must be free of dust and odours. Short-circuiting with the exit air must be avoided. The compact ventilation unit contains fans, dust and pollen filters, and a heat exchanger (and sometimes heat recovery). Each dwelling should have a central controller and a bypass of the heat recovery system.

Local ventilation

Local ventilation systems are usually considered for refurbishment projects because they have lower installation costs than central ventilation systems. Depending on the specific products, the efficiency of a heat recovery system can be higher than that of a comfort ventilation system. This applies in particular to heat recovery systems with thermal stores (see section on waste heat – heat recovery p. 135).

Heat distribution heating and cooling

Water-based heating systems use a closed water circuit. An alternative is an air-based heat and cold transport system. The lower heat storage capacity of air compared to water means air is a much less efficient medium. Moving energy using water is much more effective. Ventilation systems must be designed to achieve the air changes necessary for hygiene. If this is adequate for conditioning the building, then an air-based heating and cooling system offers a good solution because the costs of additional heat-transfer systems are saved. Possible uses in an Aktivhaus building include situations that require high rates of air change. This is necessary, for example, in offices, schools and seminar rooms where room occupancy densities are high.

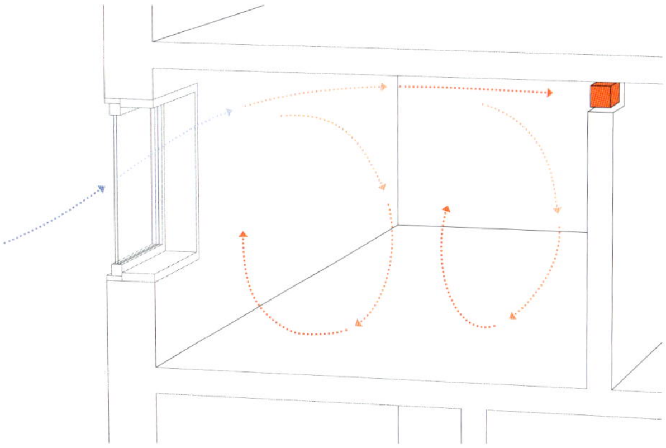

Principle of heat transfer – night ventilation

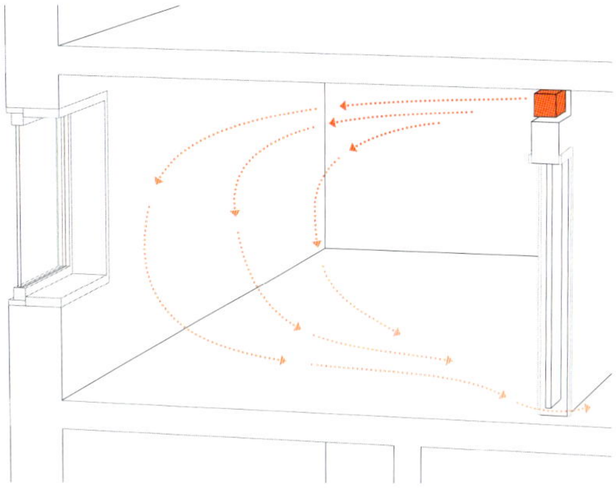

Principle of heat transfer – comfort ventilation

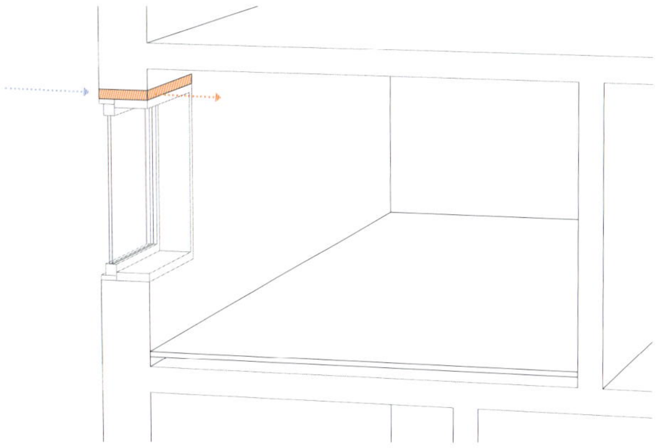

Principle of heat transfer – local ventilation, integrated into window area

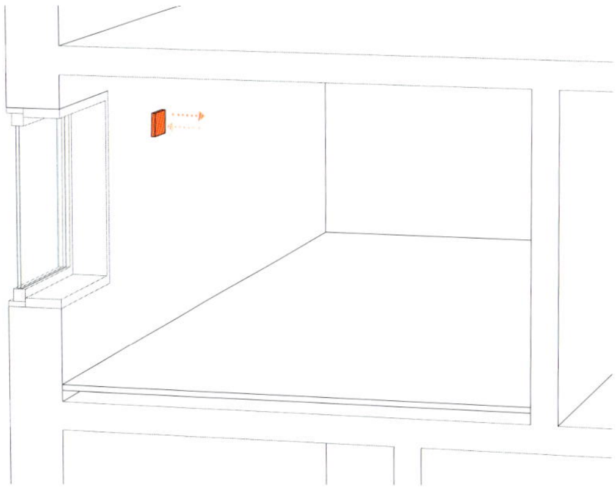

Principle of heat transfer – local ventilation, integrated into wall

Control and regulation

The technologies described for energy generation, distribution, storage and transmission must be connected together by control and regulation technology to improve user comfort and save operating energy.

Control is the unilateral process of influencing technical systems. In contrast, regulation is a two-way communication process involving a feedback loop. In the regulation process, a measured actual value is compared with a specified target value. If the two values are different, the system will seek to correct this, as long as the technical system is operating, until the actual and target values are equal.

In many ordinary buildings today, technical systems such as lighting, solar shading or heating regulate themselves automatically. This communication and interaction with the building plays an increasingly important role in saving energy and ensuring comfort. The regulating processes required for this are like those used in building management systems.

Building management technology includes all devices for the control, autonomous regulation and monitoring of building engineering plant and the capture of operational data. Building technical services, domestic appliances and multimedia devices can be interlinked in a smart house. In principle, everything powered by electricity can be automated. The user can change the automation settings by time switches, various sensors and control pads. Functions such as indoor and outdoor lighting, shading, heating, ventilation and air-conditioning, mechanically opening windows, door intercom, door locking and alarm systems, the operation of many domestic appliances and entertainment systems can be controlled in a similar way too.

A building automation system must react to different, often competing requirements. For example, if a user opens a window for ventilation, this is detected by a sensor, which alerts the ventilation and/or heating systems and regulates them. If the user wants to switch on a light, the building automation system can check whether there is enough natural light available and open the shades rather than switch on the artificial light.

It is critical to check how far automation should go. Automation should always give the user a basic understanding of the effects of each process and how it can be manually overridden. For ventilation control, for example, there are two possibilities: either the main control could be designed to have four settings (purge ventilation, standard, basic ventilation, no user presence), or it could be regulated according to need. In detached or multiple occupancy residential buildings, the relatively constant number of people allows a single central control with a simple, manually set, multistage switch is adequate. Comfort can deteriorate as a result of incorrect manual control; over-regulation can lead to dry air. On the other hand, demand-based ventilation can be regulated automatically in response to the CO_2 content of the air. This type of control in situations where room occupancy varies requires CO_2 sensors in every room. A cost-efficient alternative is the use of movement sensors. In the design of the regulation and control units, decisions should be carefully considered and made on the specific characteristics of the project.

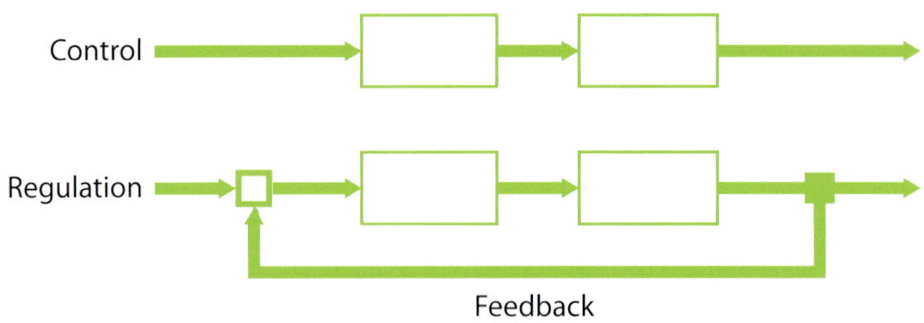

Difference between control and regulation

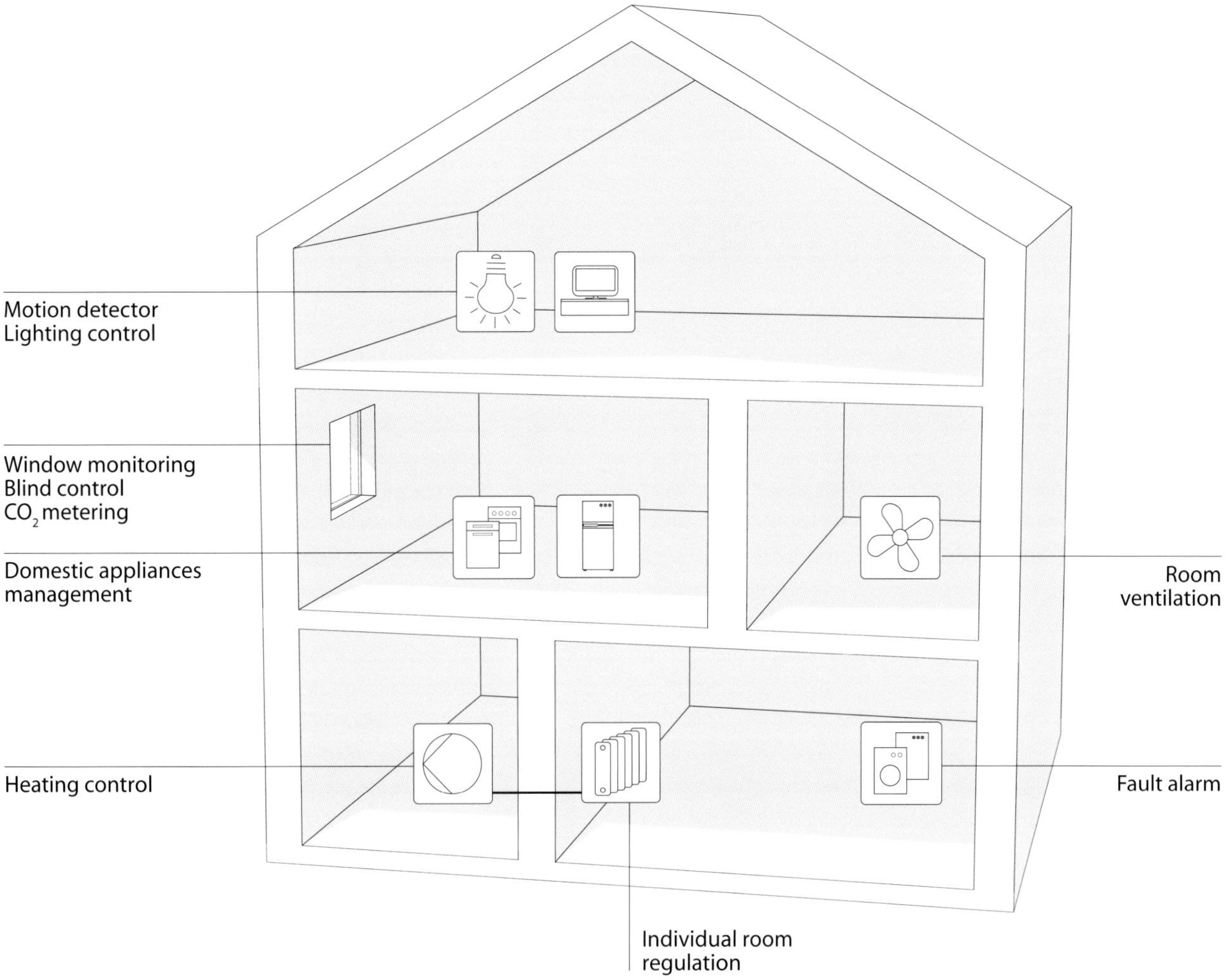

Schematic for a smart house

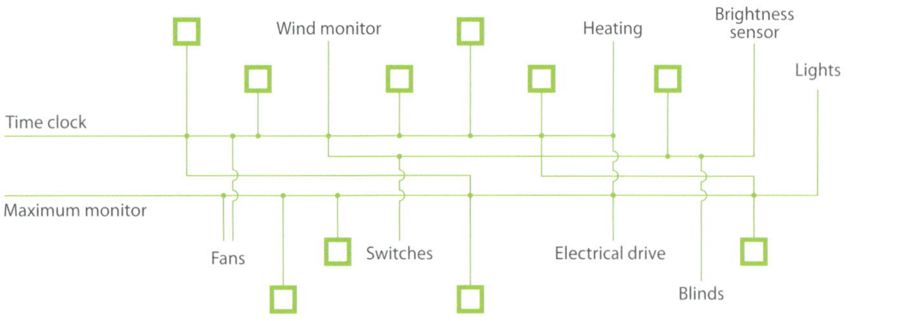

Schematic for a conventional electrical installation

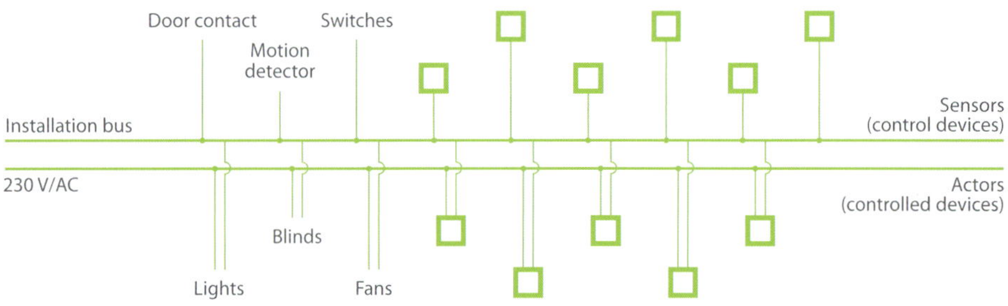

Schematic for an electrical installation with a bus system

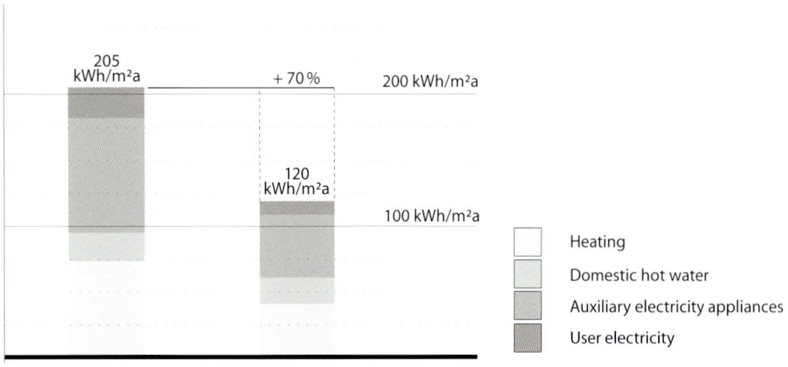

Saving on annual energy of an office building from building automation

Installation systems

The traditional task of electrical installations is to make electric power available. If the electrical circuit is broken the consumer is switched on or off. Because of ever higher requirements and the increase in number of devices, these installation systems are reaching their limits. Cabling is expensive and complicated, the fire load is greater, costs increase because labour and materials are becoming more and more expensive.

Bus systems are an attractive alternative. In these systems, all the consumers (actors) along with all the command devices (sensors) are connected by a single cable with resulting savings in cabling costs. The devices used for this system must be bus-compatible, i.e. fitted with programmable control electronics. They must also have the same interface language (e.g. KNX or EIB). A bus system is often worthwhile because of the increasing user demands for comfort, security, energy savings and reduced costs.

An Aktivhaus is normally fitted with a building management system. The system controls the regenerative energy sources, automates ventilation, makes lighting dependent on daylight levels and therefore saves energy and cost. Energy provision adjusts to match the actual use (occupation level) and user behaviour (e.g. window ventilation).

A smart building management system can start electrical consumers when enough energy is being produced by the building's photovoltaic systems. This reduces the consumption of external energy from the public grid; the proportion of self-produced regenerative electricity increases, as is required by renewable energy legislation in many countries. Where smart meters become part of the building automation system, the own-use of energy generated by or at the Aktivhaus will be able to the track the dynamic electricity price in countries where this is offered by energy providers.

With a bus system in place, the user can control the building, even when absent, via the Internet or a smartphone.

User intervention

User behaviour greatly influences energy consumption; this can cause the actual use to deviate from the designer's estimated value. Thus it is important to inform the residents about their behaviour, to clarify details, and to strengthen their energy awareness. A touch-panel or pad display lets the users know their energy consumption, regenerative energy creation and, if applicable, the price of electricity. By revealing this otherwise hidden information, smart meters raise the users' consciousness of energy savings and they can modify their behaviour accordingly. Through a simple to operate user interface, the user can explore and better relate to the issues of energy and the building's technology. It will only be accepted if it is user-friendly and intuitive to operate. One recommendation is a combination of an ordinary switch system that can be operated manually and has a conventional switch for the basic public electricity supply, and a user interface. The user interface must be graphical and easy to understand.

Users normally want to control only those processes and technologies that influence their individual needs and feelings of comfort (sun shading or anti-glare protection and room temperature). The other processes operating in the background do not have to be controllable or visible. Users do not like their actions to be dictated or their lifestyles to be limited.

Up to now, a user interface has never been a part of the basic energy supply technology fitted as standard in a house or an office. However, they will probably be increasingly requested as people become more aware of the link between user behaviour and energy use. The electricity necessary to power this must be taken into account.

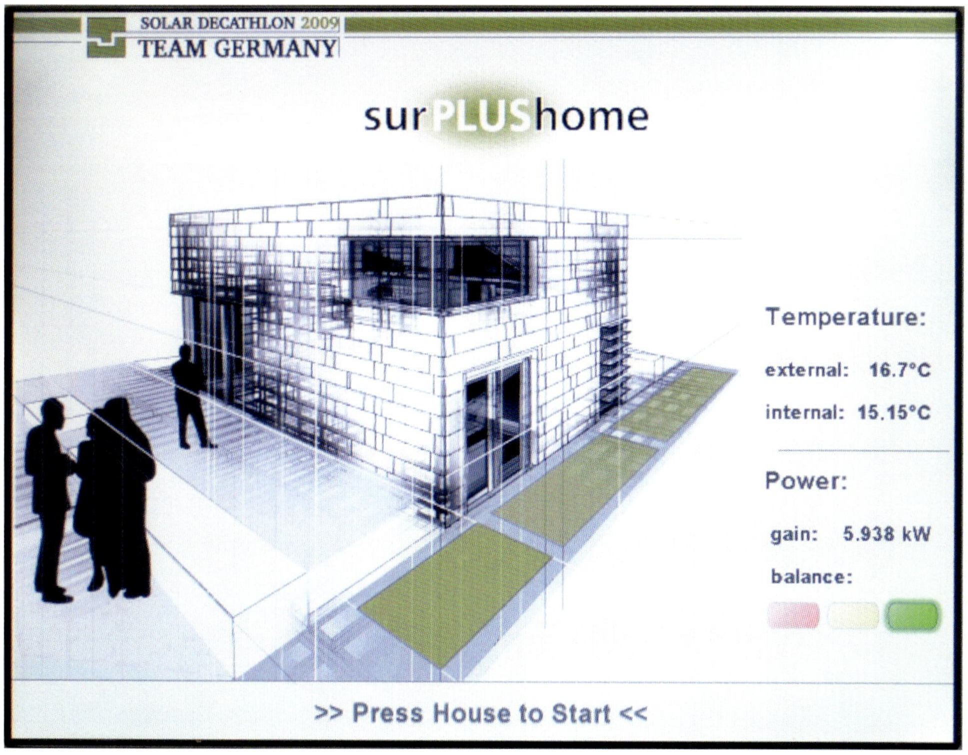

Example of a user interface with traffic light indicator from surPLUShome, Solar Decathlon 2009, TU Darmstadt (DE)

For a user interface to achieve the desired energy savings, it should be capable of displaying the following information:

- Energy consumption split accordingly to show energy services and consumers
- Energy generation
- Traffic light indicator to display the current energy balance
- Temperature display
- Recommendations for user actions, for example, the use of high-consumption appliances during periods of high production of self-generated electricity from regenerative sources
- Weather information and forecast
  It may have Internet access if necessary

The objectives of energy management using building automation and a user interface are:

- Optimisation of energy consumption
- Sensitising the user to energy savings
- Raising the proportion of own-use of self-created, regenerative energy (by recommendations, building automation, load management)
- Simple data collection and calculation system
- Opportunity to "play" with the technology and identification with the house

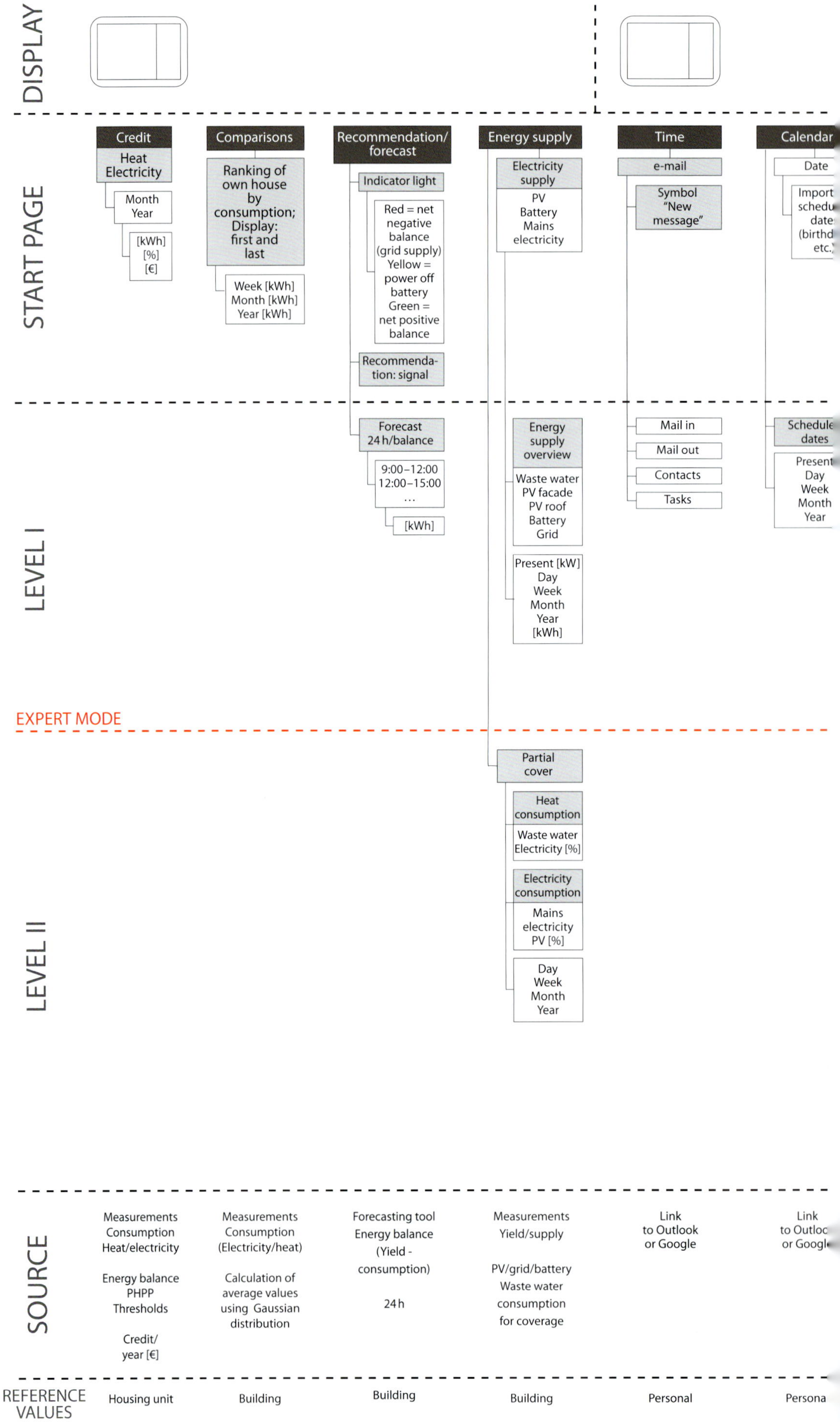

Diagram showing the different levels of the user interface from the Aktiv-Stadthaus project for investigating energy management for the user at the human-technology interface. A display like this is a complex system with many subordinate levels. The user can access different target groups via the various levels, and see at a glance on the home page all the important and current information. The first level down allows the user to see forecasts and processes, while the second level is the expert mode. The latter gives detailed information on various items and is designed for users who are interested in the technology.

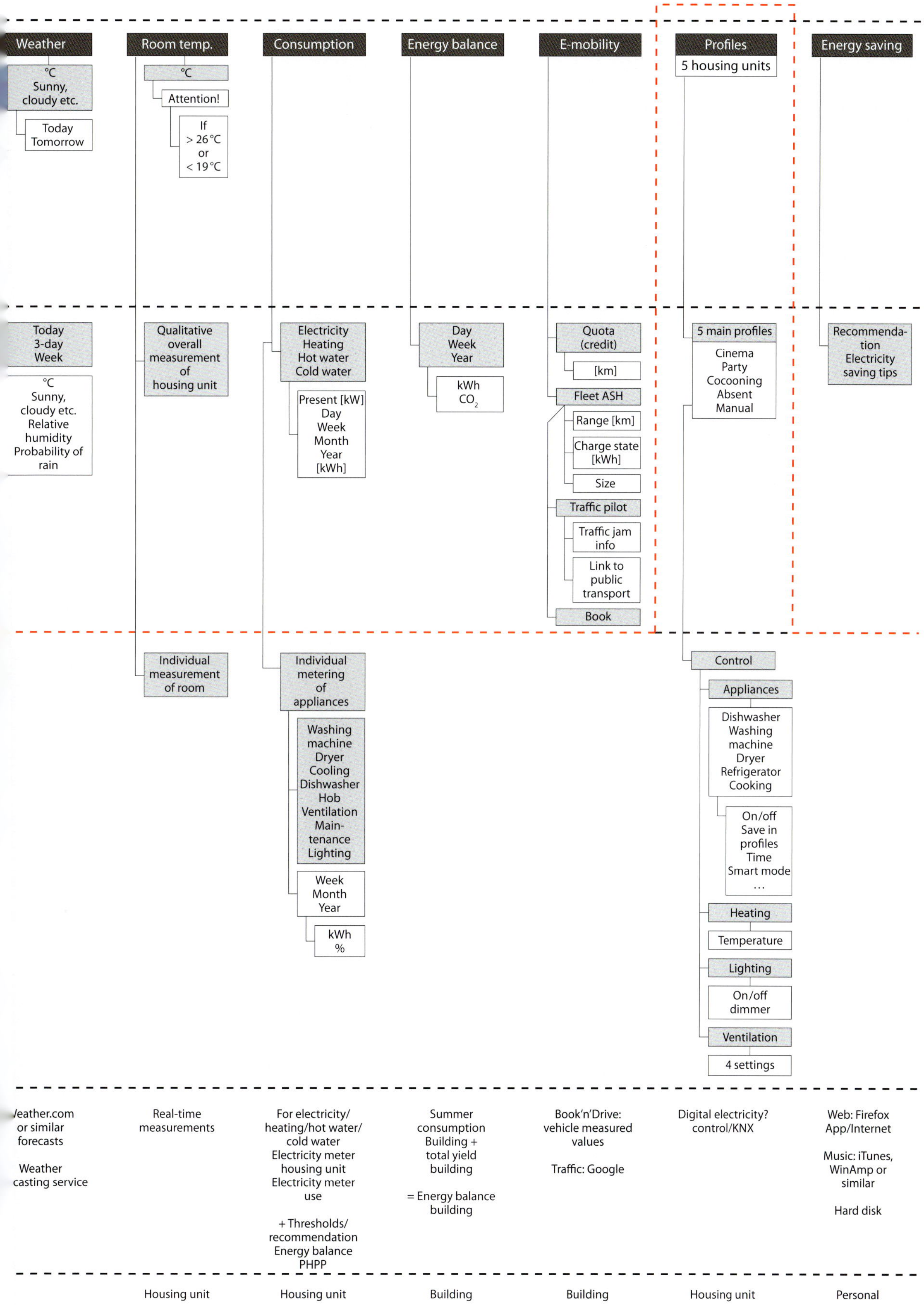
Weather
°C
Sunny, cloudy etc.
Today
Tomorrow
Today
3-day
Week
°C
Sunny, cloudy etc.
Relative humidity
Probability of rain
/eather.com or similar forecasts
Weather casting service
Room temp.
°C
Attention!
If > 26 °C or < 19 °C
Qualitative overall measurement of housing unit
Individual measurement of room
Real-time measurements
Housing unit
Consumption
Electricity
Heating
Hot water
Cold water
Present [kW]
Day
Week
Month
Year
[kWh]
Individual metering of appliances
Washing machine
Dryer
Cooling
Dishwasher
Hob
Ventilation
Main-tenance
Lighting
Week
Month
Year
kWh
%
For electricity/heating/hot water/cold water
Electricity meter housing unit
Electricity meter use
+ Thresholds/recommendation
Energy balance
PHPP
Housing unit
Energy balance
Day
Week
Year
kWh
CO2
Summer consumption
Building + total yield building
= Energy balance building
Building
E-mobility
Quota (credit)
[km]
Fleet ASH
Range [km]
Charge state [kWh]
Size
Traffic pilot
Traffic jam info
Link to public transport
Book
Book'n'Drive: vehicle measured values
Traffic: Google
Building
Profiles
5 housing units
5 main profiles
Cinema
Party
Cocooning
Absent
Manual
Control
Appliances
Dishwasher
Washing machine
Dryer
Refrigerator
Cooking
On/off
Save in profiles
Time
Smart mode
...
Heating
Temperature
Lighting
On/off
dimmer
Ventilation
4 settings
Digital electricity? control/KNX
Housing unit
Energy saving
Recommenda-tion
Electricity saving tips
Web: Firefox
App/Internet
Music: iTunes, WinAmp or similar
Hard disk
Personal

Load management, smart grid

Not only the building itself, but also the electricity networks it uses need ever more complex controls. In earlier times, energy generation was adjusted to cover the demand, but increasingly demand is adjusted to suit generation. This process, known as load management, involves arranging the demand to take place preferably when most energy is the cheapest.

In contrast to energy generated from fossil fuels, energy from renewable sources is not available continuously. To create a functioning system requires a smart grid. It integrates all consumers and energy suppliers, both centralised and decentralised electricity generators. The smart grid allows the generators and consumers not in balance to be controlled. Smart grids can optimise the utilisation of the grid and avoid expensive peak load times.

Combined with smart metering, which energy suppliers have been obliged to install in new buildings in Germany since 1 January 2010, it is now possible to measure consumption down to the second. The consuming devices are linked to and communicate with the grid. They can be controlled based on the detailed electricity cost analysis, taking into account changes in electricity prices. Non-time-critical processes, such as washing and drying laundry, and dishwashing, can take place preferably when there is plenty of power available at correspondingly cheap prices.

Load management supports a generator-oriented system. It allows energy to be consumed at the same time as it is generated and therefore supplies it at cheaper prices. This time shift helps avoid overloads on the grid. The electricity grid coordinates the balance between generation and consumption. The Aktivhaus, as an energy-generating element, becomes part of the grid. Ultimately, the ability of the building to store energy can be integrated into the system.

In the future, electric vehicles could be used to store discontinuously generated electricity and, for example, feed it back into the grid or to the user overnight. The integration of electric vehicles into the energy concept can therefore lead to a higher proportion of on-site use of self-generated electricity. In addition to the possibility of acting as an electricity store, vehicles can also play a role in stabilising the network and increasing the security of supply, and are therefore part of load management.

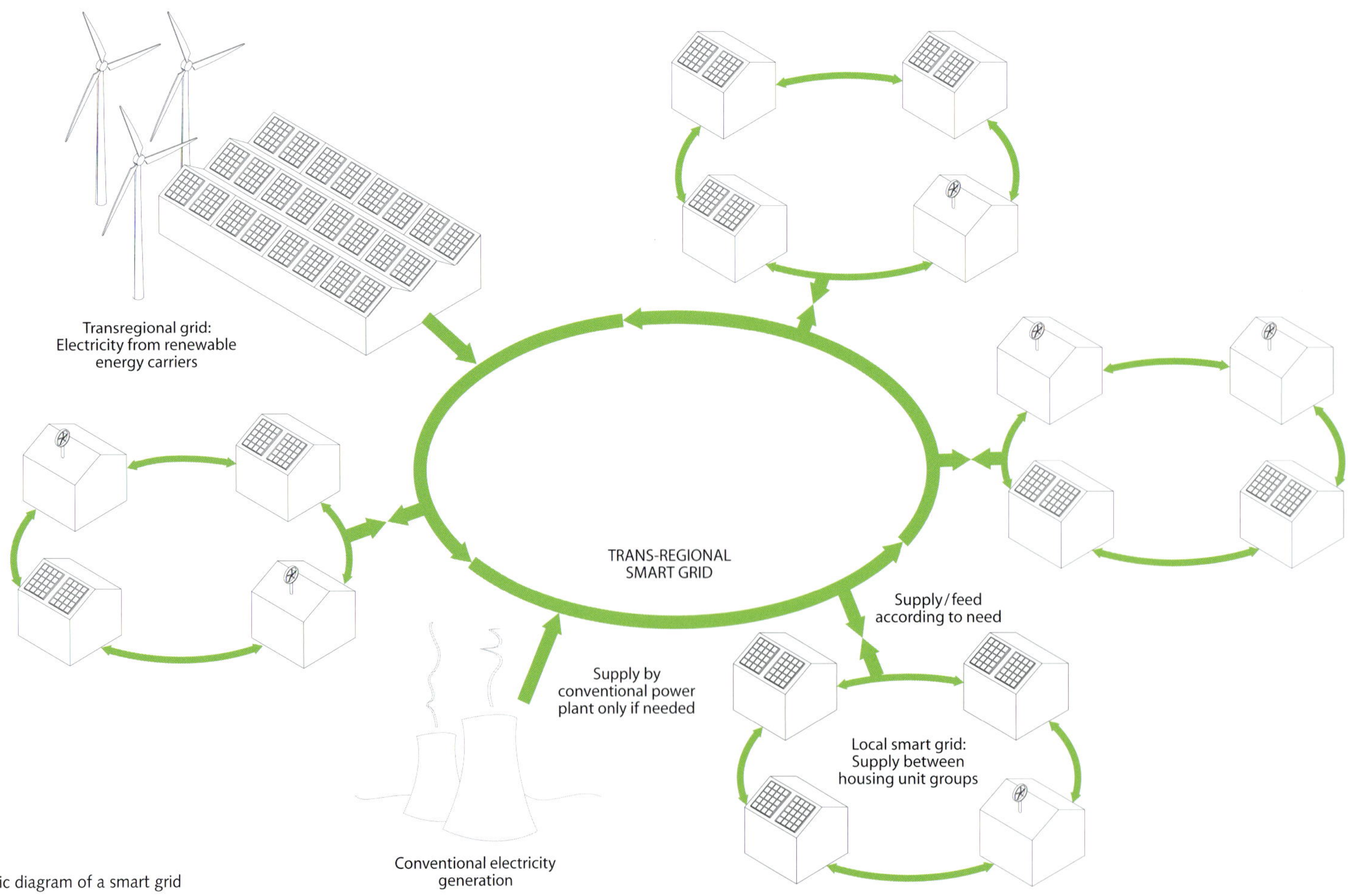

Schematic diagram of a smart grid

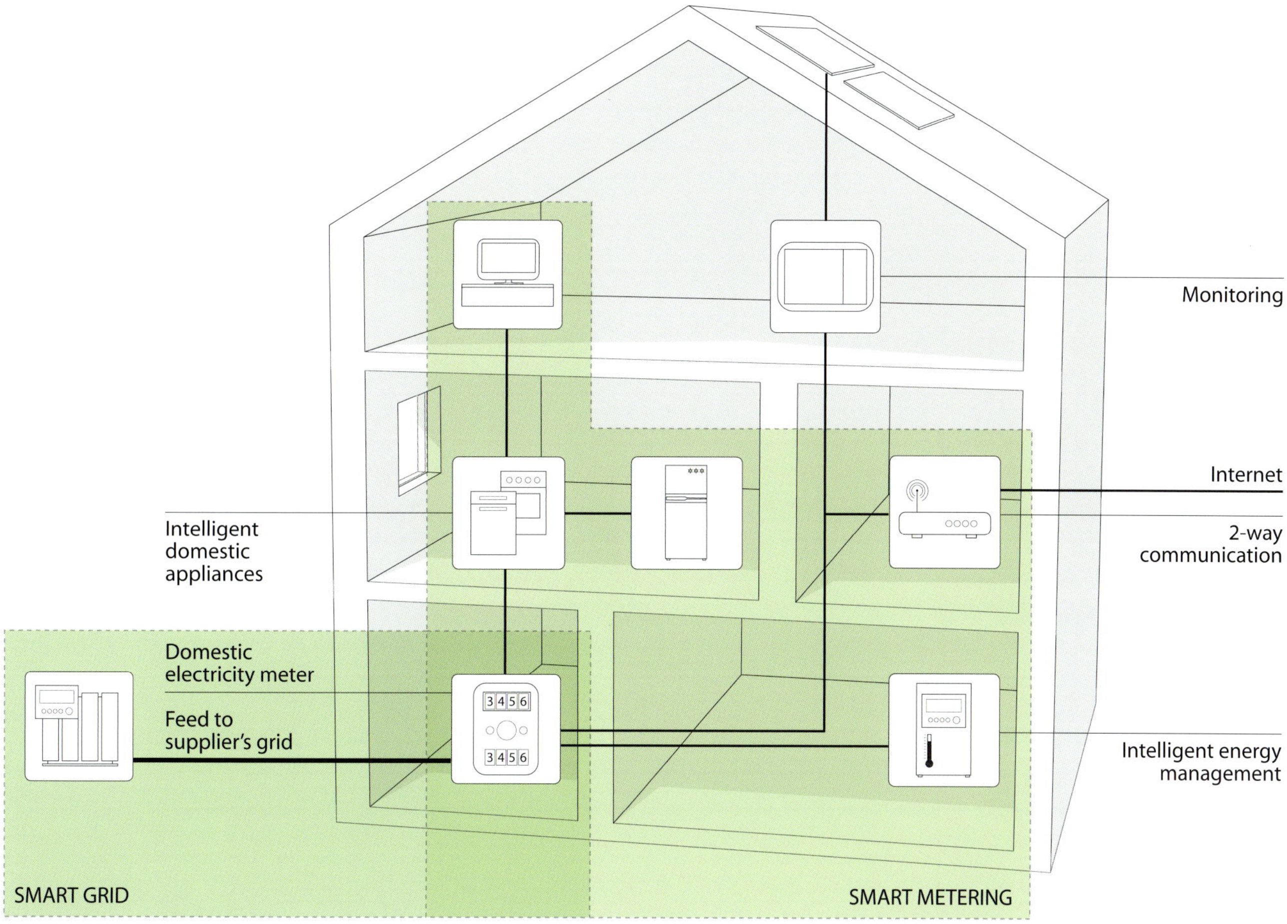

Schematic for a three-level smart house, smart metering and smart grid to allow better regulation of the energy consumption of private households

Problems of load management by automated processes:

- Legal and insurance issues (damage caused by devices starting automatically)
- Between neighbours (noise pollution caused by automated processes)
- Personal (the user must allow some control by an outside agent)
- Data protection (consumer data are passed to the electricity supplier and analysed for achieving better grid utilisation)

Advantages of load management:

- Energy savings
- Maximising on-site use of self-generated electricity, electricity purchased from the grid is minimised
- Energy costs minimised
- More transparent, precise and consumer-friendly electricity bills

Monitoring

Monitoring provides a means of checking a system regulation and control concept, optimising a system, and establishing whether the design objectives are being attained in operation. Before monitoring starts, it is important to define the benchmarks for describing the specified-actual comparisons. During monitoring the stored and consumed data are captured and saved at regular intervals, before being analysed and compared with previous data for technical and commercial purposes. Studying the data can reveal faults and allow processes to be optimised. If the building services systems are running smoothly, they can continue to be monitored in operation. Monitoring needs to take place over at least two years. The monitored building must be used according to need during this period. This recommended period is based on the assumption that some adjustment of the system takes place during the first year of monitoring and a further year of data is required to allow the check to be made. In addition, a longer monitoring period of several years can compensate for seasonal extremes of weather, such as a very hard winter or an extraordinarily hot summer.

The development of an energy monitoring concept should take into account the wider context of the building. The monitoring points and sensors should be selected based on the energy and monitoring concepts.

Monitoring should seek to give transparency to the building operation not only for the operator and the owner but also, where this is relevant, for the user and the public. Once users are aware of their behaviour and understand its effects, they can modify it in the knowledge of energy and cost implications.

A qualitative evaluation can supplement these kinds of quantitative measurements and analyses. The users can express their satisfaction and feelings of wellbeing in regular meetings. This yields knowledge of the users' subjective perceptions, which can be considered in conjunction with the data on seasonal or daily processes. The results can also show whether a changed consciousness of energy and the environment emerges, and whether living and user satisfaction and other subjective indicators of living value also change. By equating quantitative monitoring to qualitative (sociological) monitoring, building services technology can be better matched and optimised to user behaviour. Monitoring also allows the installed technology to be checked, adjusted and in some cases modified.

Preset measuring positions on systems used

System	Required measurement positions
Active use of environmental energy	
Ground loops/ collectors/ piles	Electricity consumption Supplied heat Supplied cold
Extraction wells	Electricity consumption (circulatory pump(s)) Supplied heat Supplied cold
Ground heat exchanger	Electricity consumption (fan(s)) Supplied heat Supplied cold
Mechanical night ventilation	Electricity consumption (fan(s)) Supplied cold
Feedback sensor	Electricity consumption (circulatory pump(s), fans, any spray pumps and open tank heating) Supplied heat Supplied cold
Solar thermal system	Solar radiation Electricity consumption (circulatory pump(s)) Supplied heat
Photovoltaics	Solar radiation Supplied electricity
CHP plants	
(Bio-)gas CHPP	Biogas consumption Heat generation (incl. waste gas exchanger) Electricity generation (after subtraction of self-generated electricity consumption)
(Bio-)oil CHPP	Bio-oil consumption Heat generation Electricity generation (after subtraction of self-generated electricity consumption)
Wood CHPP	Wood consumption Heat generation Electricity generation (after subtraction of self-generated electricity consumption)
Fuel cell	Biogas consumption Heat generation Electricity generation
Heating plant / waste heat use	
(Bio-)gas boiler	Biogas consumption Electricity generation Heat production
(Bio-)oil boiler	Bio-oil consumption Electricity generation Heat production
Wood boiler	Wood consumption Electricity generation Heat generation
District heating	District heat supply
(Bio-)gas HP	Biogas consumption Electricity generation (excluding from connected heat source)* Heat generation
Electrical heat pump (may be reversible)	Electricity generation (excluding from connected heat source)* Heat generation with reversible heat HP: Refrigeration
Direct electrical heating DHW	Electricity consumption Heat generation
Closed-cycle system	Electricity consumption (circulatory pump(s)) Heat gain
Waste air HP	Electricity generation (excluding from connected heat source)* Heat generation
Cold generator	
Vapour-compression refrigeration machine	Electricity consumption (excluding recooling)* Refrigeration
Absorption refrigeration machine (ARM)	Electricity consumption (excluding recooling)* Heat consumption Refrigeration
(Bio-)gas ARM	Electricity consumption (excluding recooling)* Biogas consumption Refrigeration
Remote cooling grid	District cooling grid
Desiccant cooling system Desiccant wheel	Electricity consumption (circulatory pump(s), fans, drive for desiccant wheel, water treatment) Water consumption Heat consumption Refrigeration
Desiccant cooling fluid	Electricity consumption (circulatory pump(s), fans, + brine, drive for fans regeneration) Water consumption Heat consumption Refrigeration
Storage (residential buildings only)	
Service water storage	Heat to storage Heat from storage Service water useful heat Circulation heat losses
Buffer storage	Heat to storage Heat from storage
Useful energy building technical services	
Lighting	Electricity consumption, if nec./useful by measuring operating times and performance data separately for example zones from energy balance DIN V 18599 (e.g. offices, circulation areas etc.)
Pumps Distribution	Electricity consumption, if nec./useful by measuring operating times and performance data
Air handling	Electricity consumption, if nec./useful by measuring operating times and performance data, air volume flows, air flow temperature
Operating circumstances	
Heating	Heating flow temperature Heating return temperature Heating circuit temperature
Ventilation	Supply air temperature Temperature before heat recovery Temperature after heat recovery Exit air temperature
Cooling	Flow temperature Return temperature

* Energy consumption for recooling and connected heat source are recorded separately.

Projects

As described, the Aktivhaus is a contemporary result of the continued development of existing building energy standards. It builds upon the principles of the minimisation of energy losses and internal energy consumption, and the passive use of solar radiation. An additional aspect is the active use of renewable energy sources in the house and on the property. The Aktivhaus integrates components that make use of renewable energy sources, such as solar-activated facades and roof surfaces, into the architecture. By actively using the building envelope to produce energy, new architectural images emerge that could ultimately give rise to a new building culture.

The buildings shown here as examples have taken up this challenge. They demonstrate the current status of this rapidly progressing developmental process, which is nonetheless only at an initial stage.

Not all the presented buildings attain the standard of an Energy-Plus building. Nevertheless, all are characterised by an intelligent combination of passive and active measures on the path to that goal, and consequently each one qualifies as an Aktivhaus. Refurbishment not only offers the chance to optimise existing conditions to upgrade energy efficiency, but also gives the opportunity to improve usability and the architectural appearance.

The projects shown here range from small single-family houses to multifamily dwellings and non-residential buildings, including factory buildings and office buildings as well as community facilities. In addition to new buildings, examples of refurbishments and renovations are given. They show that, even when confronting the difficulties of problematic existing buildings, it is possible to successfully generate a surplus of energy.

The degree to which implementation of the Aktivhaus concept is feasible differs depends on the building task, building form, density of the surrounding built fabric, and many other factors. The range of examples, however, shows that even under the most adverse conditions it is possible to implement an Aktivhaus – and even one that meets the Energy-Plus standard.

Energy-Plus House Luchliweg, Münsingen (CH)

NEW DETACHED SINGLE-FAMILY HOUSE (NEIGHBOURHOOD DENSIFICATION)

Project information

Architects:	dadarchitekten GmbH, Bern (CH)
Consultants / Energy concept:	Beer Holzbau AG, CTA AG, 3S Photovoltaics, Ökobaumarkt Bern (CH)
Client:	Private owner
Completion:	2010
Standard:	MINERGIE, Energy-Plus, CO_2-neutral
Living area:	160 m²
Final energy demand (heat and electricity)/m² living area:	43.65 kWh/m²a
Final energy production (renewable heat and electricity)/m² living area:	52.50 kWh/m²a

①

Balance scope per standard

- Heating
- Domestic hot water
- Cooling
- Auxiliary energy (pumps, ventilation)
- Lighting
- Equipment (household appliances)
- Electromobility

②

Careful treatment of the earth and raw materials and energy-saving construction were basic requirements for this single-family house in Canton Bern. This largely prefabricated wood building was erected in the middle of an existing residential estate from the first half of the 20th century. The task was to fit the building into the existing context as carefully as possible. After all, sustainable and ecological building means not only conserving resources and paying attention to energy efficiency, but also dealing carefully with the environment.

Additional requirements of the clients were to enable living and working in one building and to keep the spaces as flexible as possible so they could be adapted to possible changes of use. The single-family house is built on a slope and is organised clearly. The private area for the family is on the upper floor. The rest of the rooms are open to one another, thus establishing multiple visual connections. The building is oriented to the southwest, and the ancillary spaces such as the technical room and laundry room are nestled into the slope on the lower level.

According to calculations, the building produces 24 % more energy than it uses for heating, hot water, and all electricity needs, including auxiliary and household electricity. It is thus CO_2-neutral in operation.

Site plan, scale 1:2,000

The energy needed for heating and providing hot water is generated entirely by the heat pump and wood-burning stove.
The electricity required for auxiliary energy – the energy for the heat pump, the household appliances and the lighting – is supplied by the photovoltaic system. A calculated annual surplus of 24 % is the result.

169

③

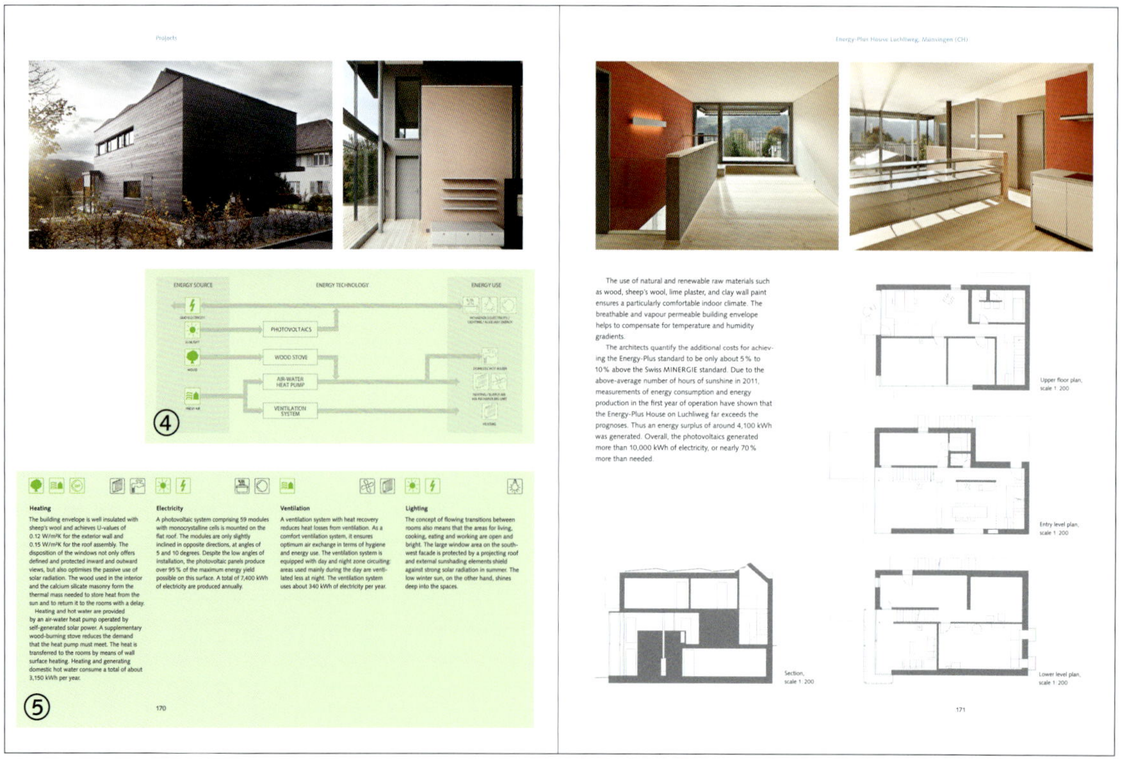

Projects

Energy-Plus House Luchliweg, Münsingen (CH)

④

Heating

The building envelope is well insulated with sheep's wool and achieves U-values of 0.12 W/m²K for the exterior wall and 0.15 W/m²K for the roof assembly. The disposition of the windows not only offers defined and protected inward and outward views, but also optimises the passive use of solar radiation. The wood used in the interior and the calcium silicate masonry form the thermal mass needed to store heat from the sun and to return it to the rooms with a delay.

Heating and hot water are provided by an air-water heat pump operated by self-generated solar power. A supplementary wood-burning stove reduces the demand that the heat pump must meet. The heat is transferred to the rooms by means of wall surface heating. Heating and generating domestic hot water consume a total of about 3,150 kWh per year.

Electricity

A photovoltaic system comprising 59 modules with monocrystalline cells is mounted on the flat roof. The modules are only slightly inclined in opposite directions, at angles of 5 and 10 degrees. Despite the low angles of installation, the photovoltaic panels produce over 95 % of the maximum energy yield possible on this surface. A total of 7,400 kWh of electricity are produced annually.

Ventilation

A ventilation system with heat recovery reduces heat losses from ventilation. As a comfort ventilation system, it ensures optimum air exchange in terms of hygiene and energy use. The ventilation system is equipped with day and night zone circuiting: areas used mainly during the day are ventilated less at night. The ventilation system uses about 340 kWh of electricity per year.

Lighting

The concept of flowing transitions between rooms also means that the areas for living, cooking, eating and working are open and bright. The large window area on the southwest facade is protected by a projecting roof and external sunshading elements shield against strong solar radiation in summer. The low winter sun, on the other hand, shines deep into the spaces.

⑤

170

The use of natural and renewable raw materials such as wood, sheep's wool, lime plaster, and clay wall paint ensures a particularly comfortable indoor climate. The breathable and vapour permeable building envelope helps to compensate for temperature and humidity gradients.

The architects quantify the additional costs for achieving the Energy-Plus standard to be only about 5 % to 10 % above the Swiss MINERGIE standard. Due to the above-average number of hours of sunshine in 2011, measurements of energy consumption and energy production in the first year of operation have shown that the Energy-Plus House on Luchliweg far exceeds the prognoses. Thus an energy surplus of around 4,100 kWh was generated. Overall, the photovoltaics generated more than 10,000 kWh of electricity, or nearly 70 % more than needed.

Upper floor plan, scale 1:200

Entry level plan, scale 1:200

Section, scale 1:200

Lower level plan, scale 1:200

171

Presentation of the projects

In the upper part of the page, all the project information is summarised in short form ①. In addition to the names of the architects and project consultants, this includes initial characteristic values. The specified and attained standard is identified along with the energy demand and production, in order to be able to compare the projects with one another.

As the selected examples come from different countries, primary energy factors, accounting methods, balance scopes, and energy reference areas can differ, so it is nearly impossible to make a comparison of the primary energy used. To enable comparison of the projects nevertheless, the absolute annual figures for the final energy demand are shown (in the right margin ③). The resulting annual figures for final energy demand and final energy production from renewable energy sources are calculated on this basis and reported (in the short summary ①) in relation to the heated living area or usable area.

The pictograms portray the balance scope of the project ②. Depending on the building's use and the standard applied for the ecological balance, different energy consumptions are incorporated into the analysis (shown in green).

The graphic comparison of final energy demand and final energy production in the right margin of the introductory page ③ enables immediate identification of a surplus or residual demand.

A distinction is also made between heat and electricity. The surplus or residual demand, as the case may be, is also expressed as a percentage, thus clearly indicating the degree of coverage by renewable electricity and renewable heat.

The generated surplus of electricity is converted to kilometres, even for projects that do not use it for powering electric vehicles, so as to illustrate the potential represented therein. The conversions are based on a consumption of 14 kWh per 100 km.

On the second double-page spread, the energy concept is explained in detail. This purpose is served in part by a diagram ④ depicting the energy flow from the power source to its use. Additionally, the concept description is divided to separately address the five essential energy services of heating, cooling, electricity, ventilation and lighting ⑤.

BALANCE SCOPES

Efficiency House (EnEV)	DE
Passivhaus	DE
Efficiency House Plus	DE
Nearly Zero-Energy House/Zero-Energy House	EU
MINERGIE (basic standard)	CH
MINERGIE-P	CH
MINERGIE-A	CH

Energy-Plus House Luchliweg, Münsingen (CH)

NEW DETACHED SINGLE-FAMILY HOUSE (NEIGHBOURHOOD DENSIFICATION)

Project information

Architects:	dadarchitekten GmbH, Bern (CH)
Consultants/Energy concept:	Beer Holzbau AG, CTA AG, 3S Photovoltaics, Ökobaumarkt Bern (CH)
Client:	Private owner
Completion:	2010
Standard:	MINERGIE, Energy-Plus, CO_2-neutral
Living area:	160 m²
Final energy demand (heat and electricity)/m² living area:	43.65 kWh/m²a
Final energy production (renewable heat and electricity)/m² living area:	52.50 kWh/m²a

Balance scope per standard

- Heating
- Domestic hot water
- Cooling
- Auxiliary energy (pumps, ventilation)
- Lighting
- Equipment (household appliances)
- Electromobility

Careful treatment of the earth and raw materials and energy-saving construction were basic requirements for this single-family house in Canton Bern. This largely prefabricated wood building was erected in the middle of an existing residential estate from the first half of the 20th century. The task was to fit the building into the existing context as carefully as possible. After all, sustainable and ecological building means not only conserving resources and paying attention to energy efficiency, but also dealing carefully with the environment.

Additional requirements of the clients were to enable living and working in one building and to keep the spaces as flexible as possible so they could be adapted to possible changes of use. The single-family house is built on a slope and is organised clearly. The private area for the family is on the upper floor. The rest of the rooms are open to one another, thus establishing multiple visual connections. The building is oriented to the southwest, and the ancillary spaces such as the technical room and laundry room are nestled into the slope on the lower level.

According to calculations, the building produces 24 % more energy than it uses for heating, hot water, and all electricity needs, including auxiliary and household electricity. It is thus CO_2-neutral in operation.

Site plan, scale 1:2,000

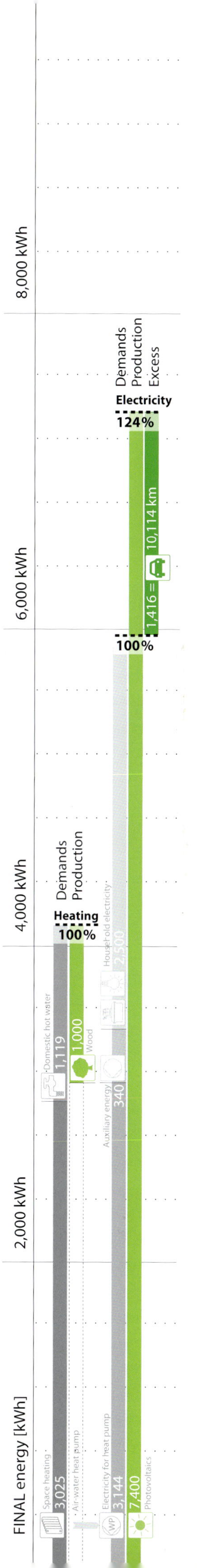

The energy needed for heating and providing hot water is generated entirely by the heat pump and wood-burning stove.

The electricity required for auxiliary energy – the energy for the heat pump, the household appliances and the lighting – is supplied by the photovoltaic system. A calculated annual surplus of 24 % is the result.

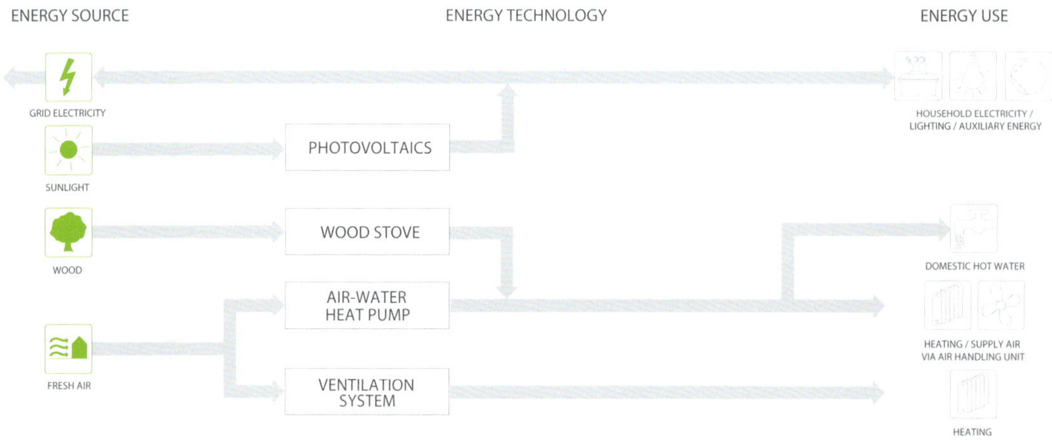

Heating

The building envelope is well insulated with sheep's wool and achieves U-values of 0.12 W/m²K for the exterior wall and 0.15 W/m²K for the roof assembly. The disposition of the windows not only offers defined and protected inward and outward views, but also optimises the passive use of solar radiation. The wood used in the interior and the calcium silicate masonry form the thermal mass needed to store heat from the sun and to return it to the rooms with a delay.

Heating and hot water are provided by an air-water heat pump operated by self-generated solar power. A supplementary wood-burning stove reduces the demand that the heat pump must meet. The heat is transferred to the rooms by means of wall surface heating. Heating and generating domestic hot water consume a total of about 3,150 kWh per year.

Electricity

A photovoltaic system comprising 59 modules with monocrystalline cells is mounted on the flat roof. The modules are only slightly inclined in opposite directions, at angles of 5 and 10 degrees. Despite the low angles of installation, the photovoltaic panels produce over 95 % of the maximum energy yield possible on this surface. A total of 7,400 kWh of electricity are produced annually.

Ventilation

A ventilation system with heat recovery reduces heat losses from ventilation. As a comfort ventilation system, it ensures optimum air exchange in terms of hygiene and energy use. The ventilation system is equipped with day and night zone circuiting: areas used mainly during the day are ventilated less at night. The ventilation system uses about 340 kWh of electricity per year.

Lighting

The concept of flowing transitions between rooms also means that the areas for living, cooking, eating and working are open and bright. The large window area on the southwest facade is protected by a projecting roof and external sunshading elements shield against strong solar radiation in summer. The low winter sun, on the other hand, shines deep into the spaces.

The use of natural and renewable raw materials such as wood, sheep's wool, lime plaster, and clay wall paint ensures a particularly comfortable indoor climate. The breathable and vapour permeable building envelope helps to compensate for temperature and humidity gradients.

The architects quantify the additional costs for achieving the Energy-Plus standard to be only about 5% to 10% above the Swiss MINERGIE standard. Due to the above-average number of hours of sunshine in 2011, measurements of energy consumption and energy production in the first year of operation have shown that the Energy-Plus House on Luchliweg far exceeds the prognoses. Thus an energy surplus of around 4,100 kWh was generated. Overall, the photovoltaics generated more than 10,000 kWh of electricity, or nearly 70% more than needed.

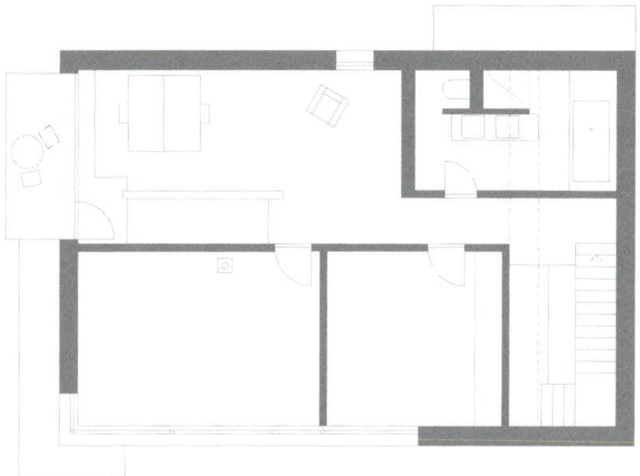

Upper floor plan, scale 1:200

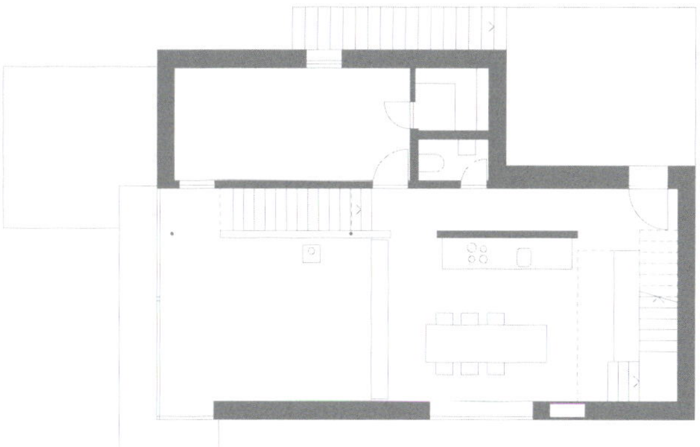

Entry level plan, scale 1:200

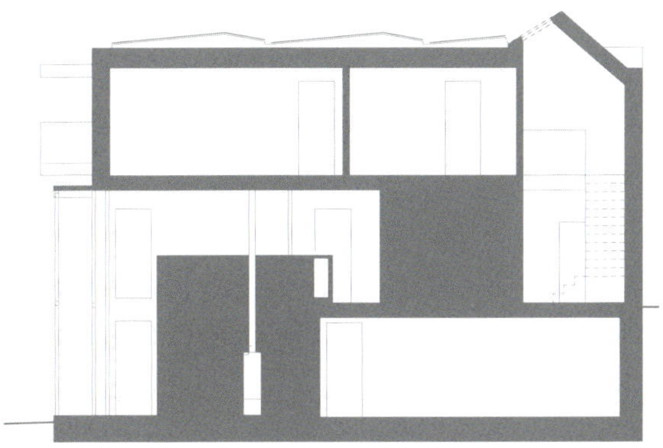

Section, scale 1:200

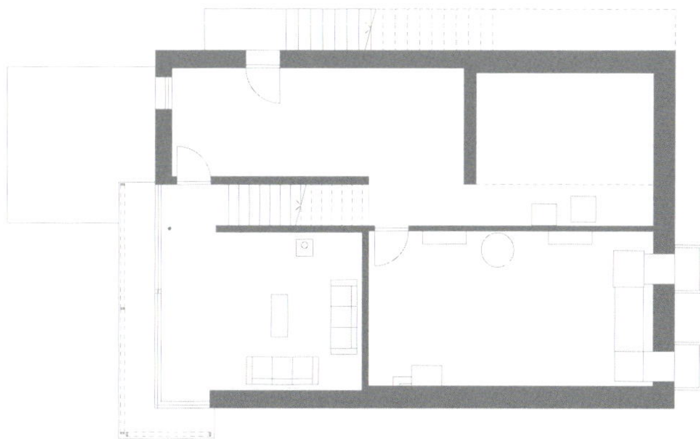

Lower level plan, scale 1:200

LichtAktiv Haus, Hamburg (DE)

REFURBISHED SEMI-DETACHED HOUSE

Project information

Architects:	TU Darmstadt, Department of Architecture, Energy-Efficient Building Design Unit, Prof. Manfred Hegger/Katharina Frey and Ostermann Architekten, Hamburg (DE)
Lighting Designer:	Prof. Peter Anders, Hamburg (DE)
Client:	VELUX Deutschland GmbH
Completion:	1954/2010
Standard:	Zero-energy house, CO_2-neutral in operation
Living area:	189 m²
Final energy demand (heat and electricity)/m² living area:	51.53 kWh/m²a
Final energy production (renewable heat and electricity)/m² living area:	59.67 kWh/m²a

Balance scope per standard

- Heating
- Domestic hot water
- Cooling
- Auxiliary energy (pumps, ventilation)
- Lighting
- Equipment (household appliances)
- Electromobility

The LichtAktiv Haus is the result of modernising a typical semi-detached house originally built in 1954. It is part of the International Building Exhibition IBA Hamburg and serves as a pilot project that helps enable the IBA site on the Elbe island of Wilhelmsburg to become a climate-neutral district. The so-called "settler's house" is one of six experimental homes that have been built throughout Europe as part of the Model Home 2020 programme sponsored by VELUX. The goal of this programme is to develop new ways of living and working in a pleasant indoor climate, with plenty of natural light and optimum energy efficiency.

The initial design idea originates from Katharina Fey and was created as part of a student competition at the TU Darmstadt. The winning design served as the basis for further development of the project. The original building no longer met current demands for either comfort or useable space. Hence as part of the renovation the house was completely altered and also expanded with an addition. The habitable area was thereby increased by a total of nearly 40% to 132 m². The new part of the building accommodates a living and dining area, kitchen, and the technical equipment room. The more private family areas – master bedroom, bathroom, and children's rooms – are located in the old house. The new addition is also used for energy production. As a result of the refurbishment, the total annual final energy demand could be reduced by more than a half.

Site plan, scale 1:2,000

The energy needed for heating and providing domestic hot water is supplied by the solar thermal system and the heat pump. The photovoltaic system covers almost all the electricity needed to operate the building (household electricity incl. lighting, plus auxiliary electricity incl. power for the heat pump).

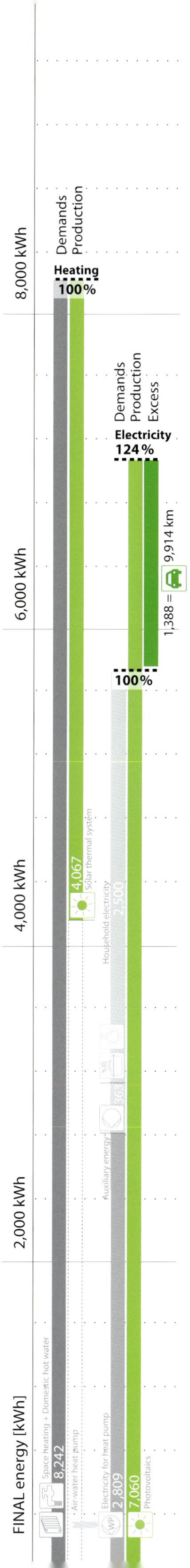

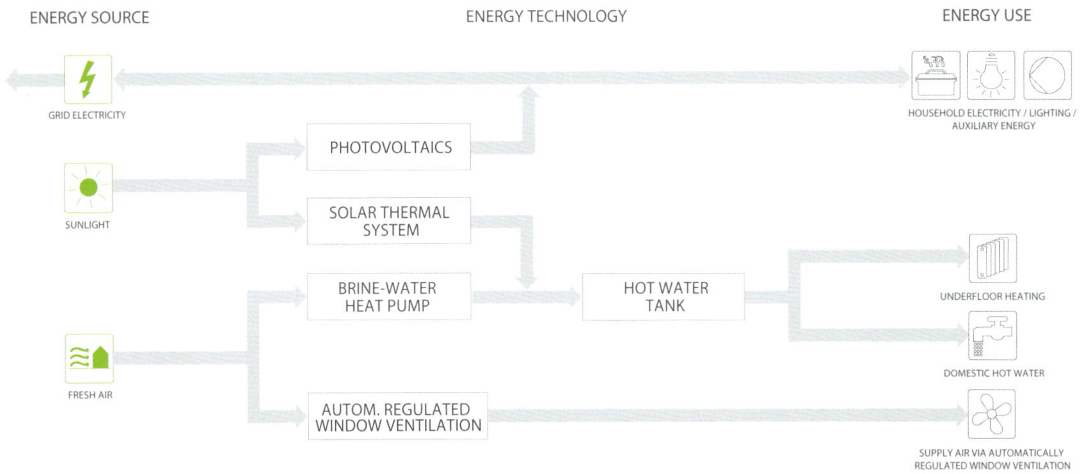

Heating

A combination of solar thermal energy with an brine-water heat pump, which extracts heat by geothermal probes, supplies the thermal energy. The 21.7 m² collector surface of the solar thermal system is located on the roof of the extension. The hot water is stored in a 1,000 l tank. The rooms are heated primarily with underfloor heating. The solar energy gains from the windows contribute to heating the interior on cold days. To diminish heat islands, the roof was covered with light-coloured reflective fibre cement tiles.

Electricity

A photovoltaic system composed of polycrystalline solar cells and measuring over 75 m² was integrated into the roof of the new building. It generates more than 7,000 kWh of electricity per year. Some of the photovoltaic cells used are glass-glass PV modules that create a beautiful play of light and shadow on the patio and the carport. With an electricity consumption of about 2,500 kWh per year for the household and roughly 3,200 kWh for the production of heating, including auxiliary electricity, over the year, the entire energy demand over the course of a year is fully covered. Surpluses power the electric car or they are fed into the local grid.

Ventilation

The house is ventilated by means of automatically controlled window ventilation, which provides for natural ventilation and ensures the minimum air exchange. The windows open by themselves as needed. Sensors measure the room temperature, the CO_2 content, and the indoor humidity. Unlike with mechanical ventilation, no ventilation shafts are necessary, which avoids potentially uneconomical and costly retrofitting needed for the former.

Lighting

The name LichtAktiv Haus speaks for itself: expansive window openings and a central, multistorey circulation space-cum-library provide plenty of daylight. With a total of 90 m², the window area has more than quadrupled. The proportion of windows in the existing building has doubled, and almost 60 m² of window area has been added in the newly built annex. As a result, the house now boasts daylight-flooded rooms, meaning that artificial lighting can usually be dispensed with even on cloudy days.

Thanks to intelligent building technology combined with sophisticated concepts for daylighting, natural ventilation, and shading as well as well-conceived spatial planning, the total energy demand for heating, hot water, and electricity can be offset year-round by renewable energy. Thus CO_2-neutral operation is made possible in the annual balance. The surplus of energy from renewable sources means that the emissions for the production, maintenance and disposal of the LichtAktiv Haus will gradually diminish. In purely mathematical terms, after about 26 years the building thus fully offsets the energy used for its renovation and achieves a neutral global warming potential. During the two-and-a-half-year-long phase of use by a test family, the energy consumption and the energy gain will be measured and the indoor climate will be monitored and documented. The results will yield insights for future modernisation projects. In addition to the quantitative measurements, sociologists are conducting a qualitative study: the residents answer questions on housing satisfaction, living feel and comfort. Thus it is possible to determine how much user behaviour changes and, at the same time, awareness for energy and comfort increases. The quantitative and qualitative surveys are not always consistent with one another; for example, the family perceives the interior temperature as varying with the seasons while the actual measured temperature is constant. An important finding is that no one wishes to do without the automatic functions of the building envelope, with its natural ventilation and air-conditioning, or forgo command of the system controls and the performance data about the house shown on display screens.

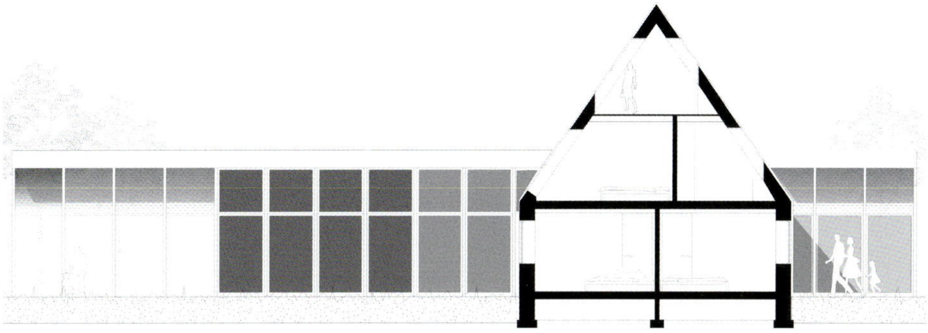

Section, scale 1:200

Entry level plan, scale 1:200

Zero-Energy House, Driebergen (NL)

REFURBISHMENT OF AN OLD VILLA AS AN ENERGY-NEUTRAL LISTED HISTORIC BUILDING

Project information

Architects:	Zecc Architects, Utrecht (NL)
Energy concept:	OPAi – oneplanetarchitecture institute, Amsterdam
Client:	Private owner (NL)
Completion:	2010
Location:	Driebergen (NL)
Standard:	Zero-energy house
Living area:	150 m²
Final energy demand (heat and electricity)/m² living area:	41.32 kWh/m²a
Final energy production (renewable heat and electricity)/m² living area:	39.00 kWh/m²a

Balance scope per standard

- Heating
- Domestic hot water
- Cooling
- Auxiliary energy (pumps, ventilation)
- Lighting
- Equipment
- Electromobility

Through refurbishment, this family house in the province of Utrecht, which was originally built the 1920s, now meets zero-energy standards for heating and hot water, Renewable sources cover the entire energy demand, including the auxiliary electricity needed for pumps, thus making it the first energy-neutral listed historic building in the Netherlands.

The red brick facade and white, wood-framed windows are a familiar sight in the region. Renovation of the historic brick building was therefore carried out with due care. The energy upgrade measures are not visible from the street, as most of the changes to the building took place within it or in the new addition at the rear. The brick villa gets its energy from this extension and from the portion of the roof of the main house that faces away from the street. The technical room is in the basement of the new addition and the rooftop solar collectors used for generating energy are on the roof. Seen from the outside, the three sides of the house that are visible from the street remain unchanged. The expansive glazing and modern cubic appearance of the new addition contrasts with the old building. A reveal on the side of the landmark-protected facade separates the old and new parts of the house.

Site plan, scale 1:2,000

The heating and domestic hot water are supplied entirely by solar thermal collectors and a heat pump. Photovoltaics supply 90% of the electricity required for auxiliary energy, including the energy needed for the heat pump. Electricity for household appliances and lighting is not taken into consideration. No surplus is generated.

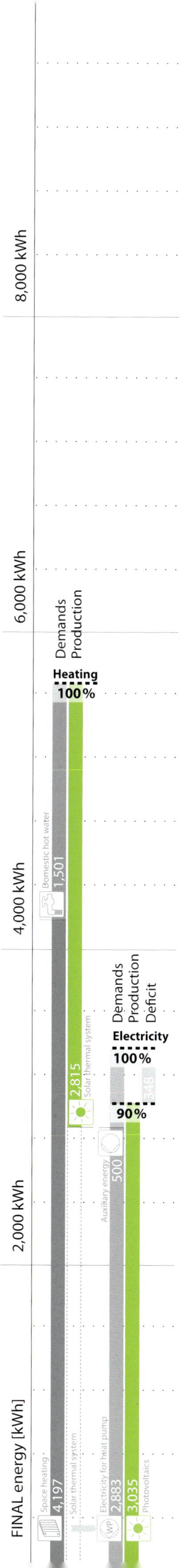

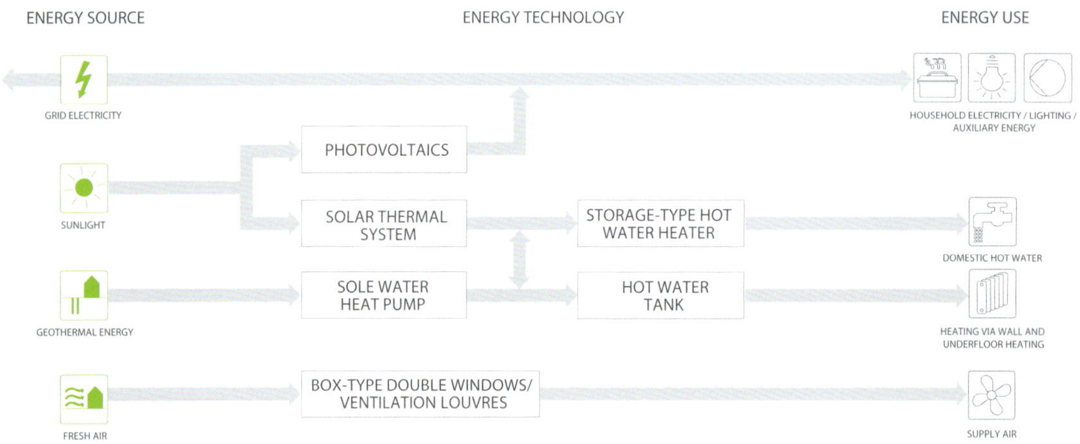

Heating

To reduce thermal losses and increase comfort, the three publicly visible facades were insulated from inside with wood fibreboards coated with earthen plaster. In order to preserve the old windows and their details, new insulating glass panes were added in front of the old windows on the inside. Because the new panes are dimensioned slightly larger, they have the appearance of display windows that highlight the old details. The rear of the building was insulated from the outside by a second wall. Its wood frame construction is insulated with flax between the uprights and an additional layer of flexible wood fibreboards against the existing exterior wall. In the second facade layer, the windows were treated in a manner similar to the interior. The new insulating glass panes set in front are so large that the old masonry and the original frames remain visible. The roof was insulated with flax, and the crawl space beneath the ground floor was filled with non-toxic granulate material made of recycled glass.

The building services are accommodated in the basement of the new addition, where the heat pump, the storage tank for heating water, and the storage-type domestic hot water heater are located. These storage tanks are fed by two heat sources: a solar thermal system, used mainly in the summer, which was installed on the roof of the extension so as to conceal it from view. There, three vacuum tube collectors generate energy to supply the heating water storage tank and the domestic hot water storage tank. The solar thermal system provides half of the annual energy needs for domestic hot water heating and a fifth of the energy requirements for heating. A second system, a brine-water heat pump, is employed mainly in the winter. It is provided with geothermal energy from a ground heat exchanger. The heat pump covers the remaining energy demand for domestic hot water heating and almost the entire heating demand. Thermal energy is transferred to the rooms by means of wall surface heating and in some places by underfloor heating.

Electricity

Without a photovoltaic system the concept of an energy-neutral historic monument would not have been feasible. The PV system was installed on the south-facing roof of the old brick building. Since that is the side facing the garden, the listed view from the street is not impaired by the PV modules. A total of 17 modules of polycrystalline cells yielding an output of 3.74 kW peak were installed. The modules have an efficiency of about 14 %. In the annual balance, the self-generated electricity is sufficient to operate the heat pump. On sunny days a surplus is generated, which is fed into the public grid. The electricity for household uses and lighting is calculated as coming from the grid.

Ventilation

The double wall on the inside not only provides thermal insulation but is also part of the ventilation system. Fresh outside air enters through ventilation louvres into the space between the old and new windows. There the cool air is warmed up and rises. The warm air then flows into the living space through ventilation slits above the second window layer. Exhaust air is extracted centrally via the bathroom. The advantage of this concept over a conventional ventilation system is that the routing and lengths of ducts remain manageable and no additional measures are necessary. This natural ventilation system was not implemented on the garden facade.

The materials used for renovating the old brick building were chosen with special care. With regard to aspects of sustainability, particular attention was given to environmental and natural considerations and recyclability. Safe materials were therefore used, such as flax for insulation and clay for the interior plaster. The aggregate used for the plaster on the new addition consists of crushed bricks from the old extension that was demolished. To conserve resources further, rainwater is recycled for greywater use.

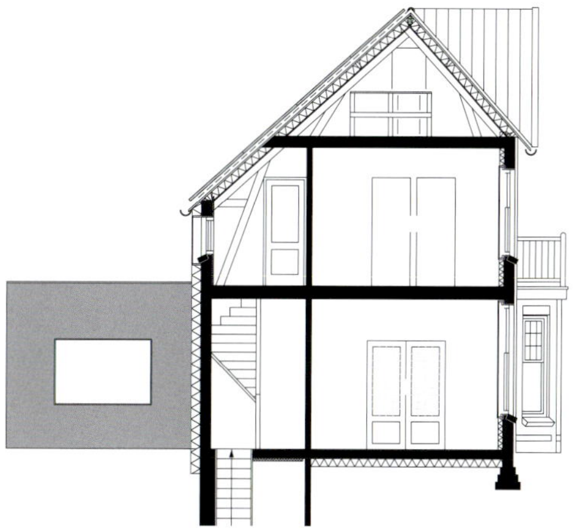

Section, scale 1:200

Lighting

The combined kitchen, dining and living area in the extension is suffused with light because it is almost fully glazed. In addition, the glazed corner of the extension is equipped with sliding glass doors that can open completely. This allows the dining area to expand onto the terrace outside.

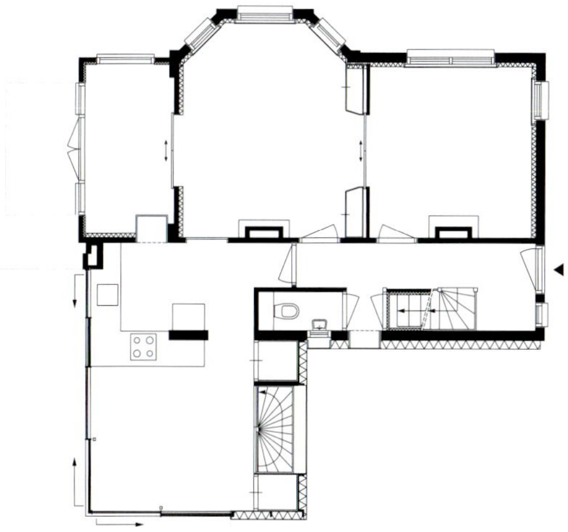

Entry level plan, scale 1:200

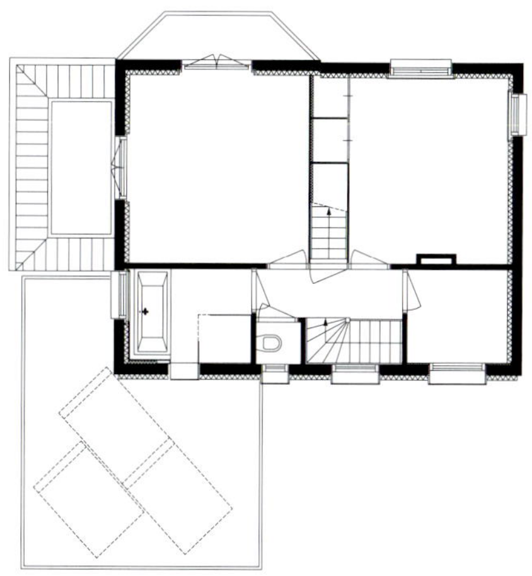

Upper floor plan, scale 1:200

14

Energy Flex House, Taastrup, Copenhagen (DK)

TWO NEW SINGLE-FAMILY HOUSES

Project information

Architects:	Henning Larsen Architects
Consultants/Energy concept:	Engineer: Enemærke & Petersen A/S
Client:	Danish Technological Institute
Completion:	2009
Location:	Taastrup, Copenhagen (DK)
Standard:	Energy-Plus, energy neutral
Usable area:	216 m² (EnergyFlexFamily)
Final energy demand (heat and electricity)/m² living area:	30 kWh/m²a

Balance scope per standard

- Heating
- Domestic hot water
- Cooling
- Auxiliary energy (pumps, ventilation)
- Lighting
- Equipment
- Electromobility

The EnergyFlexHouse project, which was developed by the Danish Technological Institute in collaboration with Henning Larsen Architects, is located 20 km west of Copenhagen. This case-study house makes it possible to test various energy-efficient constructions, components, and control technologies. The goal is to develop cost-effective energy technologies for sustainable building.

The term "Flex" refers to the building's flexibility. In this experiment, the building controls, insulation standards, and components of energy-efficient construction can be varied, added, adapted and tested.

The project comprises two buildings: EnergyFlexLab – an uninhabited laboratory for developing comprehensive systems – and EnergyFlexFamily – a home where families live for three to five months at a time in order to test the developed systems in everyday use.

The energy consumption can be followed live on the Internet or on a control panel. The consumption levels for heating, ventilation, electricity and water, as well as the energy produced by photovoltaics and the solar thermal system are updated every 15 minutes.

In the two-storey residential building, the communal areas such as the living spaces, kitchen and dining area are upstairs. This allows them to take advantage of more ceiling height beneath the gable. The individual bedrooms, bathrooms, and building services are located on the ground floor.

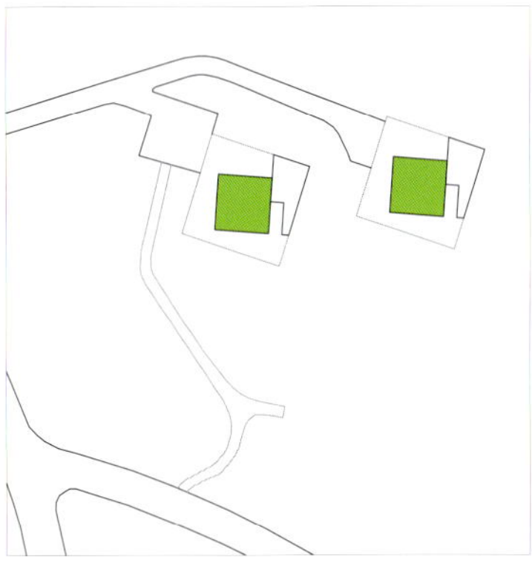

Site plan, scale 1:2,000

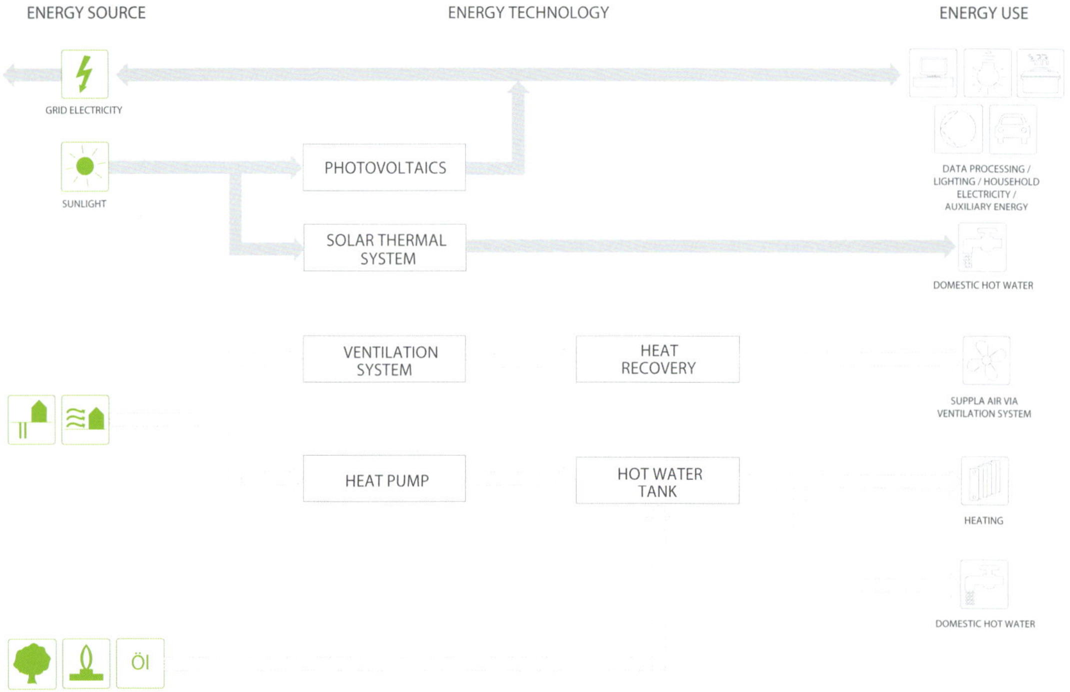

The project allows various systems to be used for heating and producing domestic hot water. The user can choose the system at will and switch as needed.

Heating

The volume of the building is kept compact in order to limit transmission losses as far as possible. Various passive measures are employed to reduce the energy demand. The roof and the walls of the highly insulated timber frame construction are fitted with 50 cm of mineral wool, thus achieving a U-value of 0.08 to 0.09 W/m²K. The foundation slab is insulated with a 40 cm thick polystyrene layer (U-value of 0.11 W/m²K). To reduce heat losses, triple-glazed windows with a U-value of 0.73–0.9 W/m²K are used.

The domestic water is heated by south-facing solar collectors with an area of about 5 m², which are integrated into the roof. In the EnergyFlexHouse, energy can be supplied by various systems: gas, oil or pellet boilers, geothermal probes, heat pumps (air-water or air-to-air), and solar thermal collectors. These alternatives can be chosen by the residents and then used as examples for testing their comfort. For the heat distribution, too, different options are available. Residents can choose from underfloor heating, radiators and a ventilation system.

To obtain a certain flexibility in the insulation standard as well, the thermal insulation was installed in several layers so its overall thickness can be varied. As a consequence, and due to the possibility of replacing windows, remedial measures can be simulated.

Furthermore, various heat storage systems are provided (phase change material, thermal mass activation, buffer storage tanks etc.).

Electricity

Photovoltaic panels are integrated into the southern surface of the gabled roof. The 60 m² facility produces so much electricity that a family's entire electricity demand (indoor climate, hot water, lighting, household appliances, entertainment media) can be covered, including an electric car. The high-performance monocrystalline modules employed here generate nearly 10,000 kWh per year. The electricity can be stored in batteries, used directly to charge the electric car, or fed into the public grid.

Ventilation

The ventilation system is located in a small room beneath the roof. The building can be ventilated using mechanical ventilation with optional heat recovery via a heat exchanger, using natural ventilation, or a combination of both methods.

The diverse architectural and technical systems in the EnergyFlexHouse yield numerous possible combinations. In addition, the Danish Technological Institute is continually developing new energy supply systems that are tested here. During their stays in the house, the families have evaluated the technical facilities with respect to their ease of use, efficiency and viability for daily use, as well as the resulting living comfort. The interaction between users and technology is a particular concern of this experiment. The house is a result of the comprehensive application of Aktivhaus principles and the use of passive, active and creative measures.

Lighting

Since the upstairs living areas are directly under the roof, the roof windows ensure good natural lighting. The window surfaces cover a total of 43 m², which is approximately 20% of the entire roof area. Of that, approximately one third of the surface area faces south, slightly more than a quarter faces north, and the remaining two fifths are oriented in equal parts to the east and west. Exterior blinds on the south-facing windows protect against overheating. Energy-saving measures provide artificial lighting exclusively with LEDs.

Home for Life, Lystrup (DK)

NEW SINGLE-FAMILY HOUSE

Project information

Architects:	Aart Architects
Consultants/Energy concept:	Esebensen Consulting Engineers
Client:	VELUX Group and Velfac
Completion:	2009
Location:	Lystrup (DK)
Standard:	Energy-Plus
Usable area:	190 m²
Final energy demand (heat and electricity)/m² living area:	53.2 kWh/m²
Final energy production (renewable heat and electricity)/m² living area:	62.6 kWh/m²

Balance scope per standard

- Heating
- Domestic hot water
- Cooling
- Auxiliary energy (pumps, ventilation)
- Lighting
- Equipment
- Electromobility

The "Home for Life" is the Danish contribution to the experiment, initiated by VELUX, to construct a series of demonstration buildings throughout Europe. The sponsor's intention is to show that generating surplus energy and creating a healthy living environment are compatible goals. The "Home for Life" project features a single-family house in Lystrup near Aarhus, on the Baltic coast of the Danish mainland. It is the first of the buildings to be completed in this showcase project. All the buildings are lived in and evaluated through monitoring over several years to determine their utility and whether they have achieved their set targets. In the "Home for Life," this has yielded valuable insights about suitable products, solutions and housing forms. One important conclusion was that technology, building materials, architectural concepts, and residential demands must be carefully brought into sync to achieve the desired effects, which were: an Energy-Plus House with a high degree of living comfort.

Site plan, scale 1:2,000

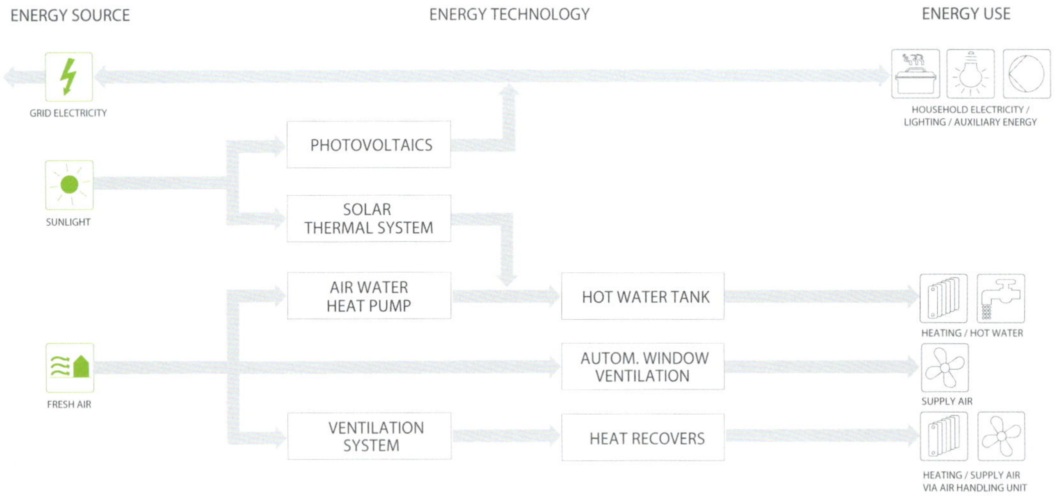

Heating

Through effective thermal insulation of the walls, roof and foundation slab, as well as triple-glazed windows, the heat losses of the building envelope could be reduced to a minimum. Thus the U-values are 0.1 W/m²K for the walls insulated with rock wool and 0.07 W/m²K for the roof and foundation slab.

In addition, approximately 50% of the required space heating can be provided by passive solar gains from the windows.

The heat is transferred through a combination of underfloor heating and conventional radiators. Hot water is provided by a nearly 7 m² of solar thermal collectors integrated into the roof. The remaining energy demand is met by an air-water heat pump, whose level of efficiency was increased by 25% by coupling it with solar thermal technology. The heat pump is used primarily to supply the heating circuits and only secondarily for the hot water tank.

Electricity

The roof, whose south-facing side extends down lower, provides enough space for 50 m² of photovoltaic panels. These are modules made of polycrystalline cells with an efficiency level of 12.5% and a total installed capacity of 6 kWpeak. This surface annually produces about 5,500 kWh of electricity and thus covers the entire electricity demand for lighting as well as household and auxiliary energy, such as operating the heat pump, over the course of a year in the single-family house. In addition, the large window surfaces reduce the electricity demand for lighting. Surplus electricity is fed into the public grid.

Ventilation

Ventilation of the "Home for Life" is provided by different systems in winter and summer. In winter, a mechanical ventilation system with heat recovery takes over the provision of preheated supply air. In summer, sensor-controlled windows ensure the direct intake of outside air.

An adequate supply of fresh air is ensured with both systems. Sensors are used to measure the temperature, humidity, and CO_2 concentration in the rooms and thus to automatically to ventilate and heat only those spaces that require it, thereby reducing energy consumption. The automatic ventilation also ensures that the indoor air humidity remains low and the formation of mould and mildew is prevented.

Lighting

The window surfaces are an essential aspect in the design and hence an important part of the energy concept. For one thing, they enclose 75 m², or almost 40% of the floor area, and for another, the windows have been placed to align with the position of the sun. Natural lighting is optimised by the orientation of the windows and with extremely narrow frame profiles, which admit daylight deep into the room.

In unused rooms, sensors automatically turn off the artificial lighting, thereby reducing the consumption of electricity.

There are two systems for shading, which give the residents the ability to regulate the brightness in a room to meet their own needs. The sun shading elements are mounted both inside and outside the windows. They not only regulate the daylight, but also ensure a targeted use of solar radiation and thus of the heat that infiltrates the building. A sensor system coupled to an intelligent control

The "Home for Life" was put to practical test for one year, during which a family lived in the house. The evaluation showed that the simulated energy balance was calculated very conservatively. On sunny days, the generation of electricity exceeded the demand by a multiple factor and it was possible to feed electricity into the public grid.

It is assumed that the energy demand can be kept even lower in the future, since the control systems needed for monitoring require more energy than will be necessary in normal operation. On the basis of the monitoring, it was possible to design changes that make it easier for the users to operate the control functions.

The survey of the test family showed that the indoor climate in the rooms was perceived as most comfortable with natural climate control. In these rooms, the CO_2 concentration was considerably lower than in the areas with mechanical climate control.

If we use the surplus energy that the "Home for Life" produces to offset the energy thal was used for the building materials, the building willcan be considered CO_2-neutral after 40 years.

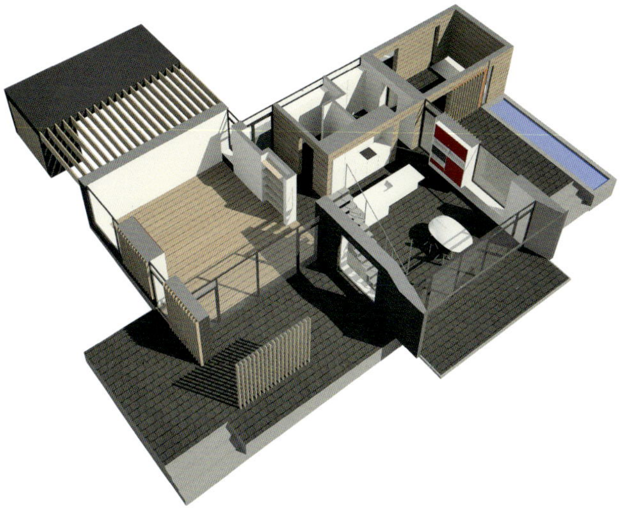

system automatically regulates the elements. Optimised from the standpoint of energy efficiency, it protects against overheating in summer and enables use of the low winter sun to heat the rooms. The exterior shading along the south is controlled automatically based on the indoor room temperature as well as the brightness. The shading along the east and west is controlled manually by the residents. In addition, a roof overhang nearly one metre deep on the southern side protects against the steep summer sun.

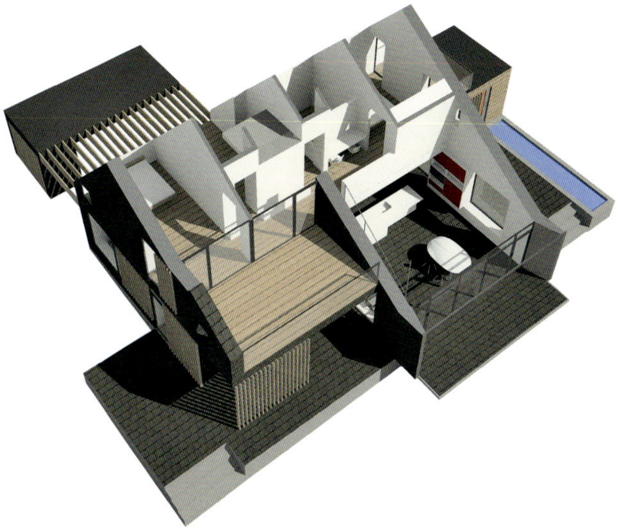

Residential and office buildings, Zurich (CH)

NEW CONSTRUCTION OF TWO RESIDENTIAL AND OFFICE BUILDINGS IN THE CITY CENTRE

Project information

Architects:	kämpfen für architektur ag, Zurich (CH)
Consultants/Energy concept:	Building services: Planforum, Winterthur (CH); Building physics: Amstein & Walthert AG, Zurich (CH)
Client:	Private owner
Completion:	2012
Standard:	MINERGIE-P-ECO
Living area:	3,370 m² and 2,150 m²
Final energy demand (heat and electricity)/m² living area:	21.63 kWh/m²a
Final energy production (renewable heat and electricity)/m² living area:	24.73 kWh/m²a

Balance scope per standard

- Heating
- Domestic hot water
- Cooling
- Auxiliary energy (pumps, ventilation)
- Lighting
- Equipment
- Electromobility

The two 5- to 6-storey buildings are located in the densely built city centre of Zurich. The gap left by demolishing the extant building was filled by a replacement building on Mühlebachstrasse and the perimeter block was supplemented by a building on Hufgasse. The result was a quiet, green inner courtyard.

The new building on Mühlebachstrasse is a retail building, which, thanks to its flexible floor plans, can also be used as a residential building. Due to its peaceful location, the rear building on Hufgasse is suitable for purely residential use. A total of 15 apartments and 6 office units were created.

The two buildings were constructed of wood, and the loadbearing exterior walls are made of large wooden frame elements with laminated wood columns. The stairwells and the basement walls are made of exposed recycled concrete. The dark slate cladding of the street facades is in stark contrast to the light framed windows with sliding shutters. Dark solar collectors integrated into the facade and light yellow facade panels dominate the walls along the courtyard.

Thanks to an energy concept based mainly on renewable energy along with an ecological choice of materials, the entire building complex achieves the MINERGIE-P-ECO standard.

Site plan, scale 1:2,000

The space heating and hot water needs are met by a pellet and solar thermal system. The photovoltaic systems provide more than twice the electricity required for auxiliary energy. Electricity for household appliances and lighting are not taken into consideration.

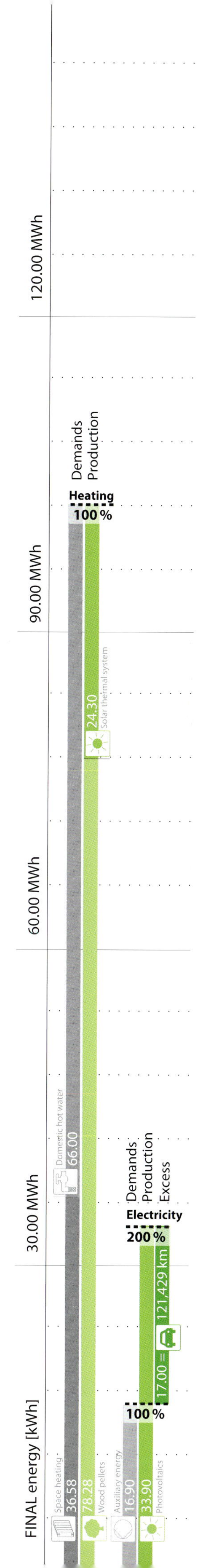

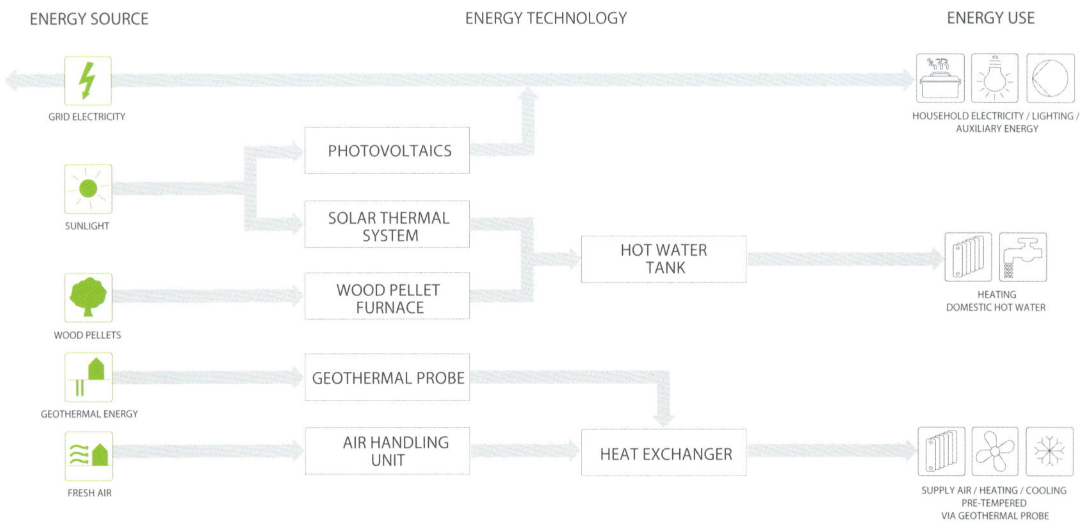

Heating

Triple-glazed wood/metal windows enable maximum passive solar gains due to their high g-value. The exterior walls and the roof were insulated with 24 cm of mineral wool. Around balconies and terraces, highly efficient vacuum insulated panels were used to reduce installation thicknesses and thermal bridges.

The heat required for domestic hot water and heating is provided by combining a pellet system with solar collectors. A two-storey stockpile of pellets is stored in the basement of the commercial and residential building on Mühlebachstrasse.

Two solar collector systems were employed. A total 95 m² of flat plate collectors are integrated into the southwest-facing courtyard facade. They fit well into the overall facade. Much more efficient vacuum tube collectors are installed on the roof of the Hufgasse building. Although the total collector surface in the facade is nearly five times as large as the 20 m² of vacuum tube collectors on the roof, the facade generates just 50 % more yield.

In order to optimise operations, hot water tanks are installed in both buildings. In addition to a central storage tank with a volume of 7,700 l in the building on Mühlebachstrasse, there is another tank with 3,000 l capacity on Hufgasse. Stratified storage tanks are used as heat accumulators due to the different temperatures required.

Each of the two solar thermal systems feeds its own storage tank. The hot water taps and underfloor heating are supplied from there.

Even though the building on Hufgasse has less area, its domestic hot water demand is higher than on Mühlebachstrasse. This is explained by the different uses. Hufgasse is largely a residential building whereas Mühlebachstrasse is an office building.

Cooling

Shading elements in the form of external sliding shutters prevent the rooms from overheating in summer. The ventilation system also provides the cooling, using geothermal probes to temper the air in summer.

Electricity

In order to generate the majority of the required electricity on site, photovoltaic systems are located on both roofs. Approximately 34,000 kWh are generated annually, enough to meet the electricity demand for the ventilation systems and the auxiliary electricity system of both buildings and to produce a surplus. The electricity needed for lighting and household use is covered by power from the public grid.

Ventilation

In order to simplify the controls, each of the two buildings has its own ventilation system. Using geothermal probes and a heat exchanger, the air is either preheated in winter or cooled in summer. Thus the air quantities for the residential and office units can be controlled separately. With a switch the user can select the desired quantity of exchange air. A bus system then operates a damper opening to regulate the volume flow.

Lighting

Large windows in the east facade and floor-to-ceiling windows in the southwest facades provide the rooms with consistent natural light.

From the very beginning of the planning, the goal was not only to minimise the energy required for operations but also to minimise the embodied energy introduced by the building materials.

The entire operating energy demand is covered by renewable sources, with self-generated electricity (excluding household electricity and lighting) and self-generated heat.

In inner-city locations, facade and roof surfaces are often shaded by the dense arrangement of buildings or they are too small due to closed perimeter block development. This makes it difficult to produce sufficient renewable heat and renewable electricity. This project shows that, even on such inner-city sites and with multistorey construction, it is still possible to meet the energy demand of buildings on the property by activating the building envelope.

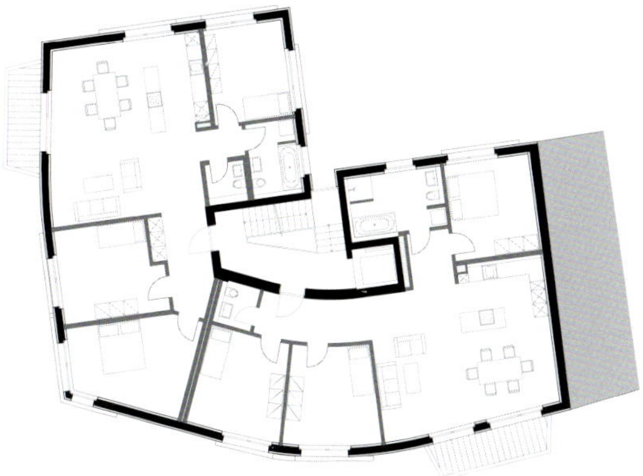

Hufgasse
Upper floor plan,
scale 1:400

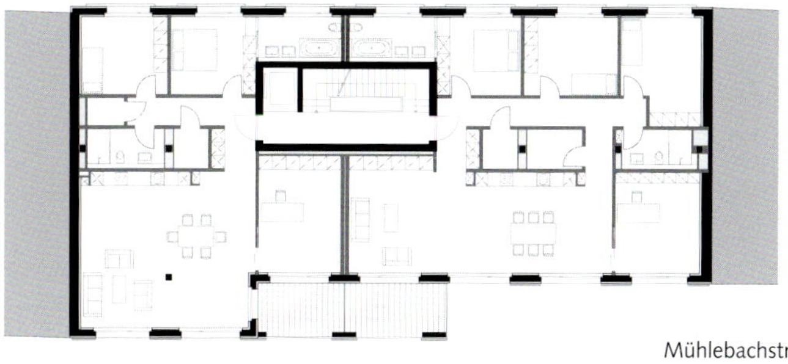

Mühlebachstrasse
5th floor plan,
scale 1:400

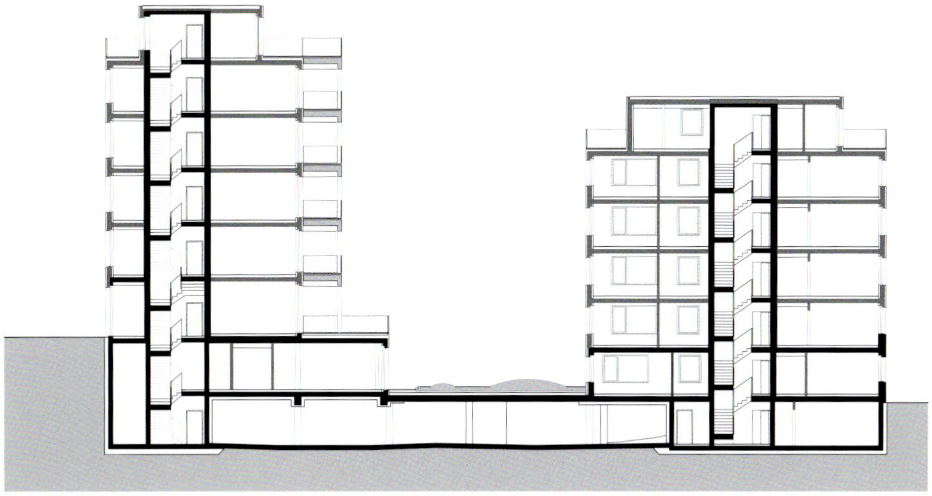

Section, scale 1:400

Kraftwerk B, Bennau (CH)

NEW MULTIFAMILY DWELLING

Project information

Architects:	grab architekten ag, Altendorf (CH)
Consultants / Energy concept:	Mechanical design: Amena, Winterthur (CH); Building services: Planforum, Winterthur (CH); Building physics: Intep, Zurich (CH)
Client:	Sanjo Immobilien, Altendorf (CH)
Completion:	2009
Standard:	MINERGIE-P-ECO
Living area:	1,380 m²
Final energy demand (heat and electricity)/m² living area:	41.67 kWh/m²a
Final energy production (renewable heat and electricity)/m² living area:	54.35 kWh/m²a

Balance scope per standard

- Heating
- Domestic hot water
- Cooling
- Auxiliary energy (pumps, ventilation)
- Lighting
- Equipment
- Electromobility

Kraftwerk B is an apartment building with seven dwelling units, situated close to Lake Zurich. It meets the strict rules of the Swiss MINERGIE-P-ECO standard. Energy consumption is reduced so far through passive measures and efficient technology that more than the total annual energy demand can be met by the active solar systems. Kraftwerk B generates almost 25% more energy on average than its residents consume in a year.

It was possible to achieve the MINERGIE-P-ECO standard by applying integrated planning from the beginning. The starting points were simple planning measures, such as creating a compact design and aligning the rooms towards the sun. The living spaces are located to the southwest while the ancillary spaces, such as the bathrooms and the stairway, are oriented to the northeast. Mainly natural building materials and easily recyclable elements were selected in order to meet the high standards for ecological quality.

For the outward appearance, special attention was paid to the integration of those building elements that generate solar gains. In addition, two 20,000 l tanks collect rainwater for irrigating the planted areas and flushing the toilets.

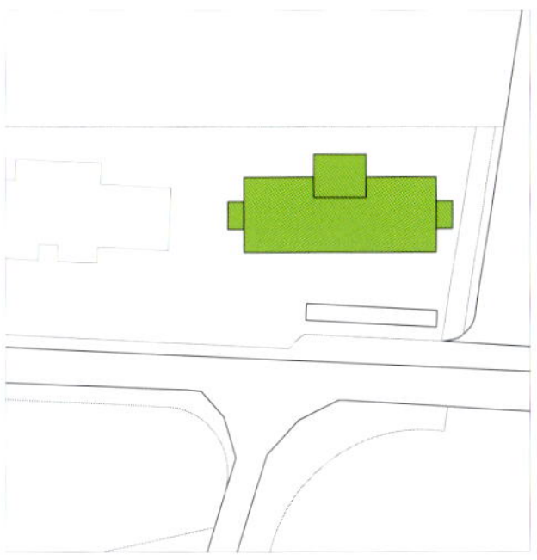

Site plan, scale 1:2,000

The space heating and domestic hot water are provided using different technologies (wood-burning stoves, heat pumps, waste-water heat recovery, and a solar thermal system). The surplus produced (about 30%) is used to supply the neighbouring buildings. The photovoltaic system, too, produces 30% more electricity than needed in a year for auxiliary electricity, household electricity and lighting. The surplus is fed into the public grid.

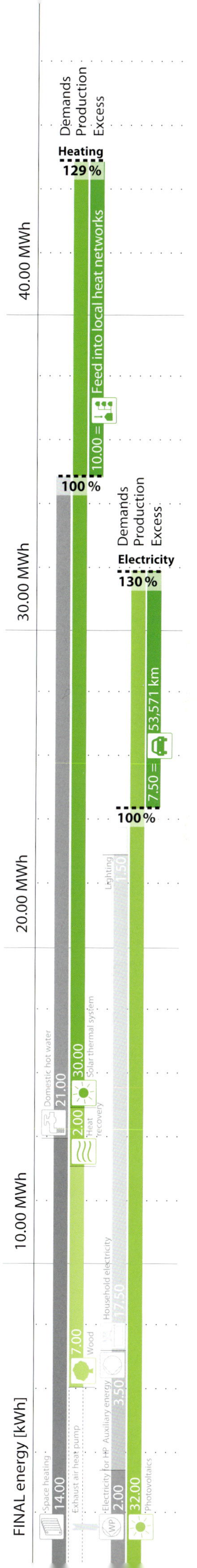

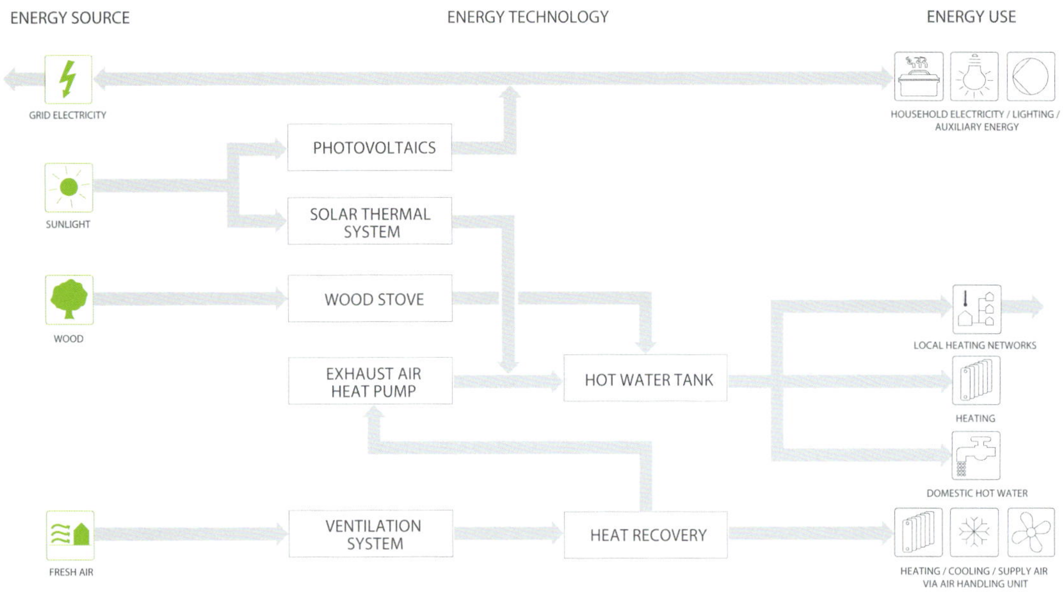

Heating

Kraftwerk B was constructed as a hybrid building. The combination of wood and concrete combines the advantages of both building materials. Prefabricated and highly insulated wood elements for the facade and roof are used as a curtain wall in front of the reinforced concrete structure. The concrete core of the building not only assumes a structural function but, together with the clay-plastered walls, also serves as thermal storage mass. It thus forms a heat and moisture buffer that regulates fluctuations in room temperature and indoor humidity.

Much emphasis was given to eliminating thermal bridges and creating an airtight construction. With a 43 cm thick layer of cellulose insulation, the facade achieves a U-value of 0.11 W/m²K. Large openings on the southwest facade collect solar gains while small windows on the northeast facade minimise heat losses. All the window are triple-glazed.

A total of 150 m² on the southwest facade consists of storey-high flat-plate collectors, which also provide weather protection and alternate with floor-to-ceiling windows. They produce heat for room heating and hot water. With the aid of a 24,000 l seasonal heat accumulator, the system provides 60 % of the heat needed over the course of a year. In summer, the system generates large surpluses that are passed on via a local heating pipeline to a neighbouring building.

Additional comfort is provided by compact wood-fired storage heating stoves that are equipped with a water-filled heat absorber to remove 50 % of the heat from the exhaust gases. They are used to supply the towel radiators in the bathrooms, heat domestic hot water, and feed a 3,000 l storage tank. In addition, heat is removed from the wastewater. The heat pump, wood stoves, and wastewater heat exchanger together generate 15,000 kWh per year, and the solar thermal system generates another 30,000 kWh of heat. That results in an annual surplus of 10,000 kWh of heat energy. The rooms are heated with underfloor heating. If the heat supply fails, an air-water heat pump acts as backup.

Cooling

Exterior louvres shade all the windows and thus prevent overheating in summer. There are no technical measures provided for cooling. The available structural thermal storage mass is sufficient to create a comfortable indoor climate.

Electricity

On the southwest side of the building, 260 m² of photovoltaic panels have been integrated into the main roof and the pavilion roof. They also form the water-draining layer. The installed system covers all the building- and household-related electrical demand and generates 32,000 kWh of electricity per year, which is 7,500 kWh more than required. The surplus is fed into the public grid. All household appliances meet efficiency class A+ or A++ and therefore use very little electricity. In addition, the washers and dryers are equipped with a heat recovery system. The dishwashers and washing machines are connected to the hot water system so the hot water they use is provided efficiently from the solar collectors instead from the power grid.

Ventilation

Controlled room ventilation with heat recovery reduces the heating demand. The central ventilation equipment is located in the basement, and fresh air is preheated via an earth tube heat exchanger. A counterflow heat exchanger minimises heat losses through the exhaust air. The windows are made as swinging casements, so they have no tilt function and therefore only allow intense air exchange. That significantly reduces the ventilation heat losses without limiting the ability for window ventilation.

Lighting

The upper apartments receive daylight from large windows at the gable ends and from the stairwell, which is fully glazed in this area.

A project's success always also depends on the user behaviour of its residents. For this reason, the tenants of Kraftwerk B can always check and manage their energy consumption on individual displays. A bonus-malus system strengthens the tenants' energy awareness through positive incentive and spurs them to alter their behaviour.

The individual components are so well coordinated with one another that more than just the Swiss MINERGIE-P-ECO standard was achieved. In addition to meeting the Energy-Plus standard, a high quality of living was established and, through sophisticated integration of photovoltaics and solar thermal technology, an attractive architectural appearance was created.

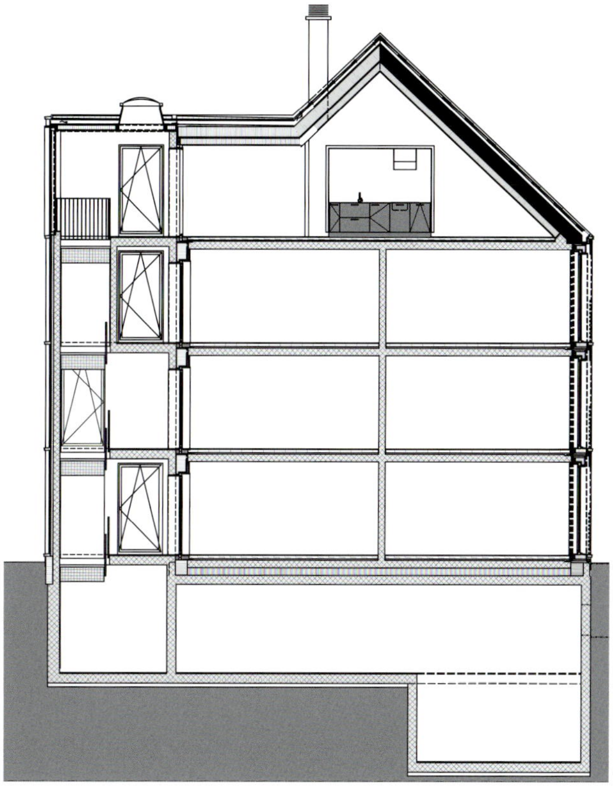

Section, scale 1:200

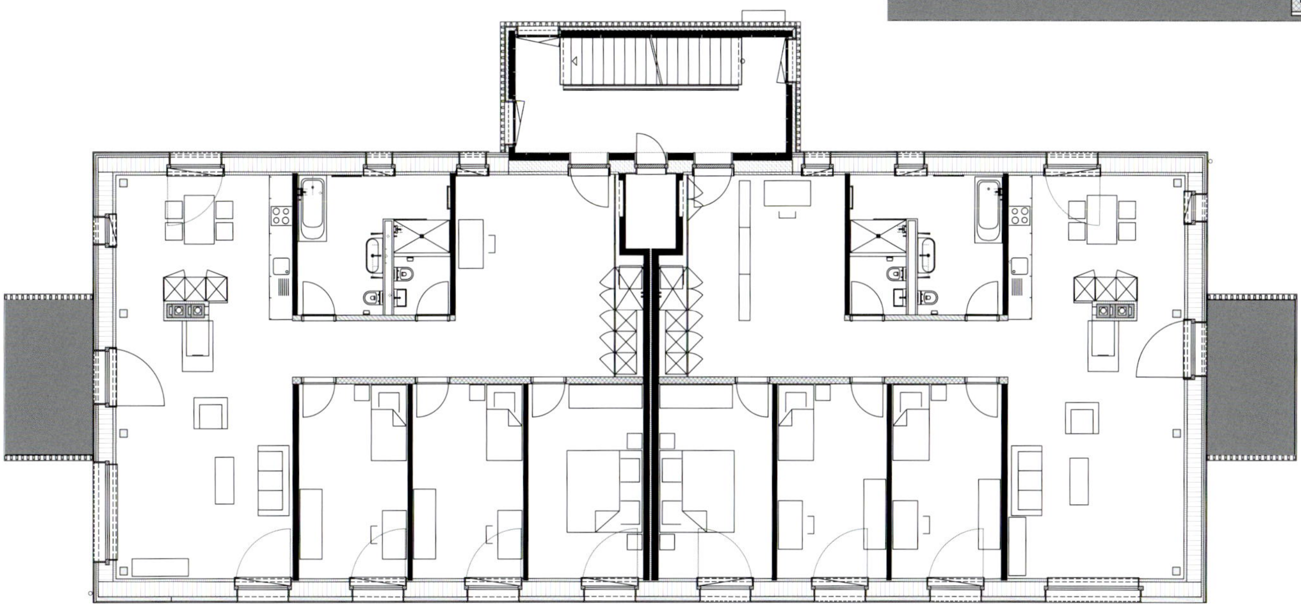

Upper floor plan, scale 1:200

Multifamily dwelling, Dübendorf (CH)

NEW MULTIFAMILY DWELLING

Project information

Architects:	kämpfen für architektur ag, Zurich (CH)
Consultants/Energy concept:	Neaf Energietechnik, Zurich (CH)
Client:	Private owner
Completion:	2008
Standard:	Zero-energy, MINERGIE-P-ECO
Living area:	727 m²
Final energy demand (heat and electricity)/m² living area:	53.96 kWh/m²a
Final energy production (renewable heat and electricity)/m² living area:	39.46 kWh/m²a

Balance scope per standard

- Heating
- Domestic hot water
- Cooling
- Auxiliary energy (pumps, ventilation)
- Lighting
- Equipment
- Electromobility

This apartment building in Dübendorf has six flats. In 2008 it was one of the first buildings in Canton Zurich to be certified according to the Swiss standard MINERGIE-P-ECO. To achieve this standard, not only did the energy demand need to be reduced to a minimum, but the material also had to be selected from an ecological viewpoint and particular attention had to be paid to creating a good indoor climate and the conditions for healthy living.

The building opens towards the south with large window surfaces and balconies in order to take advantage of passive solar gains. The ancillary spaces are oriented to the north and towards the street, where the facade appears more closed.

Site plan, scale 1:2,000

The energy needed for heating and domestic hot water is supplied by the solar thermal system and a heat pump.

Half of the electricity required for auxiliary energy (including energy for the heat pump), lighting, and household use is covered by the photovoltaic system.

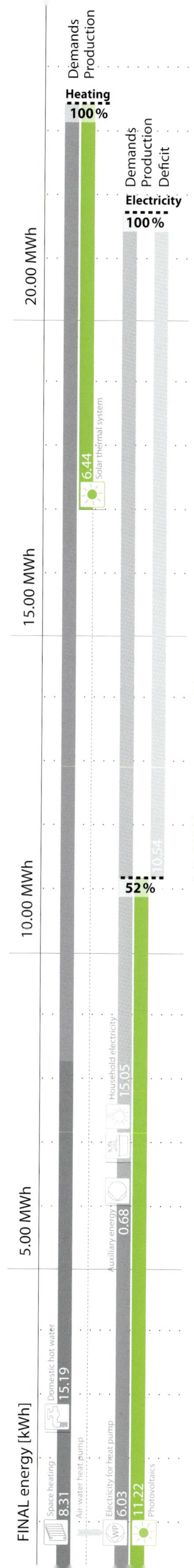

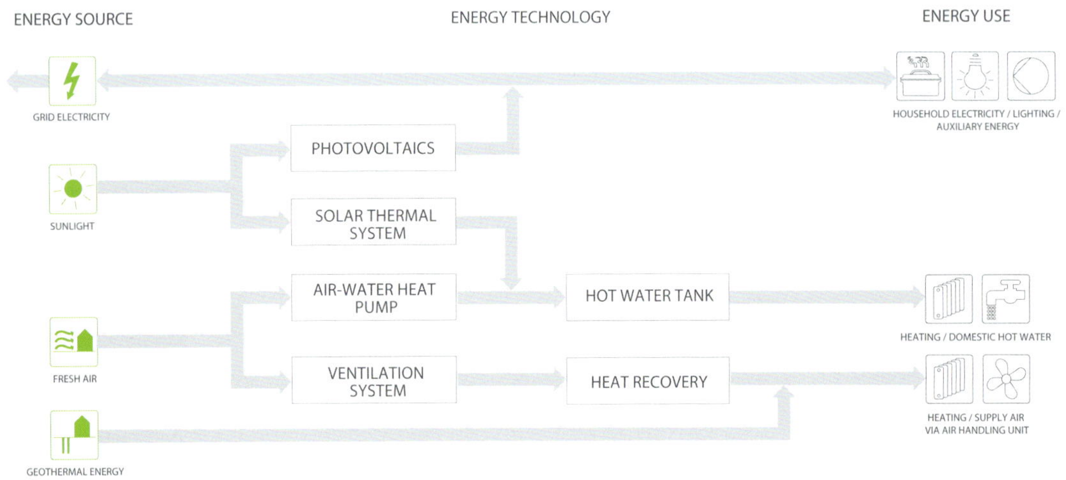

Heating

Heat is generated using an air-water heat pump and vacuum tube collectors, which are located above the glass roof of the staircase. Over an area of 14 m², the solar thermal system produces more than 6,000 kWh per year. In relation to the drive current, the heat pump produces nearly four times as much heat. For an electrical power consumption of 3.3 kW, it produces a thermal output of 11.7 kW. The vacuum tube collectors and heat pump feed a hot water tank with a capacity of about 1,800 l. The rooms are heated by underfloor heating, and also by small radiators in the bathrooms.

The fully glazed stairwell divides the building like a joint between the rented units and the private part belonging to the building owners. It functions like a winter garden. The solar thermal vacuum tube collectors above it heat the domestic hot water and act as shading elements to prevent the stairwell from overheating in the summer. They also create an alluring play of light and shadow. The timber and concrete composite floors together with the cement screed make up the building's thermal storage mass. In addition, the black natural slate tiles afford particularly good heat retention. In order to obtain a well insulated enclosure that is as airtight as possible, wood windows with triple glazing and a U-value of 0.7 W/m²K are used. The frames are additionally covered with insulation on the outside. Care was taken to eliminate thermal bridges and the unheated basement level was thermally decoupled from the areas above.

Electricity

The gabled roof has a pitch of 45 degrees and its southern-facing side has an integrated photovoltaic system composed of monocrystalline solar cells. The 94.5 m² photovoltaic surface has an output of 14 kWp and produces 11,220 kWh per year. This meets the electricity demand for the heat pump used to generate heating and hot water as well as the electricity demand for the ventilation system. The remaining electricity demand is only partially met by renewable sources. In addition to providing energy, the photovoltaic surface also assumes the functions of the building envelope.

Ventilation

The ventilation system is equipped with exhaust air heat recovery and supply air is preheated through an earth tube heat exchanger. Fresh supply air is introduced into the living rooms and bedrooms and spent air is extracted from the kitchens and bathrooms. The two parts of the building each have a separate ventilation unit with a heat recovery rate of 90%. The mechanical ventilation averts heat losses and increases air quality.

Lighting

In order to keep the electricity demand for lighting to a minimum, the stairwell was equipped with motion detectors, daylight control, and LED lighting. The large window openings to the south and the corner windows bring much light inside the building.

To shorten construction time, the recycled concrete elements used for the basement as well as the wood structure were prefabricated. In contrast to comparable solid construction, these materials contain very little embodied energy.

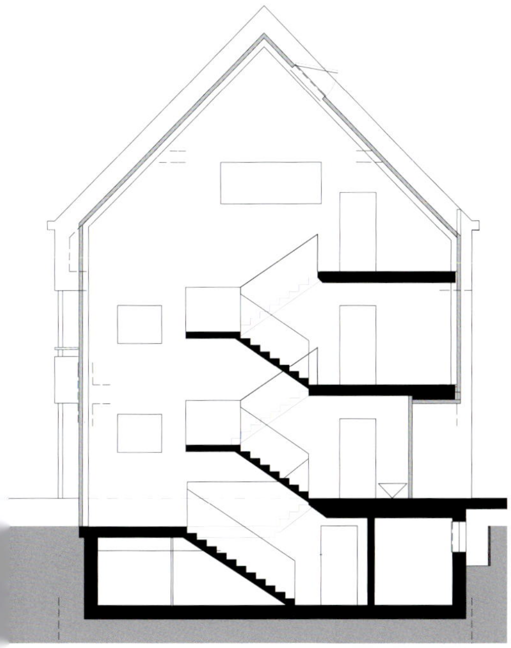

Section, scale 1:200

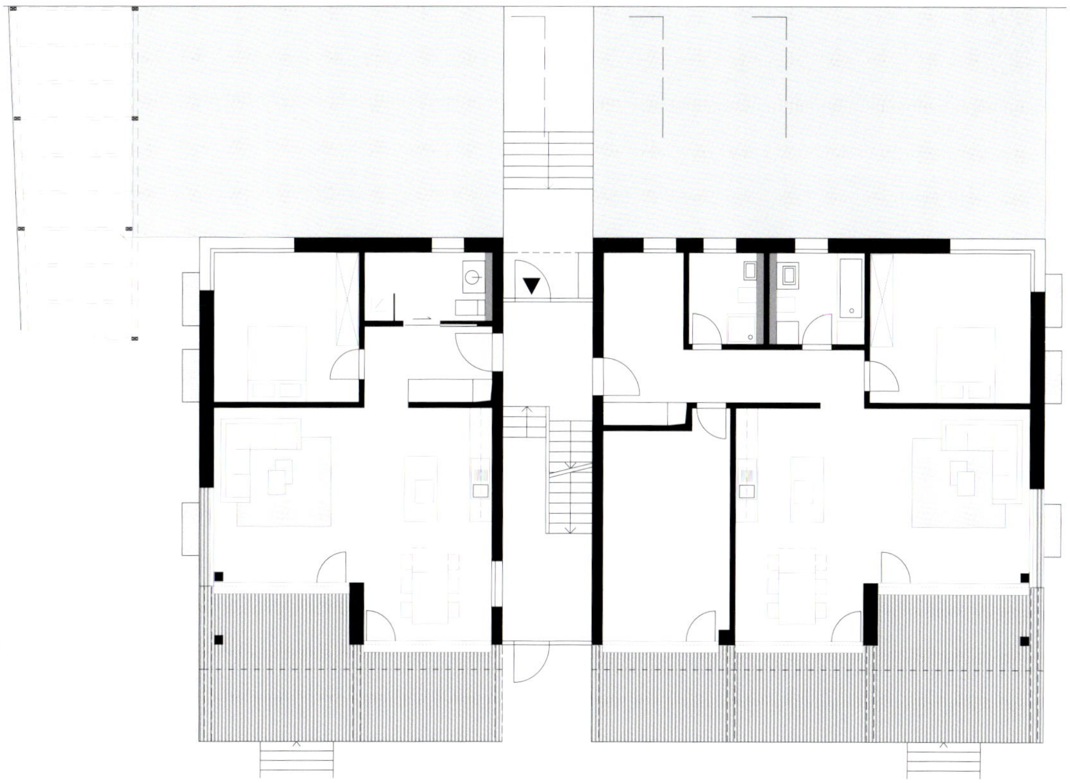

Entry level plan, scale 1:200

Solar-Werk 01, Kassel (DE)

NEW PRODUCTION HALL

Project information

Architects:	HHS Planer + Architekten, Kassel (DE)
Consultants / Energy concept:	IB Hausladen, deNET e.V., Kassel (DE) EGS-Plan, Stuttgart (DE)
Client:	SMA Solar Technology, Kassel (DE)
Completion:	2009
Standard:	CO_2-neutral in operation
Usable area:	25,700 m² GFA
Final energy demand (heat and electricity)/m² living area:	407.04 kWh/m²a
Final energy production (renewable heat and electricity)/m² living area:	229.07 kWh/m²a

Balance scope per standard

- Heating
- Domestic hot water
- Cooling
- Auxiliary energy (pumps, ventilation)
- Lighting
- Equipment
- Electromobility

With the Solar-Werk 01 production hall, SMA Solar Technology has created 450 jobs and the world's largest inverter factory – with the goal of achieving CO_2-neutral operation. Unlike with residential buildings, in industrial buildings the energy demand for cooling and ventilation is much greater than for heating. At the same time, production facilities have the highest energy consumption of electric power. Thus it is a great challenge not only to provide the energy needed for operating the building but also to make it available in a CO_2-neutral way.

To this end, the energy demand for building operations as well as for production first needed to be minimised. Then the residual energy demand needed to be met by renewable sources.

In practice, this concept meant the building envelope of the production hall was to be constructed in accordance with the low-energy standard. The architectural form helps minimise energy losses and optimise energy gains.

The entire demand for space heating and cooling is met by heat from the factory's own biogas-powered cogeneration (CHP) unit, and the residual demand is covered by district heating. The photovoltaic system and the cogeneration unit produce 36% of all the electricity. Green electricity covers the remaining electricity demand.

Site plan, scale 1:5,000

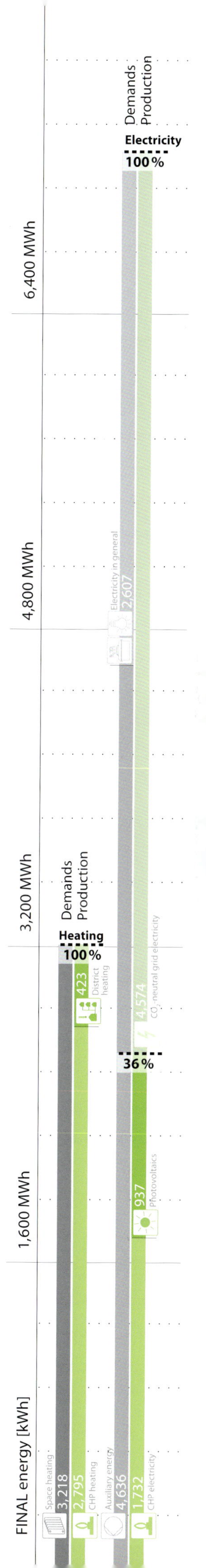

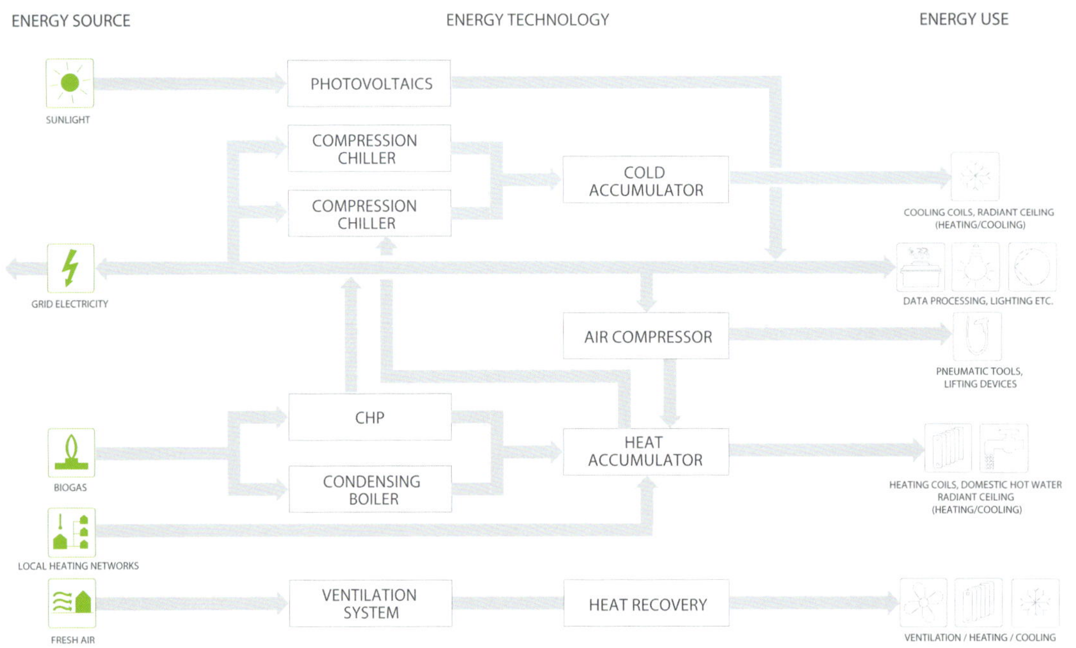

Heating

The building envelope reaches a U-value of 0.42 W/m²K on average. The final energy demand of the office and production building is around 410 kWh/m²a. Compared to a comparable conventional factory building, consumption could be reduced to 2,300,000 kWh/a through committed use of the waste heat from compressed air, exhaust air and test chambers.

A cogeneration unit run with biogas and dedicated specially to the production hall produces 2,795 MWh, which is most of the heat required. When the power plant has to be serviced, a condensing boiler that also runs on biogas takes over. The remaining heat demand of 423 MWh is met by district heating. To have thermal energy constantly available, all heat sources are connected to a water storage tank. This way, shortages can be overcome, generation production peaks can be used later, and the lifespan of the CHP unit can be increased.

Cooling

An absorption chiller for building air conditioning feeds a buffer storage tank. The chiller is powered by heat from the cogeneration unit. When there is additional demand, assistance is provided by an electric compression chiller that is also connected to the storage tank. Heating/cooling sails transfer the generated energy to the offices by means of radiant heating or cooling.

Electricity

The total amount of electricity consumed by production and by operation of the building is 7,243 MWh per year, and 1,133 MWh/a was saved in advance through the use of innovative technologies. To meet the electricity demand, the building was fitted with multiple photovoltaic units composed of polycrystalline cells. Some of these are integrated into the roof and some are installed as roof-mounted units. In total, the PV units produce 937 MWh per year.

The glass-glass photovoltaic modules integrated into the skylights on the factory building not only generate electricity, but also provide both natural lighting and shading for the hall. The roof over the logistics yard and the canopies are also made of glass-glass modules. In addition to providing protection from the weather, they also produce electricity.

The electricity generated from renewable sources is fed directly into the public grid and thus has a positive effect on the factory's CO_2 balance. The biogas-powered cogeneration unit supplies heat as well as 24% of the required electricity. For the residual demand of approximately 4,500 kWh, green electricity originating from hydropower is obtained from the local public utility.

Ventilation

The factory hall is divided into several zones and is equipped with air inlet and outlet devices. They condition the outside air and serve to heat, cool and mix air currents. The airflow can be regulated as required. In addition to the mechanical ventilation, the offices can be naturally ventilated via tiltable windows.

A good interplay of all the components is achieved by networking different renewable energy sources, thus meeting all the industrial production demands for heating, cooling, compressed air, and electricity. They various sources can be regulated according to demand and thus synergies can be exploited. As a result, CO_2-neutral operation is possible and even below the standard set by EnEV 2007 for new buildings by 36 %.

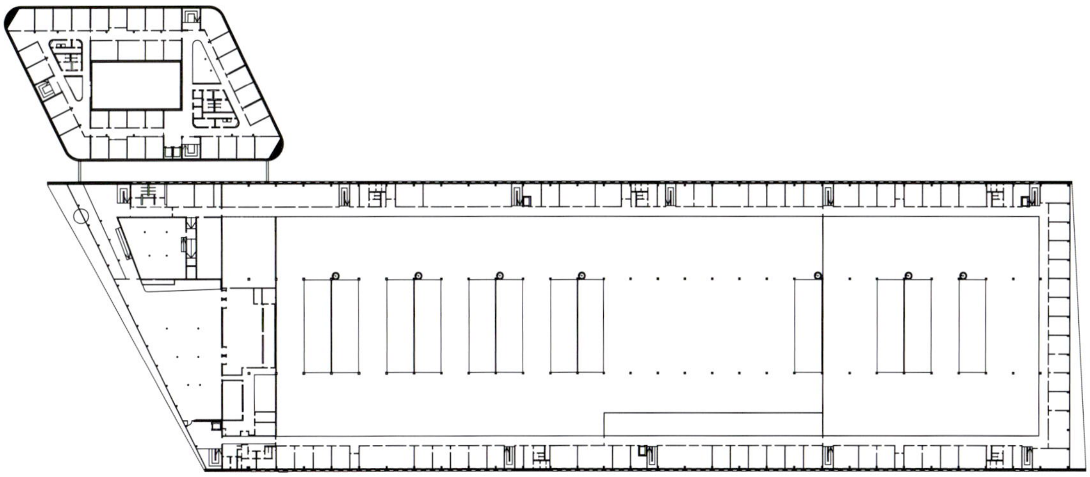

Plan, scale 1:2,000

Lighting

A high illuminance of 1,000 lux is required for the manufacture of electronic components. Luminaires with high efficiency and widespread use of daylighting minimises the electricity demand here, too. Simple measures such as skylights in the hall roof and floor-to-ceiling windows in the offices minimise the need for artificial lighting. The illumination concept for artificial lighting stipulates integration of the light fixtures into an EIB/KNX bus. This means they can be regulated with high efficiency using presence detectors.

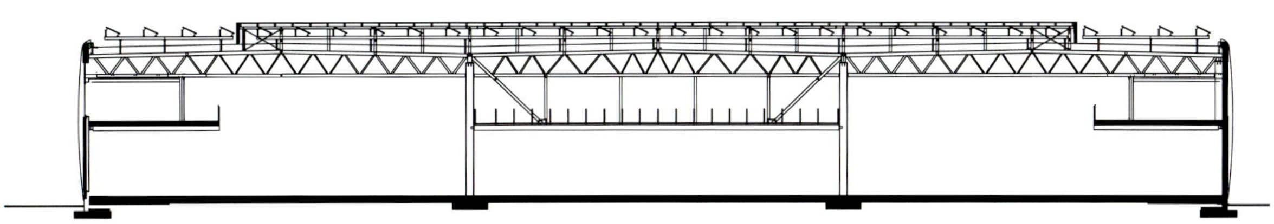

Cross section, scale 1:500

SINGLE-FAMILY HOUSES

	Münsingen, CH	Hamburg, DE	Driebergen, NL
AREA [m² living or usable area] (▪ represents 50 m²)	160	189	150
HEAT DEMAND (end energy) [kWh per m² and year] [1] (Domestic hot water / Space heating)	25.90	43.61	37.99
REGENERATIVE HEAT GENERATION (end energy) [Coverage ratio in % per m² and year]	100 %	100 %	100 %
UNDER CONSIDERATION			
ELECTRICITY DEMAND (end energy) [kWh per m² and year] [1]	37.40	36.84	35.27
REGENERATIVE ELECTRICITY GENERATION (end energy) [Coverage ratio in % per m² and year]	124 %	101 %	90 %
SURPLUS ELECTRICITY IN KM E-MOBILITY (represents 1,000 km)	10,114	700	0
ANNUAL PRIMARY ENERGY DEMAND as per guidelines [kWh per m² and year]	38.00	167.30	not specified

[1] m² living or usable area

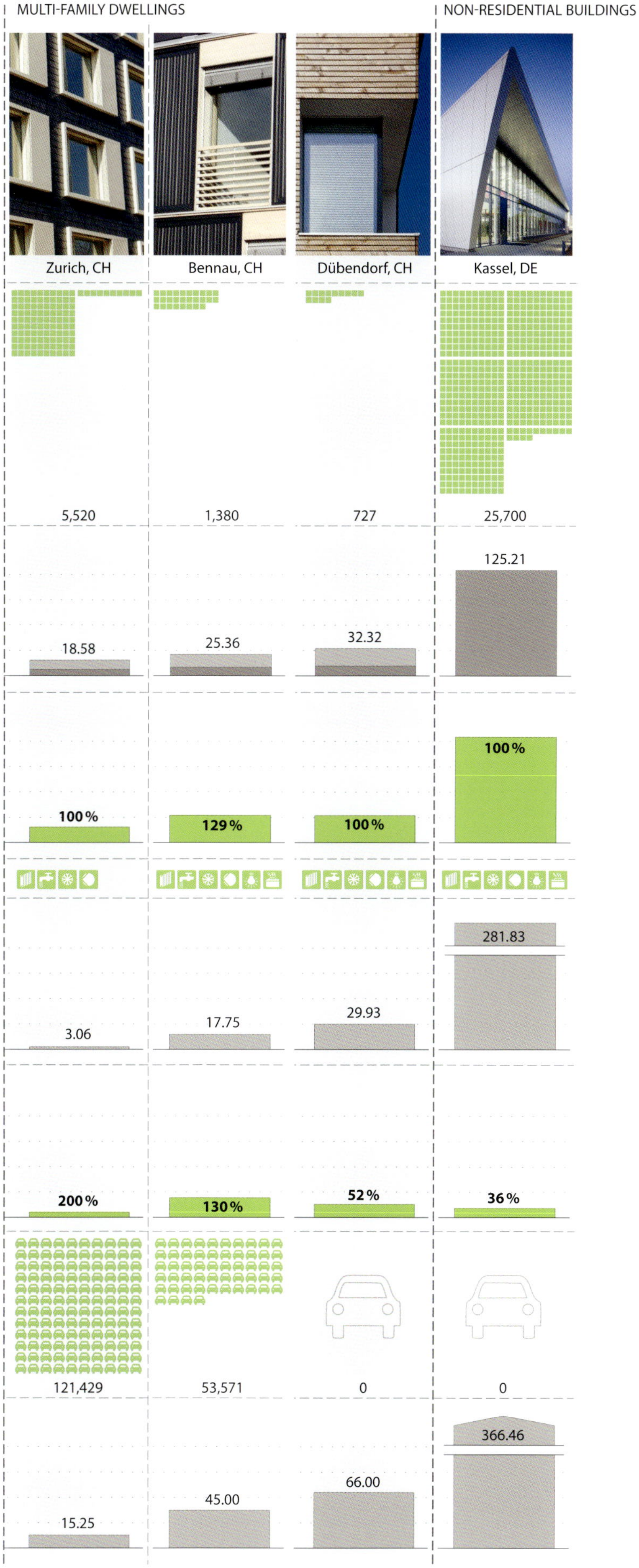
MULTI-FAMILY DWELLINGS
NON-RESIDENTIAL BUILDINGS
Zurich, CH
Bennau, CH
Dübendorf, CH
Kassel, DE
5,520
1,380
727
25,700
18.58
25.36
32.32
125.21
100 %
129 %
100 %
100 %
3.06
17.75
29.93
281.83
200 %
130 %
52 %
36 %
121,429
53,571
0
0
15.25
45.00
66.00
366.46

Positions and Perspectives

Positions

Which is the more sustainable solution: the passive or the active building concept?

INTERVIEW WITH DR. WINFRIED HEUSLER

One thing is certain: sustainable buildings are elementary for a healthy built environment. To answer the question of whether passive or active buildings are more appropriate, we have to consider the needs of people across the entire life cycle of the building. The first step forward is a holistic design approach that takes into account environmental, economical, social and cultural aspects. A really sustainable concept diminishes or even eliminates the unnecessary and useless consumption of energy, materials, time and money. Its acceptance is closely linked to the quality of comfort, as well as the wellbeing of people within the living and working environment. Thus, the importance of comfortable temperature, fresh air and daylight cannot be overestimated. In addition to practical aspects we have to consider the regional culture of building ("Baukultur"), with its specific formal and symbolic aspects. Formal aesthetic functions distinguish two antagonistic principles: order versus complexity and reduction of stimuli versus richness of stimuli in terms of shapes, colour, texture and material. The observer's imagination will associate symbolic functions with objects in the particular context. Accordingly, buildings and facades are bearers of meaning, as well as having practical functions. We are convinced that our world needs sustainable buildings. The question is: which is the more sustainable solution, the passive or the active building concept?

Do we need passive or dynamic facades?

The facade is mainly relevant for the external appearance of the building, its fitness for purpose and durability, and for protecting people and property. Thus, the optimisation of facades should not stop at reducing thermal losses. Transparent and translucent surfaces collect solar gains. In buildings with high internal loads and large glass areas, solar radiation causes overheating if no additional measures are taken. External shading systems reduce the solar gains greatly. Daylight systems, on the other hand, have the role of evenly distributing the entering daylight within a room and optimising the daylight quality.

Passive facades seal off the interior from external factors as far as possible. They are highly thermally insulated and airtight. This requires an overall optimisation of frame, glazing and non-transparent areas of the facade. Thermal bridging within the component and where individual components are joined is also significant. Conversely, dynamic facades respond specifically to changing external (weather) and internal conditions (e.g. presence of users). The focus is on natural ventilation as well as on the use of solar energy and daylight. A facade that is really optimised for energy and comfort should be able to react to the comfort needs of the users and the changing outside conditions.

If a facade is optimised with regard to its energy efficiency, it is recommendable to consider renewable energy sources to compensate for the remaining energy consumption. In the case of facades, there are mainly two active-solar energy technologies to be considered: electric and thermal. Their performance can be increased with the addition of storage systems.

Do we need passive or active building concepts?

In a passive building concept, passive facade components seal off the interior from external factors as far as possible. Contemporary mechanical systems ensure a comfortable interior environment. Conversely, in active building concepts, dynamic facade components respond specifically to changing internal and external conditions. Following the bioclimatic design strategy, the aim of this concept is to minimise the use of mechanical systems by means of natural ventilation as well as solar energy and daylight.

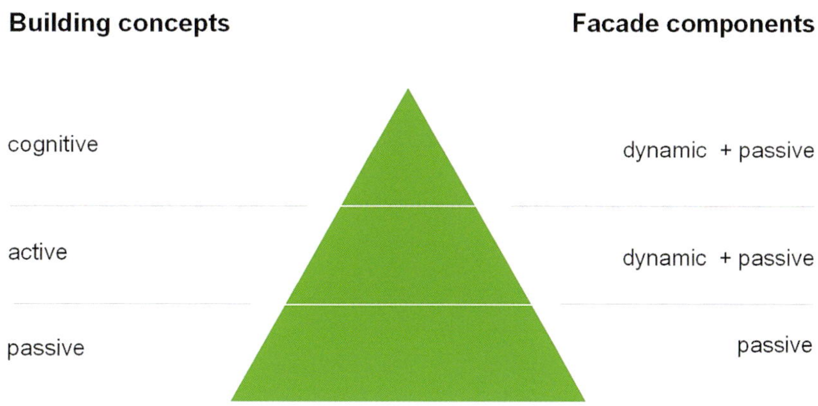

Passive, active and cognitive building concepts, a first step forward to an advanced building operating process

In our experience, purely passive building concepts are advantageous only if the location, the height or the use of the building excludes natural ventilation, solar energy and daylight for at least two thirds of the year.

In many moderate climatic zones, optimum energy efficiency is provided by cognitive building concepts. Their facade and mechanical-system components with dynamically adjustable functions are connected to each other through an intelligent building automation system. This enables the cognitive building to react to variable conditions that are in many instances predictable and can be calculated, such as annual or diurnal swings in meteorological conditions (i.e. solar altitude angle) or the times of a building's operation. However, unpredictable weather and operational aspects – such as variations in cloud cover, or the spontaneous presence of users – should be included by means of appropriate sensors or from internet weather forecasts. The expansion of the system boundaries by interconnecting the buildings of a city district offers additional potential.

In the end, we have to remember the fact that building techniques and methods should be tailored to the people for whom the building is intended rather than matching the tenants to the newest possible techniques and methods. Many megacities (with high growth rates) are in regions with tropical climates. One solution for that specific climatic zone can be the passive building concept. The alternative is to return to the traditional cooling method of natural ventilation and to create building zones with different levels of comfort in accordance

with the onion-skin principle. The core zone has to be sealed off from the surrounding buffer zones (airtight and well insulated). Courtyards, atria, loggias and sky gardens can be part of these concepts. The outer layer should not be glass but a rigid, partially transparent solar shading installation that allows air to permeate. Movable and in particular motor-driven components in such regions are only suited to buildings whose owners have a positive attitude to maintenance.

The appropriate tools and processes

In current architecture there is a desire for optimum comfort and maximum design freedom on the one hand, but a need for cost and resource efficiency on the other. To find a solution for this challenge we have to optimise the planning, construction and operation processes. First of all, we should follow a holistic design approach that is subject to a detailed, step-by-step analysis with respect to cost-efficiency (investment, operating and maintenance costs), design considerations (practical, formal and symbolic aspects), energy requirements (heating, cooling, ventilating, lighting etc.), and room comfort (thermal, visual, acoustic etc.). The analysis should consider the entire life cycle of a building, from design, planning and construction, through operation and usage, updating and upgrading, up to demolition with possible reuse or recycling of building components or materials.

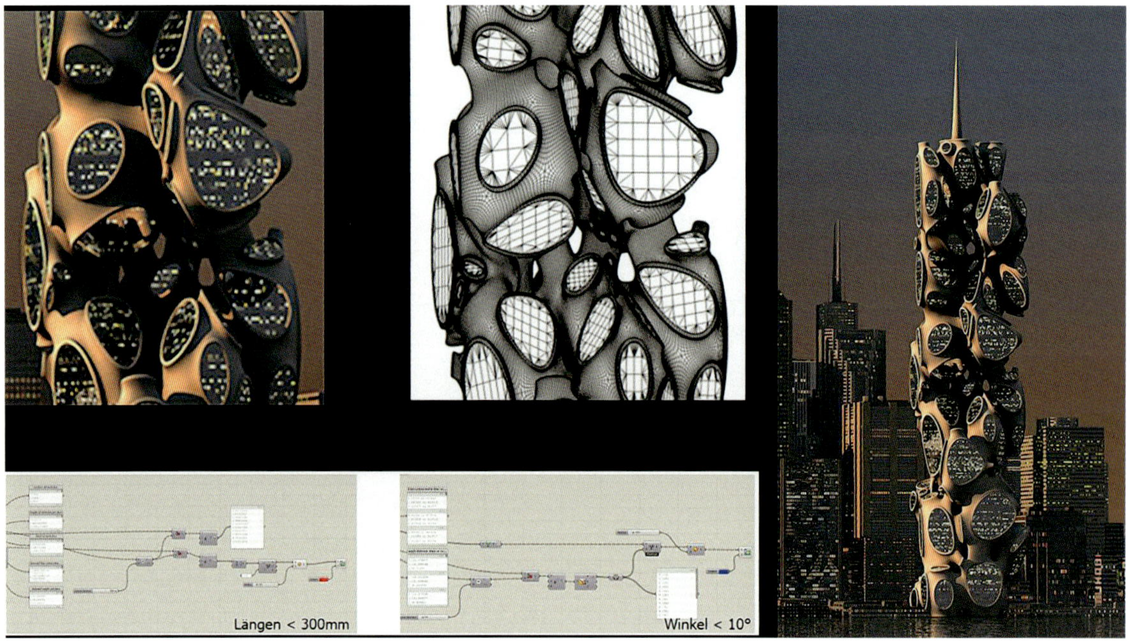

R & D concepts, the basis for long term research on advanced methods of facade design and construction

To approach a suitable solution for challenging tasks, the cooperation of individuals who share their experience offers the greatest potential. However, this requires a paradigm shift in the way people work together. In the end, a cross-disciplinary team with a spirit of openness will develop a fruitful discussion between architects and engineers. They complement each other in giving the answer to the question of whether a passive, active or cognitive concept is the most sustainable for a specific project.

Dr. Winfried Heusler,
Head of Corporate Building Excellence and the technical ambassador for Schüco worldwide

Aktivhaus Concept with a dynamic facade
incl. solar shading and photovoltaic lamellae
Head Office EDP, Porto (PT), (2011)
Design: APEL Architectura / Porto (PT)

Tomorrow's sustainable buildings, today: a holistic approach

INTERVIEW WITH MICHAEL K. RASMUSSEN

Where do you see the biggest challenges for building and living in the future?

The key challenge is climate renovation. Why? Because buildings account for 40 % of Europe's total energy consumption. And, in 2050, more than nine out of ten existing buildings in Europe will still be in use. So it is vital that any renovation proposals are reproducible across Europe.

Many existing buildings are already overdue for modernisation. Large energy savings can be achieved right now by applying current technologies, materials and know-how. The real challenge is to find sustainable solutions that combine high energy efficiency and a better indoor climate with daylight and fresh air for the occupants. When scaled up, sustainable living initiatives benefit society as a whole and ultimately contribute to better living conditions for millions of people in the growing urban centres around the world.

Spurred on by these facts, the VELUX Group initiated the Model Home 2020 programme in 2008. Their intention was to determine whether the predicted energy regulations for 2020 could already be met using solutions available today, based on masses of daylight and fresh air – without compromising a good indoor climate. Our ambition from the start was to do it solely with products, processes and technologies already available – and to raise the bar and set higher standards for liveability and comfort, health and wellbeing in future buildings in order to meet the 2020 regulations today. The result was six experimental buildings, constructed between 2009 and 2011 in five European countries. Each of them was built to CO_2-neutral and high-comfort standards and tested by real people. Their operation was closely monitored quantitatively by 18 universities and scientific institutes – and qualitatively by the people who lived in them – until 2014.

One of the six projects is the climate renovation of a typical "settler's house" in Germany from the 1950s. It is called LichtAktiv Haus (LAH) and proves positively that intelligent energy design can be successfully combined with the highest standard of liveability. Its daylight architecture not only gives the building plenty of light and fresh air, but also provides the evidence that the current building stock can also be made energy efficient.

The results are indisputable. We *can* sustainably renovate existing buildings using the materials and knowledge we have at our disposal today. Right here and now, with standard products, off the shelf. Encouraged by this, we embarked on more climate renovation projects throughout Europe to make old buildings more energy efficient and raise liveability standards for its residents. Among them are the renovation of the Poorterstraat district of Montfoort in the Netherlands and the forthcoming RenovActive project in Brussels – both climate renovations of old buildings based on Active House principles, with a focus on replicability and affordability.

The Model Home 2020 projects make a vital contribution to a more sustainable future, focusing on optimised comfort, energy efficiency and reduced environmental impact.

The VELUX Group has been developing concept houses throughout Europe for years. Why is a roof window manufacturer concerning itself with overall concepts for living in the future?

Society needs energy-efficient buildings that provide a healthy indoor climate for the people living, working or playing in them. But they must also have as little impact as possible on the environment. We need to make a transition. The VELUX Group is taking an active part in this transition by working closely with stakeholders in the building industry, conducting experiments and producing high-quality roof windows. VELUX roof windows allow people to live healthy and comfortable lives in buildings by providing daylight and fresh air, and yet maintaining a good energy balance.

We are committed to helping overcome the challenges of climate change, limited energy resources and the basic human need for health and wellbeing inside buildings. This is also one of the areas of focus of the

World map of Active Houses, built during the last 17 years. Active Houses with engagement by the VELUX Group

Active House Alliance, so the VELUX Group gives it our full support. Our goal is to contribute to the passing of European and national legislation that ensures maximum energy efficiency without compromising on liveability. Since 2005, we have been driving the agenda of sustainable living in buildings. We have done so using the knowledge and experience gained from our global network of customers, architects, engineers, contractors, researchers, building products manufacturers – and, of course, the Active House Alliance.

Healthy and climate-friendly buildings can go hand in hand. The more than 25 demonstration buildings and Active House projects that we have built together with these partners over the last decade have proved that. Each of them was designed to the basic principle of our founder Villum Kann Rasmussen – "one experiment is better than a thousand expert views". The experiments in sustainable construction, and the promotion of architecture that enhances our quality of indoor life, started in the late nineties with the climate renovation in Torzhkovskaya Street in St Petersburg, Russia.

We continued in 2005 with SOLTAG, a demonstration house designed by Rubow Architects and funded by the EU 6th Framework Programme.

The third project, Atika, was specifically designed for the Mediterranean climate and built in Bilbao in 2006. It focused on creating a healthy, low-energy indoor environment with good summer comfort and optimal year-round daylight conditions. Today, Atika has found its permanent location at Politecnico Milano, under the new name of VELUXlab.

The SOLTAG project showed particular promise and, together with Rubow Architects, we refined it further. It became the VELUX House pavilion, a CO_2-neutral modular housing unit that could be built at an attractive price. It functioned as the exit pavilion of the COP 15 Climate Conference in Copenhagen in 2009, and today serves as a VIP and visitors' pavilion in La Rochelle, France.

But brilliant and innovative design is one thing and performance another. These pavilion projects have shown that assessment studies are crucial to the understanding of any demonstration project. Monitoring and analysing the energy and indoor environment performance of these buildings are essential phases of the experiment. But so is reaching a true picture of how the living conditions are perceived by their occupants. With these pioneering

steps successfully taken, the VELUX Group launched the Model Home 2020 project.

To what extent did the chosen approach of energy-efficient and comfortable living spaces influence the planning process of the European Model Homes?

The focal point when planning a building should always be the user's wellbeing and ensuring a healthy indoor climate. We firmly believe that future-oriented buildings should meet both needs: being energy efficient and respectful of natural resources, while being persuasive comfortable, attractive living spaces with plenty of natural light and fresh air conducive to wellbeing.

For this reason, the VELUX Group is a major driver and key knowledge partner in the Active House Alliance, a network of universities, research institutions and researchers, and building contractors from all over the world. The Alliance's goal is to create healthier and more comfortable buildings that use renewable resources and have no negative impact on the climate or environment. The Alliance has set itself the target of drawing up specifications and guidelines for planning the next generation of sustainable Aktivhaus buildings. An Aktivhaus is one that integrates the three basic principles of comfort, energy and environment in the design and use of a building. It is evaluated on the interaction between those three parameters. The Alliance's ambition is for Aktivhaus specifications to become the benchmark for new buildings and renovation. This vision has already spurred various countries to establish their own Aktivhaus initiatives and associations.

The six project buildings built by the VELUX Group as part of the Model Home 2020 experiment were planned to the Aktivhaus principles outlined above. As a result, the planning process never got bogged down with issues of "either/or" but was always searching for integral solutions. This holistic approach to new buildings and climate renovation projects will provide maximum benefits to people and the environment, combining all requirements for comfort, energy and environment in an attractive package. But the individual concepts of each project building are also geared to specific national, or even regional, aspects. All the buildings were designed by local architects, engineers and universities, and individually tailored to the climatic, cultural and structural factors of the area in which they were to be built. So the traditions and lifestyle of each country are reflected in the basic architectural concept of the building.

The same also applies to a number of other projects implemented in recent years together with various partners, all to Aktivhaus principles. The projects include the Future Active House in Trondheim, Norway, the Russian Active House in Moscow, and the Solhuset kindergarten in Denmark. The Aktivhaus concept has also gained a foothold on the other side of the Atlantic, with the Smith Residence in St. Louis, Missouri, and the Great Gulf Active House in Toronto, Canada.

What do you believe are the main findings and experiences gained from the European Model Homes?

Thanks to our Model Home 2020 programme, we can conclude that nearly zero-energy buildings of the future, with ambitious requirements for indoor climate, can be built using the technology and products available today – without compromising energy savings. So we recommend that future legislation for such buildings should be based on Aktivhaus principles. And we hope that future revision of European directives on buildings and energy efficiency in buildings will incorporate what we have learned – especially with regard to natural ventilation and daylight.

The various projects of the Model Home 2020 experiment provide unprecedented insights into sustainable construction and climate renovation. In Germany,

"settler's houses" like the LichtAktiv Haus (LAH) were mainly built during the financial crisis of the 1930s, and after the Second World War up to the 1960s. There are still about 10 million of these homes – and the majority need modernising. If only one million of them were renovated into zero-energy buildings like the LAH, we could achieve annual energy savings of about 30 TWh. This corresponds roughly to the annual electricity production of three nuclear power plants. At the same time, we could cut carbon emissions by 12 million tonnes a year.

With these monitoring and evaluation results of Aktivhaus projects, it is possible to make a direct correlation between quantitative and qualitative parameters – between how successful the building is technically and how residents perceive its performance. This insight is of great importance to the construction sector, as well as to owners who are planning to modernise their house or flat. We hope that what we have learned can be integrated into European and national legislation for the renovation of our building stock. The associated social science research is particularly crucial for interdisciplinary research into energy-efficient living in buildings.

The findings of the living experiments with test families who moved into the Model Homes confirm the significance of combining both – living quality and resource conservation – as the concept for future living. Optimum indoor air quality and generous amounts of light in all rooms significantly increase living comfort and were perceived positively by all the test families. They also meet expectations in terms of sustainability: the test families demonstrated a greater awareness of the concept of energy saving and have modified their energy behaviour accordingly. These case studies allow us to identify the subjective indicators for sustainable living in energy-efficient buildings and communicate them back to architects and planners via the Active House Alliance.

Since 2005, we have been driving the agenda of sustainable living in buildings, co-creating ideas and finding solutions to problems. We can now demonstrate that the EU targets for Net Zero-Energy Buildings in 2020 can be achieved with solutions available today. All permanent projects – implemented since 2009 – have been built and evaluated to Aktivhaus principles. As stated above, that means the design process is never a matter of "either/or" but always a quest to find solutions in the design of new builds and renovation of existing buildings that maximise benefits for people and planet alike. Looking ahead, we see an imperative to scale up these well-tested solutions and to take up the challenge of the existing housing stock. 90% of the building stock that will exist in 2050 is already there, so the question arises as to how to renovate them to become healthy, sustainable and affordable.

Tomorrow's buildings are here today.

Michael K. Rasmussen,
Chief Marketing Officer, VELUX Group

Energy efficiency in the heating market

INTERVIEW WITH PROF. MARTIN VIESSMANN, VIESSMANN GROUP

Pellet boiler (photo left) for the need-based supply of heating for detached and semi-detached houses as well as business establishments.

Micro-CHP systems also make efficient heat and power cogeneration technology usable for the modernisation of small residential buildings.

Professor Viessmann, you have three times consecutively received the German Sustainability Award for your commitment to sustainability, and in an international context you have also received the renowned "Energy Globe World Award". How is this reflected in your business? How does your business show its commitment to sustainability?

In our company, sustainability is firmly rooted in the brand as well as in the organisation. It underpins the whole business. A good example of this is our strategic sustainability project. In this project, we have considerably increased efficiency of materials, work and energy at our company's headquarters in Allendorf (Eder)/Hesse. 4,200 of our 11,600 employees work here, which makes us by far the largest employer in the region. As a result, we have reduced the consumption of fossil energy by two thirds and CO_2 emissions by 80%. We have thus improved our competitiveness and made the site and its jobs more secure.

At the same time, high levels of energy efficiency, low emissions, straightforward integration of renewable fuels, and complete recyclability are important factors in the success of our products. We also take on social responsibility by sponsoring art, culture and science along with social institutions and projects. We have launched our own foundation for this purpose. With partners from business, politics and science, for the German-speaking countries in the first instance, we have created an information platform, our "Alliance for Sustainability". This platform promotes sustainability in relation to the domains of building, housing and modernisation to all interested parties and opinion leaders, while also serving as a source of inspiration for them.

What are the consequences of the European Union's energy and climate policy targets for the heating market?

The immediate consequence is the necessity to modernise as quickly as possible those existing buildings that are completely outdated with regard to energy. The EU's energy and climate policy targets can only be achieved when this modernisation backlog is cleared.

Ninety percent of approx. 120 million heating systems in the EU are outdated and urgently need to be renewed. Only a little more than 10% use efficient technologies such as condensing technology, combined heat and power generation, or renewable energies.

It has been calculated that the potential for saving energy in the European heating market would be sufficient to reduce the energy demand of the EU states by up to 11%. This means that with modernisation of the current, outdated systems, more than two thirds of the political objective of an annual 1.5% primary energy saving by 2020 could be achieved. The necessary technology is available. What we need for implementation is an appropriate political framework.

The energy efficiency of buildings can be increased by both technical and structural measures. What is the value of modernising heating systems in this context?

The European Union's timetable envisages a reduction in CO_2 emissions of 80 to 85% by 2050. A precondition for this is that buildings within the Member States should be – for the most part – climate-neutral. Because the roof or facade of a house is usually renewed only once within this timeframe, it is important that these opportunities are used to carry out insulation work at the same time.

Heating systems have shorter modernisation cycles. They are likely to be renewed twice in the period up to 2050. Therefore the first step should always be to exchange a heating system that is already outdated for one that uses high-efficiency technology. Thus, with a relatively small investment, an immediate increase in efficiency can be achieved, and the possibility of using renewable energy can be introduced. When the building's heat demand is gradually reduced by subsequent insulation measures, the rate of utilisation of modern heat generators actually increases.

Which products do you see the biggest opportunities for on the European markets? Are technologies for using fossil fuels still worthwhile?

There is a good outlook for all products that help to increase energy efficiency on the heating market and to extend the reach of renewable energies. The question of which technical solution is most suitable depends on the framework in each particular country, that is, on the available infrastructure, the availability and price structure of fuel sources, regulatory requirements and, not least, the climate. Because currently, production costs for renewable energy are still higher than those for fossil fuels, their use is mainly dependent on the subsidy programmes that are available in the individual countries.

But around 90% of all heating systems in Europe are still run on fossil gas or oil today. This state of affairs will change only very slowly, at least as far as existing buildings are concerned. Condensing technology, increasingly combined with solar heating, will therefore continue to have the greatest impact on market share over the next few decades. It is the most economic option because it is highly efficient and the relatively small investment in a condensing technology system is usually paid back after a few years. Condensing technology also has a green future because of biogenic components in oil and gas, because of concepts such as power-to-gas, or because of a combination with highly efficient micro-CHP systems, for example fuel cell heating devices.

North America, unlike Germany or the European Union, has no mandatory energy and climate policy targets. The development of unconventional energy resources through fracking and the temporarily low oil price count as an additional factor. This has led to a weakening of emerging trends towards efficiency in the USA. What potential do you see in view of this background for reducing fossil energy use and expanding renewable energies?

In North America, fossil fuel sources contribute more than 90% to heat generation. Wood, and in the USA hydropower, are the only renewable energy sources to make an appreciable contribution to the energy supply of the 350 million people in the USA and Canada. There, current systems are problematic from two points of view: the residential sector mostly uses electrically driven hot air heating, which does not even come close to the efficiency levels of hydraulic systems that use condensing technology for heat generation.

In the north of the USA and Canada, there are also water-driven systems, but their technology is usually completely outdated, too. This creates a huge modernisation backlog that the federal states are addressing only very cautiously, through individual efficiency standards for modernisation projects. There is a lack of incentives for using modern, efficient heating systems and for expanding renewable energies.

Thermal solar collectors can be employed not only for making hot water but also for supplementary heating. When combined with a highly efficient condensing boiler, energy consumption can be reduced by up to 40 percent in comparison to outdated heating technology.

Prof. Martin Viessmann,
CEO and president of the supervisory board of the Viessmann Group

Perspectives

The Aktivhaus paves the way toward climate-neutral buildings and sustainable cities. It utilises the passive properties of buildings for saving energy together with the specific characteristics that predestine every building for energy production. An Aktivhaus saves and wins: energy losses are offset by the gains it generates from renewable sources.
What additional development opportunities are linked with this strategy? This final chapter seeks to provide information on necessary and useful developments that move beyond the Aktivhaus, namely energy-generating individual buildings.

Typical energy demands
for selected types of use

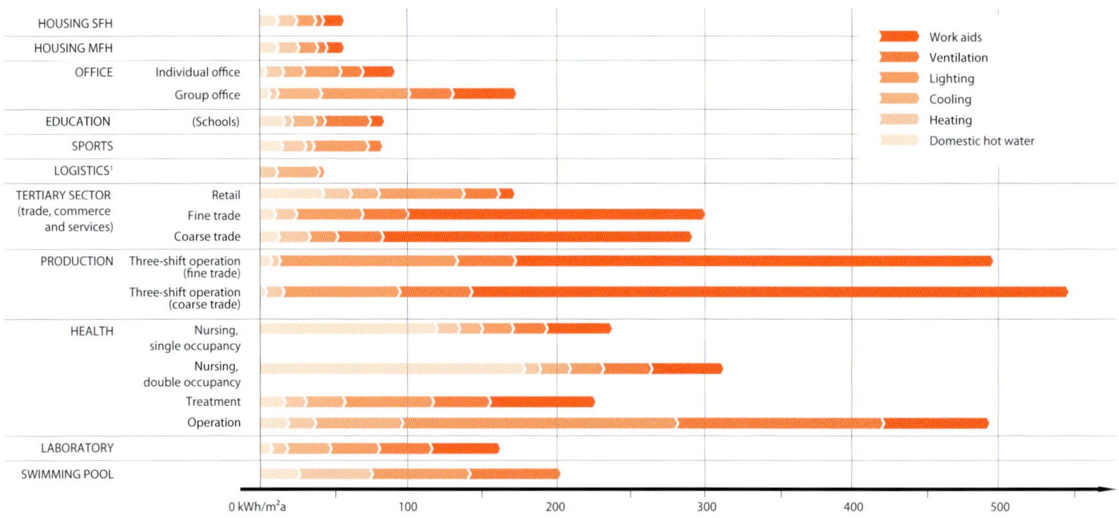

[1] without cold storage

Specific net energy demand
per year (kWh/m²a) – Passivhaus standard

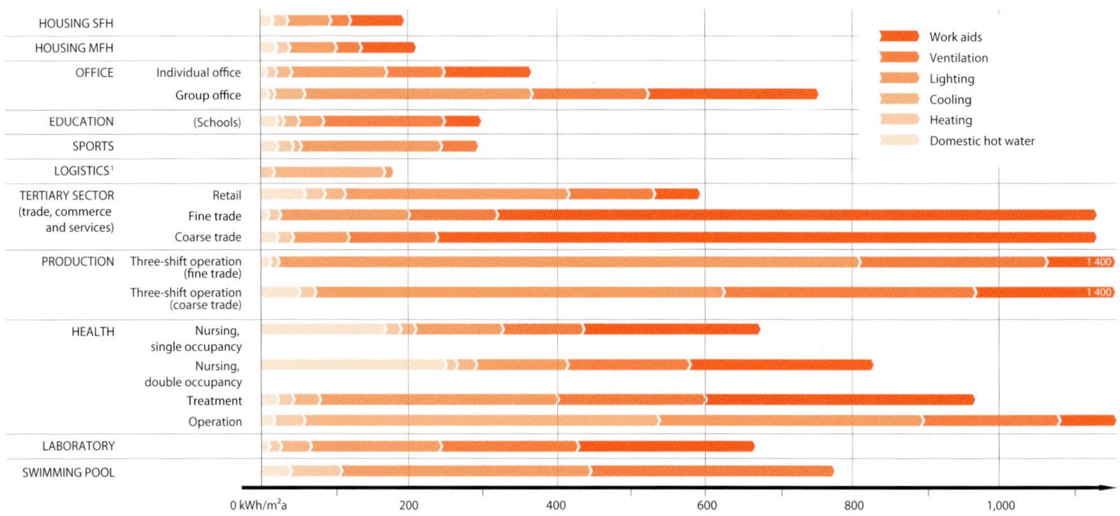

[1] without cold storage

Specific primary energy demand
per year (kWh/m²a) – Passivhaus standard

Performance

Unlike with the Passivhaus standard, for an Aktivhaus there are no stipulations pertaining to required qualities for the building envelope or maximum energy demand for the building, as already described. And in contrast to the Energy-Plus standard, there is no binding stipulation that the building must necessarily produce more energy than it consumes on balance over the year.

There are several reasons for eschewing standard parameters. Each building is unique. This concerns its geographic location, the climatic and the geological conditions, its location within the urban space, and many other criteria that make it more or less suitable for using environmental energy. Even more important for its suitability as an energy collector and for possibly producing an energy surplus, however, are two other building characteristics: the use and the building height, or rather, the number of storeys.

The use determines the energy demand for the building, whether for heating, cooling and ventilation, for auxiliary energy, lighting, or all the powered equipment in the building. Likewise, the use determines the internal loads from people and equipment. Thus, depending on the type of use, the specific energy demand per unit area is different: very low, for instance, for a logistics facility or a gymnasium; in the middle range for a school or an apartment building, and high, say, for a laboratory building.

A glance at these few examples shows that, for logical reasons, an energy standard should be defined independently of use. This is also illustrated by the table of typical energy demands for selected uses (see p. 218).

The number of storeys and thus the building height determine how much surface area a building has to collect energy from its surroundings for its use. Due to its large enclosing surface, a single-storey building is thus particularly well suited to collecting energy: through the surface in contact with the ground, its roof area, and – to a limited degree – its wall surfaces. Thus a detached, one- or two-storey single-family house or a logistics facility can generate many times its energy needs from renewable sources.

The energy performance of an Aktivhaus is contingent on its use and the number of storeys; it will usually produce more energy than it consumes.

A multistorey building has less enclosing surface per square metre of usable floor area. It has proportionately less roof surface and more wall surface. However, wall surfaces contribute less to energy collection from the environment than roof areas do. Because buildings with many storeys are also generally located in denser urban spaces, shading of the roof and wall surfaces can further reduce the yields.

The performance of an Aktivhaus will always be dependent not only on geographic criteria but also on its use and the number of storeys / building height. Many use types of use – such as schools, single-family homes, or typical-height multifamily housing – are well suited to achieving the Energy-Plus standard. In contrast, with facilities such as laboratory buildings or high-rise office buildings, it is difficult or even impossible to generate more renewable energy than they consume. The diagram below shows, for various types of use and different

The Energy-Plus standard for the Aktivhaus as a function of type of use and number of storeys

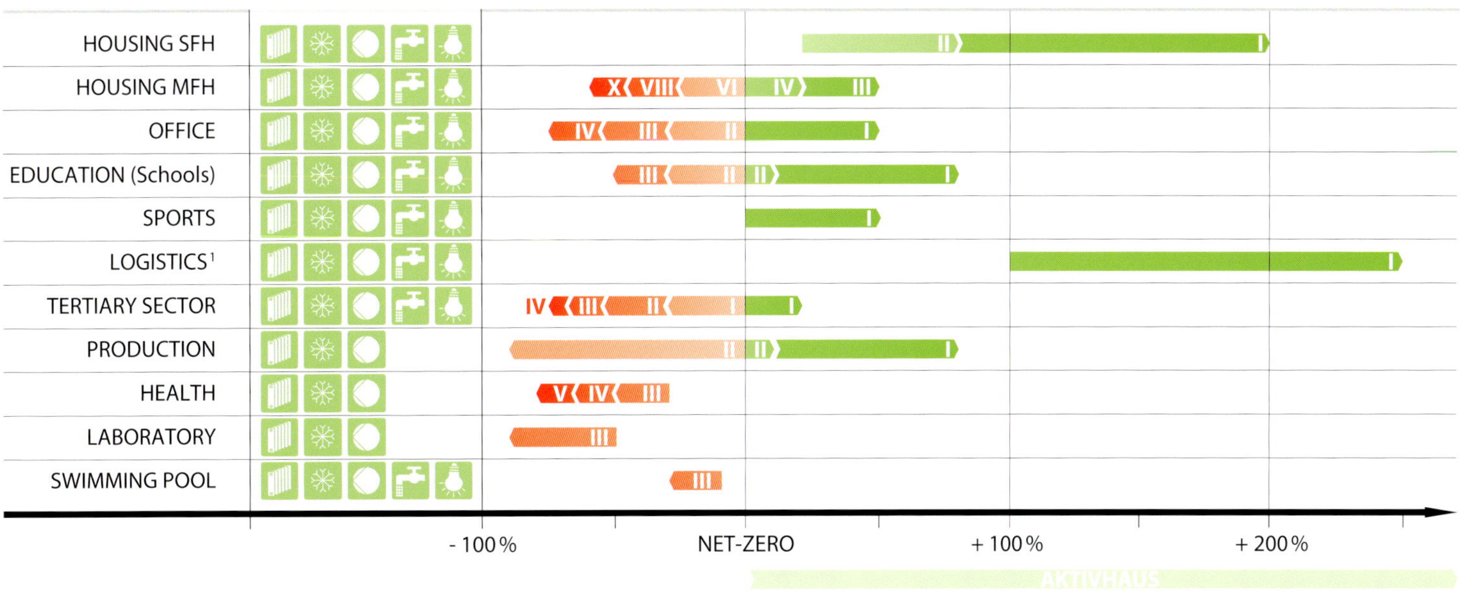

[1] without cold storage

building heights, the performance likely to be attained by an Aktivhaus under today's economic and technological conditions. Performance requirements would need to be updated along with evolving changes to the qualities of building materials and technologies for renewable energy production.

Today's state of progress in the development of opportunities to create an Aktivhaus, even under difficult conditions – that is to say, in a dense urban context, on an extremely confined site, and with eight storeys – is depicted by the study for an apartment building in Frankfurt city centre as an example.

It clearly shows that even today, many new buildings can generate a significant share of their energy needs from renewable sources, or even produce a surplus of energy. If we take the policy objectives of the EU, the US government, and many other nations seriously, Energy-Plus buildings will be the norm by about 2020 onward, although this standard would need to be differentiated according to the specified building characteristics. Clients are, in any case, well advised to follow these standards even today. The long lifespan of buildings would otherwise mean that the investments they represent will already be out of date in less than ten years and that their values will need correction.

Aktiv-Stadthaus, Frankfurt (DE)

NEW MULTIFAMILY APARTMENT BUILDING

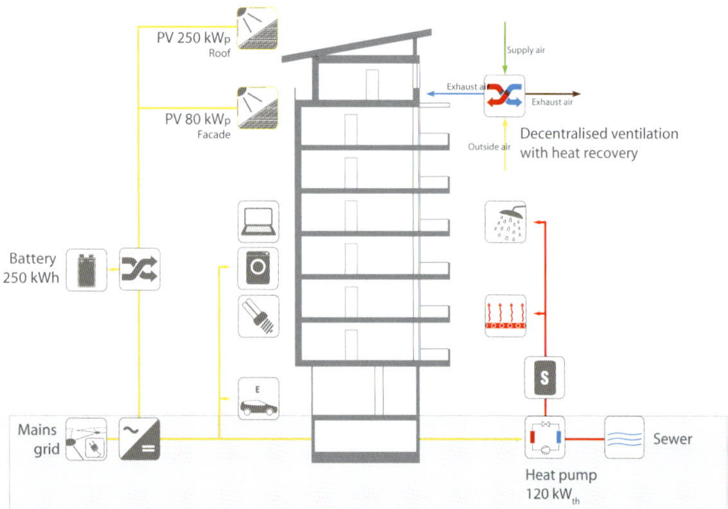

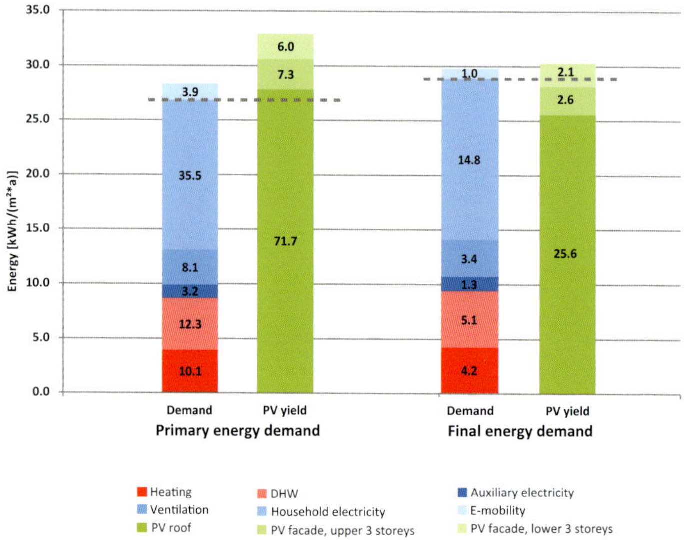

Users and operation

Renewable energies are not continuously available. As a result, production and consumption are not congruent. A better balance can be achieved if energy is mainly used when it is available from the environment. So-called user interfaces – that is, interactive information systems – can provide information about when and to what extent environmental energies are available. Then energy consumers whose operation is not necessarily tied to specific times, such as dishwashers, washers, dryers etc. could be activated accordingly. In addition, appliances that have internal storage capacities, such as deep-freezer chests, refrigerators, and hot water tanks, could also be put into operation manually or automatically. Tiered rate systems can help to reward the use of electricity generated from renewables.

The goal in employing such user interfaces is to raise awareness for saving energy and to increase the proportion of the energy produced by the building that is used on-site. They increase awareness. But there must not be an obligation or necessity to use such interfaces. Rather, such an interface between humans and technology should promote a playful approach to buildings and energy, and it should enhance a user's identification with the building and its performance.

The Aktivhaus is both producer and consumer; it sensitises its users in their dealings with the building and energy.

The on-site use of locally generated energy can be increased further by using means of storage within the building. The passive storage capacity of the building has priority, because it is available without additional structural or technical complexity, being defined by the design and construction of the building itself. In addition, heat accumulators and battery storage systems can be used to further increase the proportion of energy used on-site.

The Aktivhaus in existing contexts

In the developed countries of the world, the rate of new construction is low. In Germany, for example, the yearly rate stands at approximately 0.55% of the building stock. The greatest leverage for a climate-friendly and sustainable transformation of our built environment is therefore attained by renovations that upgrade the energy efficiency of our existing building stock.

A building renovation is often unable to attain the same energy efficiency qualities that are possible with new construction. Pre-existing built conditions are often inadequate and can only be modified with great effort and expense, as with thermal bridges or uninsulated foundation slabs, making it impossible to match the characteristics of new buildings. Improving the enclosure qualities of the building is nonetheless a prerequisite when seeking to improve user comfort and reduce energy demand.

To avoid misunderstandings: an old apartment building whose roof is covered with solar collectors, a school

Architects: HHS Planer + Architekten AG

Research/Energy concept: TU Darmstadt (DE), Energy-Efficient Building Design Unit, Prof. Manfred Hegger, and Steinbeis Transfer Center EGS, Prof. Norbert Fisch

Client: ABG FRANKFURT HOLDING Wohnungsbau- und Beteiligungsgesellschaft mbH

Completion: 2014

Location: Frankfurt (DE)

Standard: Efficiency House Plus

Energy reference area per EnEV: 8,764 m²

Final energy demand per m² of energy reference area: 28.8 kWh/m²a

Final energy supply per m² of energy reference area: 30.3 kWh/m²a

Example of a user interface for displaying energy production and consumption, Aktiv-Stadthaus, Frankfurt (DE)
HHS Planer + Architekten AG, Kassel (DE)

Office and residential building in Darmstadt (DE), c. 1900/2007
Renovation by opus Architekten, Darmstadt (DE)

building that is retrofitted with geothermal energy and a heat pump, or a factory roof equipped with photovoltaic panels is, in our view, no more an Aktivhaus than an old barn equipped with photovoltaics would be. An existing building can be converted into an Aktivhaus only if there is parity between the construction measures needed to upgrade the building and its use and the technical measures that serve to produce energy. What matters is not solely an improved energy balance, but comprehensive improvements to the building – especially in terms of its function and its relationship to the urban framework.

Thus we can say from the outset that not every building is suited for conversion into an Aktivhaus. First considerations: Is it worth the expense? Will the site still be attractive in the future? In the face of demographic changes and migration, is its continued use assured in the future? Does the building possess good functional preconditions for also meeting future needs? Is it in a technical and structural condition that gives no apparent cause to expect any hazards?

If the answers to these and other questions are positive, the building can be transformed into an Aktivhaus. For a building that is not listed – individually or as part of a protected architectural ensemble – has no special design qualities, and does not play a notable role in people's mental maps, a renovation that converts it into an Aktivhaus can be tackled boldly. The building's appearance can be altered. There are, however, prerequisites for such action that must always be met: every technical modification must also be combined with an improvement in quality for the city, under the motto "Change for the better."

The challenge is a different one with a listed building or one that plays a special role in the collective memory of the city. In that case, the distinctive characteristics of the outward form should be retained where possible. The measures for upgrading energy efficiency will then be restricted to concealed areas. But even then, a significant improvement can be attained. With measures such as new or double windows, roof insulation, possibly interior insulation, and efficient building systems, much can be done to improve the use and the building fabric. The renewable energy production will, under certain conditions, then be focused on the use of geothermal heat or solar energy that does not adversely affect appearance.

A holistic approach should always be taken towards renovation. Better energy performance should be used to improve usability and design.

In the course of extensive renovations, additional storeys can be added or buildings can be otherwise expanded. Evocative contrast and a correspondence between old and new can significantly aid in bringing both history and

Werkzeile in Munich (1954/2009)
Renovation by Koch + Partner Architekten und Stadtplaner, Munich

the present into their own, and they can put an end to the widespread despondency encountered in the face of such tasks. Taken to its logical conclusion, the present can then be an Aktivhaus, in the form of an extension that also produces energy for the adjoining, pre-existing building.

But the aim of the renovation will always be to attain high quality. Experience shows that a renovated building does not begin another renovation cycle for 30 years or more. Half-hearted renovations that barely meet today's requirements are too short-sighted. If we want to achieve the policy goal of carbon neutrality by 2050, every renovation undertaken today would already need to meet this target.

From Aktivhaus to Active City

As the name implies, the Aktivhaus considers the potentials and consumption of an individual building. The building saves energy and reduces environmental pollution while also gaining renewable energy. The goal is autonomy at the building scale.

Autonomy is absolutely necessary where central supply and disposal systems are lacking or far away: in remote locations such as islands – in so-called island mode, also known as isolated operation. Such isolated operation requires a balance of energy production and consumption at the level of the individual building.

Buildings in a neighbourhood or an urban environment are not, however, reliant on this form of autonomy. It is here, in association with other buildings, that the Aktivhaus can fully exploit its capabilities. If well planned and optimised for use, it can produce more energy than it consumes on balance over the year; depending on usage and density, it may even generate many times its own demand. If examined over time, however, there will inevitably be periods when the building produces less energy than it consumes, just as there will be those when the situation is reversed – when, in other words, it will produce more energy than needed. As much of that as possible will be compensated by the building's internal storage capacity. A connection to the mains grid makes it possible to compensate for the rest.

Since the balance limit of the building is overcome, new opportunities arise. The potentials for energy production and storage expand considerably through interconnection. Furthermore, the supply and demand can be evened out much more easily in a neighbourhood, district or city than at the level of individual buildings. The superposition of different load profiles with varied peak usage (as for homes, offices, schools, shopping centres) plays a key role in this. During the day, little-used residential buildings already pass on solar-generated electrical energy through the mains grid to offices, schools and other consumers. In addition, energy sinks for one use can become sources for another (such as using industrial waste heat, or waste heat from refrigerated display cases, to produce domestic hot water).

The Aktivhaus in the city is interconnected. It exchanges energy with its neighbours, thus further improving its energy balance.

This interaction brings the Aktivhaus together with renovated and non-renovated existing buildings; some buildings serve as energy suppliers and others purely as consumers. In addition, urban open spaces can supply nearby buildings with energy in many different ways, such as providing energy from biomass or geothermal facilities. If the regions surrounding cities and towns are also brought into the picture, the network includes considerable renewable energy potential from the undeveloped space.

Intelligently managed networks significantly increase the proportion of the energy that is used locally at the neighbourhood, district and city level. The microgrids or smart grids required for the electrical distribution already exist; ways to regulate them intelligently like virtual power plants, though, are still pending. For heating and cooling, it can make sense to establish small local heating networks. For larger networks, however, the question for the long term is whether improving the energy efficiency of the buildings that supply them could or should lead to such a substantial reduction in energy demand that, when viewed economically and technically, expansion of the heating and/or cooling network hardly makes sense any more.

Expansion of the balance limit from the building to the neighbourhood

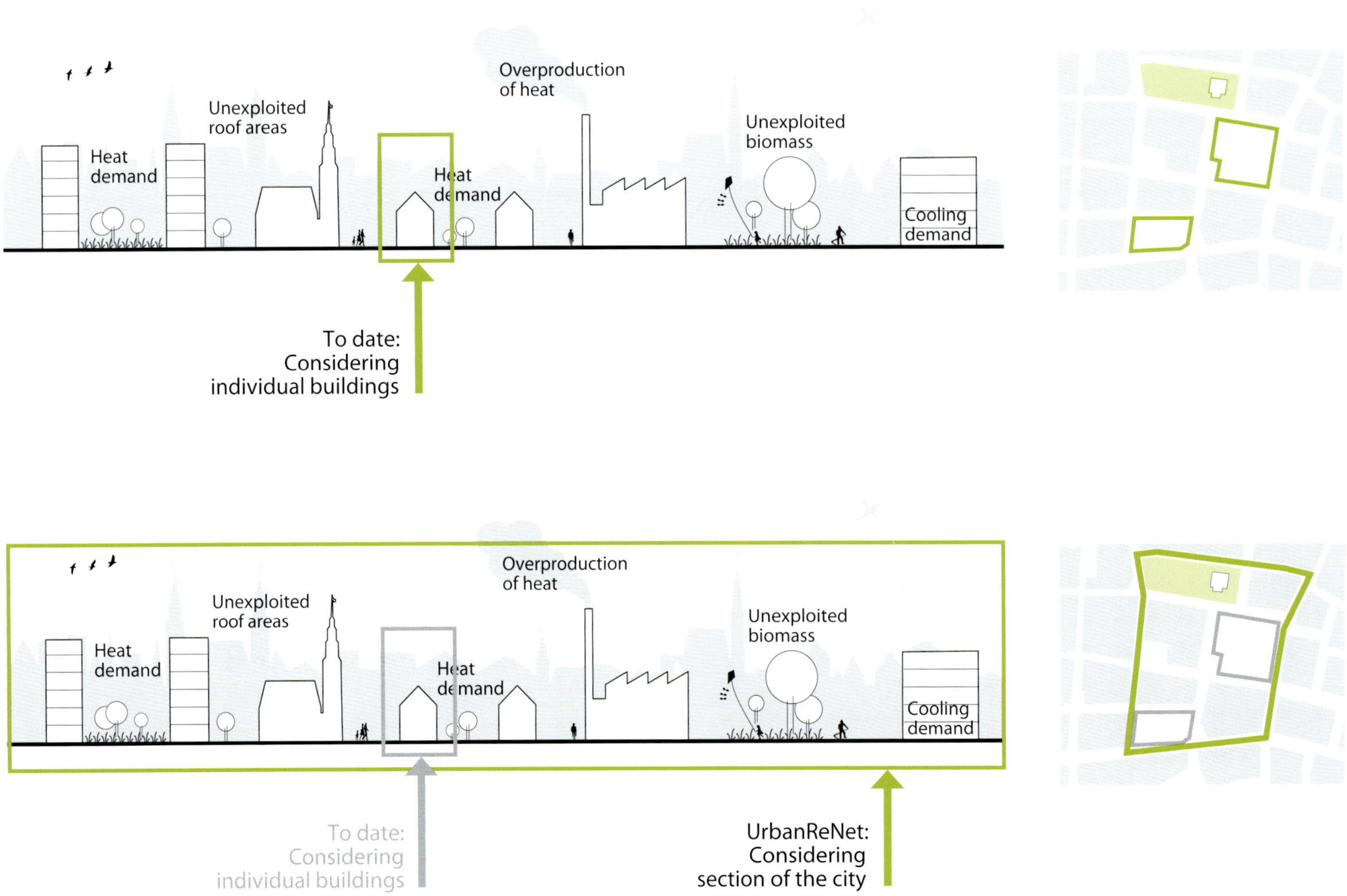

Sustainable building as model

By making full use of the potential for energy savings and the production of environmentally friendly energy, the Aktivhaus takes an important, further step towards a sustainable built environment. In view of the challenges of climate change, increasing world population, and the looming shortage of resources, however, a comprehensive look at the consumption of resources during construction is required. The matters of concern extend far beyond operation of the building.

Choice of materials

Through a conscious choice of materials, planners can significantly influence the use and the reusability of building materials and components for future generations. Are they ideal candidates for recycling? Are they particularly durable? Are they made of renewable raw materials? Especially in view of the new decisions about materials that come into play with the Aktivhaus concept, and the additional expenditures linked with their use, positive responses to these and other questions are needed. To make the necessary choices, it is helpful to take into consideration the lifespan and the ecological assessment data for the materials, so as to be able to give comprehensive consideration to the environmental impact of constructing and operating the building over its entire life cycle.

Construction

Intelligent design is always geared to minimising the costs of materials for construction. Reversible construction without adhesive bonding and sandwich panels enables easy dismantling of the building at the end of its life cycle and the most complete possible reuse of components and building materials. Maintenance-friendly constructions enable easy, non-destructive replacement of parts without damaging other, still functional compo-

Expansion of the balance limit: As part of the IBA Hamburg International Building Exhibition, the former anti-aircraft bunker in Hamburg's Wilhelmsburg district will be refurbished and transformed into an "energy bunker". The building houses both public and technical functions, including a museum and a café along with a central plant facility. The Energy Bunker will be equipped with a biomass CHP plant, solar thermal collectors, a photovoltaic system, and energy storage for supplying heat and domestic hot water to the neighbouring residential area, HHS Planer + Architekten AG, Kassel (DE)

nents in the process, which would thereby compel their premature replacement. Forgoing paints, coatings and cladding simplifies the segregated recycling of components according to different materials. Documentation of the materials and products installed facilitates maintenance and disposal as well as the reuse of materials temporarily stored in a building.

Site selection

The choice of a suitable location is of significant consequence, especially in the context of rising energy costs, demographic change, and increasing urbanisation. This raises questions such as: Will the location remain attractive in the long view, even with a declining population? Do additional expenditures for transport negate an otherwise favourable energy balance in relation to the building and thereby possibly also invalidate an environmentally friendly ecological balance? Is the technical and social infrastructure necessary for a site secure in the long term? Many areas become less dense as a result of smaller households, migration movements, or other demographic developments. The high costs of mobility, in terms of both money and time, are questionable in many cases.

Building programme

As demonstrated by the rebound effect described in the chapter "Basics" (see p. 22 ff), achievements in energy efficiency and sustainability can be outstripped by increased spatial needs. This is especially true for housing; in many countries the area requirement per person has multiplied over the past 50 years. Factors such as area economy with good spatial quality, clever built form and zoning, and layouts that are adaptable and universal can effectively avert a further increase in land requirements. This also suggests it would be prudent in the future to relate the demand for energy and resources to a person, as is the case in the "2,000-watt Society" model in Switzerland.

The Aktivhaus encourages thinking about more than just energy issues. Sustainability is the goal. Effective conservation of resources succeeds by considering the impacts of individuals.

Planning and design

The issue of successfully integrating active energy-generating systems into the architecture arises repeatedly, for both existing buildings and new buildings alike. Public acceptance of renewable energy forms has been harmed by the way many wind farms are not integrated into the surrounding landscape, the scarring of agricultural areas by barns perfunctorily covered with solar panels, and the disfigurement of schools and industrial buildings with innumerable slanting photovoltaic elements. This incidentally also applies to many early efforts towards architectural integration. It takes time and a great deal of of precision work to reach compelling solutions that are worthy of emulation. Thorough detail work is necessary to integrate new building elements and systems into the architecture and to consolidate changing requirements and new technologies into a good form.

Intensive development work and integrated planning are required. One must first cleverly apply typical design factors in the interest of sustainable and energy-efficient construction, such as a building form suitable for the urban space, the use, and the desired energy efficiency, intelligent fenestration, a sensible ratio between solid mass and transparency, and heat- and moisture-absorbing materials. The design process will always also be confronted with new requirements. But an unbiased examination of the unfamiliar also holds opportunities for something new to develop. New expressions of architecture can be found that serve the goals of sustainable and energy-efficient construction, employ new materials or bring materials together in innovative ways, and create new forms from the intelligent application of new technologies.

The energy transition will succeed when the Aktivhaus is well designed, when sustainable product design and architecture fascinate their users.

For architects and other planners, good design of energy-efficient and sustainable buildings is the greatest challenge for the coming years. For this is not just about area economy and energy efficiency. Only a building that is loved by its users and urban society as a whole is truly sustainable. Only then will it be used lastingly. And the effort needed for increasing energy efficiency and for using long-lasting materials will only be truly worthwhile when communities and individuals passionately identify with their built environment. This, and no less, is what each Aktivhaus must achieve.

Conclusion

The aforementioned perspectives show that development of the Aktivhaus and the Aktivstadt has great potential. The tasks ahead have been identified. They can only be tackled when all the stakeholders – architects and engineers, the property industry, public authorities and government institutions, banks, and building users – work together in close cooperation. These tasks require a willingness to accept the challenges facing us as a result of scarce resources and changes in energy policy. Much creativity will need to be employed to create new solutions that are both technically and artistically compelling. We believe the conditions are in place. If we use them, this will be the opportunity for a transformation that is socially desired and globally overdue.

Appendix

Glossary

A

A/V ratio
The ratio of the area of the building envelope surrounding the heated space A to the total building volume V. The A/V ratio is a measure of the compactness of the building. An A/V ratio of 1.00 means that there is one square metre of envelope for every cubic metre of volume. A detached house will usually have an A/V ratio of 0.60 to 1.20, for terraced houses the value is 0.50 to 1.00, while for multistorey, compact residential buildings, the A/V ratio may come down to 0.30.

Absolute humidity [g/m³]
Absolute humidity is the mass of water vapour that a specific volume of air can contain. It is expressed as grams of water per cubic metre of air. The upper limit to this value is determined by the maximum amount of moisture that the volume of air can contain.

Absolute humidity is a direct measure of the quantity of water vapour contained in a volume of air. It indicates the maximum amount of condensate/moisture that could precipitate from the air.

Absorber
An absorber is part of a solar collector and absorbs incoming solar radiation through a carrier medium (water + antifreeze). A higher performance can be achieved by a black absorber, or better still, through selective coatings.

Absorption
During absorption, a material takes up e.g. heat or moisture. It absorbs the heat or moisture (e.g. in absorption cooling machines etc.).

Absorption cooling systems
Absorption cooling systems are an example of active cooling. They use the system of thermal cooling and most use solar energy. Absorption chillers are based on the principle of evaporative cooling. A coolant circulating in a closed system is caused to evaporate under extremely low pressure and low temperature (evaporator).

Acidification
Acidification occurs mainly as a result of the conversion of atmospheric pollutants into acids. The result is a lowering of the pH of rain.

Acidification potential
Acidification potential is one of the earth's most important environmental indicators. The lowering of the pH of rain harms soil, watercourses, living organisms and buildings. Acidification potential is expressed in sulphur dioxide equivalent [kg SO_2-equivalent]. Secondary effects that show how acid rain damages buildings include corrosion of metals or the decomposition of natural stone.

Adiabatic cooling
Evaporative or adiabatic cooling is a process used to cool room air using the latent heat of evaporation. Rather than moistening the cooling air directly, this process moistens a second air flow. It is therefore an indirect process. In evaporative or adiabatic cooling, only air and water, i.e. regenerative energy sources, are used to create the cold. Evaporative cooling is the same in principle as the process of perspiration. When we sweat, the water evaporates and draws heat from the body.

Air change
The air change is the air volume flow for a room or building. A 3-times air change, for example, means that the air in the space is replaced three times an hour.

Air change rate [1/h]
The air change rate measured in the unit [1/h] is a number that indicates how often the air in the room / building volume is replaced in an hour. It plays an important role in the ventilation of buildings. Example: Air change rate = 15/h The air in the room/building volume is replaced 15 times in an hour.

Air collectors
Solar collectors that use air as the heat carrier.

Amortisation period [a]
The amortisation period is the time that has to pass until the actual investment is repaid by savings during operation.

The energy amortisation period is the time during which a system used to create energy must be operated until the energy used for its manufacture has been created again. While systems that are operated with renewable energy have energy amortisation periods of a few months or years, conventional power plants never achieve amortisation because they have to be continuously supplied with more primary energy. Therefore it makes no sense to give an energy amortisation period for this type of power plant.

Anergy, exergy [kWh]
Anergy is energy no longer capable of being used directly in a work process, e.g. environmental heat. It must be activated by the use of exergy.

Anergy is a measure of the maximum mechanical work that could be gained in a system in thermodynamic equilibrium with the existing environment if it could be brought into equilibrium with a new environment at absolute zero ($T = 0$ K). This energy cannot be used (is not capable of performing work), however, because this environment is not available.

The opposite of anergy is exergy, which is a measure of the maximum mechanical energy that could be gained using the environment if the system is brought into thermodynamic equilibrium with the environment.

A system that is in equilibrium with the environment is not without energy; it has no exergy but always contains its own anergy.

For systems that are above the ambient temperature and pressure: Anergy + Exergy = Energy

Annual heating demand [kWh/m²a]
The annual heating demand is the quantity of heat required annually for heating the whole building (not taking into account the heat required for hot water production).

Annual heating energy demand [kWh/m²a]
The annual heating energy demand is the quantity of energy that must be supplied to the building for heating and hot water over a year. It takes into account the losses from the heating plant and associated equipment.

Annual primary energy demand [kWh/a]
The annual primary energy demand Q_P [kWh/a] is the quantity of primary energy required to heat, ventilate and provide hot water to a building over a year. It takes into account all energy gains and losses.

Annual use efficiency
The annual use efficiency is a measure of how much of the capacity of a heating plant is utilised. A well set up and designed system works economically; poor annual use efficiencies arise, for example, through over-dimensioning.

Artificial light
Unlike daylight, artificial light is light created from artificial sources.

Atmospheric humidity [%]
Atmospheric humidity, or air humidity for short, is the proportion of water vapour in the mixture of gases forming the earth's atmosphere. When related to the proportion of moisture in the mixture of gases in a room, the term used is "humidity". Absolute humidity is the amount of water vapour in g/m³ contained in one cubic metre of air. It is often expressed in percent as relative humidity. This describes the ratio of the actual water vapour content to the maximum possible content at the same temperature and pressure.

Auxiliary energy [kWh/a]
Auxiliary energy is the energy required for operating pumps, fans, regulating equipment etc. in the heating, cooling and drinking water systems etc.

B

Biomethane production
In the production of biomethane, hydrogen is thermo-chemically synthesised with CO_2 (methanisation). The biomethane can be stored and fed into the gas network, from where it can be converted into heat as required. The efficiency of the conversion from electricity to methane is 60%, in other words 1.0 kilowatt hour [kWh] of electricity produces 0.6 kWh of the energy carrier methane.

Blower-door / differential pressure test
The blower-door test is a leakage test of the building envelope. Airtightness is an important requirement for a building according to the German Energy Saving Ordinance (EnEV). A blower-door test is not required. However, if the minimum requirements for windtightness can be demonstrated using a blower-door test, a lower minimum rate of air change and hence lower ventilation heat losses can be assumed in the calculations.

Break-even point
The economic break-even point is the point at which the income and the costs of production (or a product) are equal and therefore neither profit nor loss has been made. If the break-even point is passed, a profit is made; if it is not reached, then the result is a loss.

Brine
A solution of salt in water used as a heat carrier, e.g. in heat pumps.

Buffer store
A buffer store provides short-term storage of heat energy to cover daily fluctuations in heat demand and heat production.

Building compactness (A/V) [m²/m³]
See A / V ratio

Bulk density [t/m³ or kg/dm³]
The density of a material is the ratio of the mass to the volume and is expressed in t/m³ or kg/dm³. Bulk density is the density of porous materials including the volume of the pores (e.g. porous concrete).

C

Capillary action
Capillary action is the driving force responsible for the movement of fluids in materials with pores.

Cascade storage
Cascade storage is a multistage storage system that charges a store for the day's requirement, for example, and places the excess in a second store that can be brought into use on demand.

CIS cells (CIS, CIGS, CIGSSe)
CIGS (also CIGSSe or CIS) is a thin-film technology for solar cells. The names are acronyms and stand for the elements used; copper (Cu), indium (In), gallium (Ga), sulphur (S) and selenium (Se). In practice, various combinations of these elements are used: The most important examples are $Cu(InGa)Se_2$ (copper indium gallium diselenide) and $CuInS_2$ (copper indium disulphide).

CO_2 stores
Regenerative raw materials are described as CO_2 stores because they convert carbon dioxide into biomass by photosynthesis while they are growing. At some point in the future, the climate-damaging CO_2, which had been bound in the material until then, is released again by thermal recycling or combustion.

Coefficient of performance (COP)
The coefficient of performance for heat pumps is the ratio of heat output at the condenser in kW to the electricity used in driving the compressor in kW. It is a measure of efficiency. The electricity required to operate the system increases with increasing difference between the temperature levels at the evaporator and the condenser because higher compression is required.

Cold bridges [W/m²K]
Cold bridges are local weak spots in the normal construction of a building component and can occur at a point, or over a strip or area. They may be formed due to geometry (external corners), construction details (penetrations such as an embedded cantilever balcony slab, butt joints between components) or be material-related.

Cold bridge-free design
To keep design costs down, a simplified criterion for cold-bridge-free design was adopted for the Passivhaus concept, in which cold bridge losses UCB ≤ 0.01 W/m²K do not have to be taken into account in the calculations. However, this assumes that cold bridges are considered and designed out from the start. Cold bridges must be identified and taken into account early in the design phase. Later modifications in the completed building are certainly technically possible but are usually very time-consuming and expensive.

Cold local heating
Cold local heating is based on passing near-surface water through heat pumps, which extract the heat from the water (for heating) or put heat into the water (for cooling), and then return it to the ground.

Component-related average thermal transmittance [W/m²K]
The component-related average thermal transmittance is the average thermal transmittance for individual building component groups. This average U-value of individual components is the quotient of the sum of the thermal transmittance losses of the components and the area of the building component surface. The EnEV stipulates maximum thermal transmittance values for individual components, and in setting these requirements it distinguishes between, for example, opaque and transparent components.

Concentrator cell
In concentrator cells, all the sunlight is directed (by refraction, mirrors) onto a small multicoated cell (tandem/triple), which can then operate at very high efficiency, currently up to 40.7%.

Condensing power plant
A condensing power plant is a conventional thermal power plant in which heat is converted into electricity. It is used exclusively for the production of electrical power and makes no further use of the residual heat, merely giving it up to the environment through a condenser or a cooling tower. The steam leaving the condenser power plant turbines, which is at a low temperature and pressure, is condensed in a cooling system. The efficiency of these power plants is between 40 and 60%. The direct alternative to the condensation power plant is the combined heat and power plant.

Cooling load
The cooling load is the heat load that has to be removed from a room by convection to achieve or maintain a specified room air condition. It is divided in accordance with VDI 2078 into external and internal cooling loads.

Counterflow heat exchanger
The counterflow principle is fundamental to heat transfer. In this process, two substances of different temperatures, usually water or air, pass one another in different directions, with the result that heat from one flow direction/substance is given up to the other.

D

Daylight
Daylight or natural light is the visible light of the sun.

Daylight quotient
The daylight quotient is a means of evaluating the quality of the daylight in a room. The calculation is not a legal requirement for an energy balance (the requirements for the room uses can be taken from the German Workplace Guidelines and DIN 5034). Nevertheless, calculating the daylight quotient can be an advantage in increasing the energy efficiency of the building. The daylight quotient is always dependent on the illuminance outside the building and the actual illuminance available indoors.

Deep geothermal heat
Deep geothermal heat begins at a depth of more than 400 m and a temperature of over 20 °C. Deep geothermal heat in the true sense is reached at depths of more than 1,000 m and has temperatures higher than 60 °C.

Diffuse radiation
Diffuse radiation refers to the solar radiation that reaches us from all directions – after sunlight has been scattered by clouds, mist, mountains, buildings etc.

Diffusion
Diffusion is the mixing of two substances through the natural motion of their molecules alone; e.g. the movement of water vapour in buildings. In the same way as heat always migrates from the warmer to the cooler side, water vapour moves between areas of different moisture content (water vapour diffusion). Temperature, atmospheric pressure and relative humidity affect the speed of diffusion and the quantity of diffusing vapour.

Direct radiation
This is solar radiation that strikes the collector directly from the sun. It is more intensive than diffuse radiation but over a year, the quantities of direct and diffuse radiation striking the collector are about equal.

Dynamic building simulation
Dynamic simulations are often carried out to calculate the internal and external summer heat loads, particularly for non-residential buildings. The simulations are programmed with weather and building operating scenarios to check the use of passive and active cooling measures.

E

Efficiency [%]
Efficiency is the ratio of the power output to the power input under optimum operating conditions. The efficiency of a conversion process, e.g. in power plants or heating systems, is the ratio of the usable energy created to the energy used for the conversion process.

Efficiency of a solar cell or module
The efficiency indicates what percentage of the light striking the cell or module is converted into usable electrical energy.

Electricity net generation
The amount of electricity measured at the generator terminals of a power plant less the electricity used for operating the plant.

Energy [J] / [Wh]
Energy is a physical property and represents stored work in a predefined system, or the capability of the same system to perform work. Energy is usually measured in joules [J] or watt hours [Wh]. Energy cannot be created or destroyed; it can only be changed from one form of energy into another. Therefore, the law of conservation of energy applies in closed systems. A distinction is drawn between the following forms of energy: mechanical, thermal, electrical and magnetic energy, electro-magnetic vibration energy, chemical energy, nuclear energy. Energy is the sum of anergy and exergy.

Energy balance
Sum of all the energy gains and losses of a building.

Energy demand / final energy demand [kWh/m²a]
The energy demand of a building is, in the context of the preparation of an energy performance certificate or verification of EnEV compliance, the calculated value of how much energy a building requires under standard conditions. It is used to compare the insulation standard and the technical equipment of buildings.

Energy content
The energy content is the quantity of heat that can be obtained by complete combustion of a quantity of a fuel.

Energy Saving Ordinance 2002, 2004, 2007, 2009 (EnEV)
German regulations for energy conservation applicable to heated buildings. The main requirement parameter of the EnEV is the annual primary energy demand Q_p in relation to the compactness A / V of the building. The amalgamation of the Heating Appliance Ordnance and Thermal Insulation Ordnance extended the earlier energy balance framework in Germany.

Energy consumption [kWh/a]
Energy consumption is a measured parameter that gives the actual consumption of a building.

Energy efficiency
Energy efficiency is the evaluation of the energy quality of buildings by comparing their energy requirement parameters with reference parameters (e.g. the requirements of the EnEV) or with comparative values (e.g. the average for buildings of the same use).

Energy efficiency ratio [-]
The energy efficiency ratio expresses the energy efficiency of heat pumps. It is the result of dividing the heat produced by the electrical energy used. It is the ratio of the energy yield to the energy expended, in other words, efficiency.

Energy reference area [m²]
A different reference area is used for calculating the energy balance of non-residential buildings. The floor area (NF) from DIN 277 is used to determine the energy reference area. The energy reference area equates to 100 % of the usable floor area (HNF) and the area ancillary to the main function (NNF), and 60 % of the circulation area and ancillary area for services (VF + FF). Stairs, shafts and unheated areas are not taken into account.

Entropy
Entropy is the thermodynamic parameter with which the heat transfers and irreversible processes in thermodynamic processes can be calculated and presented in an understandable way.

EU Energy Performance of Buildings Directive 2010/31/EU
Considering that the building sector was consuming 40 % of the total energy used in the EU, and the obligations of the Kyoto Protocol on climate change issued by the United Nations, the European nations produced an EU directive to achieve the declared objectives. The Energy Performance of Buildings Directive (EPBD) 2010/31/EU is an update of an earlier directive of the same name, 2002/91/EC, issued by the European Parliament.

Eutrophication
Eutrophication is a term for nutrient enrichment. Too much nutrient in watercourses can lead to fish death or even to the sudden biological death of the watercourse. Plants in soils over-enriched with nutrients exhibit weak tissue growth and lower resistance to environmental influences. High eutrophication also leads to nitrate enrichment of the soil and drinking water, where it can result in nitrate poisoning of humans. Eutrophication potential [kg PO43 equivalent] includes other substances by comparing their effect with that of PO43.

Exergy
see Anergy

External thermal composite insulation system (ETICS)
An external thermal composite insulation system is a multilayer product used for insulating external walls. The insulation, which is attached to the wall, is designed to be used in combination with special plaster coatings. External thermal composite insulation systems are particularly suitable for refurbishing existing buildings with plastered or concrete facades.

F

Final energy [kWh]
Losses occur each time energy is converted or transported. Final energy is the amount of energy including the system and distribution losses, or the amount of energy of an energy carrier supplied to the building boundary before conversion.

Flow temperature
The temperature of the hot water in the flow pipe of a heating circuit.

Free enthalpy
Free enthalpy (G) is the driver of all chemical, biological and biochemical processes. It gives information about whether a process in which there is an exchange of energy between system and surroundings takes place is reversible or irreversible. G is the maximum usable work of a process at constant pressure and constant temperature.

Fossil energy carriers
Fossil energy comes from energy carriers with an energy content that was converted into a concentrated form a long time ago and does not renew itself in terms of human time scales. Fossil energy carriers form naturally as a result of biological and physical processes, such as changes below and on the earth's surface over long periods of time. Natural gas, crude oil, lignite and coal are based on organic carbon compounds. By burning fossil energy carriers, not only is energy released in the form of heat, but also other combustion products, such as carbon dioxide, nitrogen oxides, soot and other chemical compounds, depending on the composition and purity of the fossil fuel.
Crude oil continues to be the most important supplier of energy in the world. Around 40% of the energy we require comes from crude oil. Fossil energy carriers are finite.

Functional equivalent
The functional equivalent is the thickness of a material that will provide the same level of functionality. Only by using a functional equivalent can the environmental effects of construction materials be directly compared with one another.

G

Global irradiance
Global irradiance is the solar radiation falling upon a horizontal surface. It is composed of direct and diffuse irradiance and depends on the geographical latitude, the season, the degree of cloud cover, and particles in the atmosphere. The greater the angle of incidence, the higher the radiation density. Under a cloudy sky, only diffuse radiation reaches the surface of the earth, which is why the global irradiance in Central Europe is below 100 W/m². On clear summer days, however, it reaches approximately 700 W/m². The annual global irradiance in Germany is between 900 and 1,200 kWh/(m²).

Global warming/greenhouse effect
In everyday conversation, the global warming effect or greenhouse effect is referred to as the warming of the planet due to greenhouse gases and water vapour in the atmosphere. The term "greenhouse effect" was originally used to describe the rise in temperature observed behind the glass panes of a greenhouse when the sun is shining. Today the term is much broader and the build-up of heat in the atmosphere from the sunlight striking the earth is called the atmospheric greenhouse effect, because of the similar physical principles of both processes.

Global warming potential (GWP)
The global warming potential or greenhouse warming potential describes the emission of gases that contribute to the global warming effect. These gases reflect the infrared radiation given off by the earth's surface and turn some of it back down to earth. This naturally occurring process is strengthened by the greater concentration of these gases in the troposphere, where they are responsible for global warming.
The global warming potential takes into account all gases by relating their effects to that of carbon dioxide. As the persistence of each of these gases in the troposphere varies, the global warming potential is usually quoted with a time horizon. This is normally 100 years but can also be 50 or 20 years.
The (relative) global warming potential or CO_2-equivalent (CO_2 is always used as a comparator) is a measure of how much a specific amount of greenhouse gas contributes to the greenhouse effect.

Grey energy
Grey energy is the quantity of energy used directly or indirectly in the manufacture or supply of a product or service. It relates to a specific place of production or supply of the product or service. Grey energy differentiates between renewable and non-renewable energy.

H

Heat capacity
See Specific heat capacity

Heat carrier
Liquids or air that transport the heat from the collector to the store are called heat carriers. In solar systems, this is usually a mixture of water and antifreeze to prevent the collector from freezing in winter.

Heat load [kW]
The heat load is the maximum heat to be provided by a heating plant. The coldest days in the year, usually in January and February, determine the maximum heat load. This is the heat that must be supplied to provide comfortable indoor temperatures.

Heat output [kW]
The heat output is the usable heat provided by a heating plant over a specific period of time (e.g. an hour). It is given in kW (kilowatts). The heat output must be at least equal to the heat load.

Heat recovery
Heat recovery is a collective term for various methods of making the thermal energy of a mass flow involved in the process available for reuse. The basic objective of heat recovery is to minimise primary energy consumption, focusing on the energy and ecological requirements. Heat recovery has the characteristics of regenerative energy.

Heat recovery efficiency
This is the efficiency of the heat exchanger, for example in a ventilation system with heat recovery. The value is an important parameter in any energy analysis of all the building services.

Heat sources
Any object that can give off heat in any form (radiation, convection) is called a heat source. This can have a positive effect in winter as a heat gain, but a negative effect in summer as a heat load.

Heat transfer coefficient [W/m²K]
The heat transfer coefficient is a factor of proportionality that determines the intensity of heat transfer at a boundary surface. The heat transfer coefficient in W/(m²K) is a specific parameter of the arrangement of a material with respect to an environment. The higher the heat transfer coefficient, the worse the insulation properties of the material boundary surface. Its reciprocal is the heat transfer resistance RS in (m²K)/W.

Heat transfer resistance [m²K/W]
The heat transfer resistance R_S in (m²K)/W is the inverse of the heat transfer coefficient. The higher the heat transfer resistance, the better the insulation properties. The heat transfer resistance is given for the internal and the external sides of a component. These parameters in conjunction with the thermal resistances of the individual component layers give the thermal resistance.

Heating curve
The relationship between the outdoor temperature and the flow temperature necessary to heat the heated areas of a building is set by

the heating curve. The heating curve depends on the particular building and is usually determined by trial and error during the building's operation. The flow temperature is normally adjusted by the heating system regulator taking into account the outdoor temperature.

Heating degree day [Kd/a]
In Germany, a heating degree day is calculated based on the heating base temperature of 15 °C (specified in VDI Guideline 2067/ DIN 4108 T6) and an indoor temperature of 20 °C. For a Passivhaus, the heating base temperature is between 9.5 and 11 °C because of the inertia of the building. In the Passive House Planning Package (PHPP), an energy balance for the standard climate uses a heating degree day value of 84 kKh/a (heating period approx. April to September).

Heating degree hour [kKh]
See Heating degree day. The heating degree hour obtains a more accurate energy balance using hours instead of days.

Heating demand [kWh/a]
The quantity of heat that must be used per year for space heating. This is calculated for standard conditions and is classed as useful energy.

Heating demand (per m²) [kWh/m²a]
The heating demand (HWB) is the calculated amount of energy supplied to a building during the heating period to maintain the desired indoor temperature (e.g. by radiators).

Heating energy demand [kWh/m²a]
According to the definition in the German Energy Saving Ordinance (EnEV), the annual heating energy demand is the amount of energy that must be supplied to a building for the purposes of heating, ventilation and hot water, determined in accordance with the EnEV calculation method. It is given in kWh/(m²a) or in kWh/(m³a).

Hot water storage tank
There are various types of hot water storage tanks (e.g. stratified heat storage). What they all have in common is that they always contain hot water (in contrast to continuous flow water heaters).

I

Illuminance [lx]
Illuminance E in lux is the amount of light falling on a surface divided by the area of that surface.

Impact assessment
Preparation of an impact assessment follows the inventory in the life cycle assessment. The impact assessment assigns emissions to each material and energy conversion processes in the inventory. The different emission types are placed in groups according to their ecological effect categories (e.g. contribution to global warming effect) to arrive at a better assessment. Their effects are calculated as equivalents relative to a main pollutant to indicate the effects of all the pollutants involved. There are over 30 different main pollutants available for use as reference points. If the inventory gives no specific information about the processes, then the necessary information for the assessment can be obtained from databases. If this is the case, care must be taken to maintain traceability in the LCA.

Infrared thermography
Thermography uses a special camera (thermal imaging camera) to make visible the thermal energy being given off by a building or structure. To understand this process, it is important to know that objects with a temperature greater than absolute zero radiate heat in the infrared range. This means that even cold objects, such as ice, give off infrared radiation. Moreover, the higher the temperature, the more intensive the infrared radiation given off, and the more red the area appears in the thermal image (blue = cold, red = warm).

Insulating glazing
Insulating glazing, also called double/triple glazing or insulated glazing, usually refers to windows with at least two panes of glass. The glazing cavity is filled with a noble gas, usually argon or krypton, to increase the thermal insulation effect. The glazing may also have additional properties, e.g. solar protection, sound protection etc.

Internal heat gains
The use of electrical equipment, computers, artificial lighting, the presence of people, and activities such as cooking in a room give off heat and warm it. These internal heat gains are taken into account as energy contributors in the design of Passivhaus buildings.

Isotherms
Isotherms are calculated lines joining places of equal temperature on a building component. By visualising the thermal condition of a body, they make it easier to understand.

K

kWh
The abbreviation for kilowatt hour as a measure of energy or physical work. 1 kWh = 1,000 watts over a period of 1 hour.

L

Latent heat stores
These do not change in temperature during the charging and discharging loading process; instead the heat storage medium changes its overall state. Usually the change from solid to liquid (or the reverse) is used because hardly any change in volume occurs. The storage medium can be charged or discharged beyond its latent heat capacity, and only then does the energy flow cause the temperature of the medium to rise. Latent heat stores combine sensible and latent heat storage.

Life cycle assessment (LCA)
Life cycle assessment is a means of analysing the resource consumption and the environmental effects of a material over its life cycle. It analyses the life of a building material over the stages of raw material extraction, manufacture, processing and, where relevant, transport, use, reuse and disposal. The balance boundary is determinant for the information that can be gained from an LCA.

Life cycle impact assessment (LCIA)
Life cycle impact assessment calculates, on the basis of materials used, the effects of the manufacture and production processes of a product in terms of, for example, emissions. It is a generally applicable technique and does not relate only to building products. It can be applied to any process, for example services, production processes, or to a complete economic entity, such as a company.

Life cycle costs
Life cycle costs are the costs that a product incurs from the original idea to its withdrawal from the market. This includes the capital investment and expenditure, but not the positive returns in the form of receipts.

Lighting
Illuminating a room or building with artificial light is described as lighting. If this is done by sunlight instead of artificial light, then the word "lighting" applies only when technical means, such as mirrors, are used to direct the light.

Local climate
In energy calculations, it is often possible to take into account the location of the building. For EnEV standard calculations, the climate for Germany must be used for reasons of comparability. However, a more specific location and the associated climatic influences can often be entered for individual optimisation.

Local heating network
Local heating is the transmission/transfer of heat between buildings for the purpose of heating. In contrast to district heating, this is done over relatively short distances. Local heating networks are promoted by government policies because they offer a way of transporting locally produced heat energy to users. They are energy creation systems with high overall efficiencies and enrich the economies of the localities.

Long-term stores
Long-term stores are stores that take up heat and store it for several weeks or months. Consequently the number of charging cycles per year is relatively small.

M

Microclimate
The microclimate is the climate of the near-ground layers of air up to a height of about two metres or the climate that forms in a small, clearly delineated area (for example between buildings in a city).

Minimum air change
The minimum air change is the amount of supply air that must be supplied to the net building volume for hygiene purposes. The minimum air change for residential buildings is determined from DIN 1946-6. DIN EN 13779 provides the information for non-residential buildings.

P

Peak oil
The time of maximum global oil extraction – known as peak oil – is the high point at which half of the world's conventionally extractable crude oil will have been used.

Phase change material (PCM)
Phase change materials are materials that can reversibly change their overall state as a result of the action of outside influences, such as light, pressure, water or temperature. They have the ability to change their state from liquid to solid by crystallisation and release the energy stored in them that they absorbed previously at a higher temperature. A PCM behaves as a latent heat store.

Photochemical ozone creation potential (POCP) [kg C_2H_4-equivalent]
The photochemical ozone creation potential [kg C_2H_4-equivalent] is an estimate of the near-ground ozone creation and is based on the effects of ethylene (C_2H_4).

Photovoltaics (PV)
Photovoltaics is the direct conversion of radiation energy – mainly solar energy – into electrical energy.

Pollutants
Pollutants are substances that are harmful to the environment (people, animals and plants). These include carbon dioxide, sulphur dioxide and nitrogen oxides.

Carbon dioxide is a colourless, odourless gas that results from combustion and is partly responsible for the greenhouse effect. It can be reduced only by using less fuel.

Sulphur dioxide is a foul-smelling toxic gas that irritates the skin. It arises from the combustion of sulphur-containing fuels (coal, wood etc.). It is partly responsible for acid rain (forest dieback).

Nitrogen dioxides adversely affect breathing and cause acid rain.

Primary energy [kWh]
Primary energy is the energy made available by naturally occurring energy forms or energy carriers.

Primary energy content [MJ] or [kWh]
The primary energy content of a material is the energy input from energy carriers (resources) required for the manufacture and use of the material. It makes a distinction between non-renewable primary energy (lignite, coal, natural gas, crude oil, uranium etc.) and renewable primary energy (water and wind power, the exploitation of the sun through solar energy and biomass etc.).

Primary energy demand [kWh/m²a]
The primary energy demand, which is part of the actual energy demand of a system, is the energy demand that occurs outside the system boundaries in the upstream process chain in the extraction, conversion and distribution of the energy carriers (primary energy). It gives information about energy efficiency and resource conservation of the energy use. In calculating the energy balance, the use of primary energy in the energy demand is taken into account by multiplying it by a primary energy factor reflecting the energy carriers involved.

Primary energy factor
The losses arising from the extraction, conversion and transport of an energy carrier are taken into account using a primary energy factor and thus increase the basic demand figure in the primary energy evaluation. The primary energy factors are different for each balancing system and country.

Process heat
Process heat is the heat used for technical processes. Process heat usually arises from combustion processes or electrical current; in favourable circumstances, waste heat can be used as process heat.

R

Rate of utilisation
The rate of utilisation of a plant or device is the ratio of the energy made usable by the plant or device to the energy supplied to it over a given time period. This period may include breaks in operation, no-load running, start-up and shut-down times. In the case of plants for producing electricity using combined heat and power systems, the rate of utilisation or total rate of utilisation is the ratio of the total amount of usable energy delivered (sum of electricity and heat delivered) to the energy used. It must be differentiated from the (electrical) efficiency, in which only the electricity delivered is taken into account. Because the rate of utilisation is determined partly by the heat demand and is therefore greatly influenced by the seasons, the annual rate of utilisation is normally used to assess plants. It should be noted that the rate of utilisation for hot water provision using fossil energy carriers is particularly low. Well-insulated houses with a low demand for heating energy can therefore have a low annual rate of utilisation and this makes it worthwhile to use solar hot water.

Rebound effect
In the context of energy economics, the term "rebound" describes the situation where the savings potential due increasing efficiency are not realised, or realised only partially. A situation where the increased efficiency leads to greater consumption (this means a rebound effect of over 100%) is called backfire.

Reference building method
The EnEV from 2009 onwards stipulates the maximum permissible highest values based on the reference building method. This method, which provides maximum parameters, is based on relevant values from a building of the same construction (geometry, alignment, usable floor area, standardised building components and technical systems). The previous verification based on the A/V ratio of the building is no longer used.

Relative air humidity
The relative air humidity (relative humidity or RH for short) is expressed as a percentage and is the ratio of the actual water vapour content in the air of a system (room) to the maximum possible water vapour content.

Renewable energy
Renewable energy is energy from a source that will not be exhausted for as long as humankind exists. Solar energy from thermal collectors, photovoltaics, natural light, wind, water and bioenergy are all forms of renewable energy.

Renewable Energy Sources Act (EEG)
In 2000, the Renewable Energy Sources Act replaced the Energy Grid Feed-In Act in Germany. The EEG covers mainly the acceptance, transmission of and payment for electricity fed into the grid from renewable sources.

Renewable Energies Heat Act (EEWärmeG)
The objective of this German legislation for promoting the role of renewable energy in providing heat is to increase the proportion of renewable energy used for heating, hot water, cooling and process heat to 14% by 2020. To achieve this objective, the legislation stipulates that new buildings (building applications from 1.1.2009) of 50 m² or more must obtain a proportion of their heating energy from renewable energy carriers, such as solar heat, biomass, environmental heat or geothermal sources.

Residential building
A residential building is defined as any building used as a main residence (e.g. a house for one or more families) – hotels are excluded by this definition.

Resources
Resources are material or non-material goods. In the construction sector, they are mainly those quantities of raw materials that can be extracted with the currently available technical means.

S

Seasonal coefficient of performance
The ratio of the heat supplied in a year in kWh to the electricity in kWh used to operate the compressor, auxiliary equipment and connect to the heat sources. The higher the seasonal coefficient of performance, the lower the amount of energy required, and the more efficient is the operation of the heat pump (see Coefficient of performance).

Secondary energy [kWh]
Secondary energy is the remaining energy form after the conversion of the primary energy carriers into useful energy carriers. Secondary energy generally has one of the following properties:
— good storability (e.g. coke, refined oils)
— good transportability (e.g. electrical energy)
— high energy density (e.g. coke)
— simple/cheap to manufacture, e.g. briquettes).
One of these properties is usually preferred. The choice depends on the location and purpose. The by-products of the manufacture of secondary energy are often also usable secondary energy (e.g. gas from petrol production and heat from the production of electrical energy are by-products that can be used as process gas or district heating, respectively). These by-products are, however, not always used.

Smart grid
The term "smart grid" (intelligent electricity network) covers the networking and control of power generation plants (centralised and decentralised), energy suppliers, energy stores and electrical consumers. A smart grid optimises consumption in terms of location and time, and ensures better coordination of energy producers and consumers, whose uses of the grid are not well matched. Smart grids can optimise the utilisation of the grid and avoid expensive peak load times. The objective is to secure the energy distribution system through efficient and reliable operation.

Solar collector
This consists of an absorber made of black metal plates behind a sheet of glass. The solar collector absorbs the solar radiation and converts it into long-wave heat radiation. So that the energy is not lost, the collector is well insulated at the sides and bottom (flat-plate collector) or contained in a vacuum (vacuum tube collector). The heat is transported by a liquid (frost-proof brine) in small tubes before being transferred to a water storage tank via a heat exchanger.

Solar cooling
Solar cooling is a term applied to systems in which solar thermal energy provides the heat required to drive the absorption chiller.

Solar fraction [%]
The percentage of the total heating demand provided by the usable energy from the solar system.

Solar store
Stores into which solar energy is fed. They bridge periods of bad weather and the peaks in the daily energy demand.

Solar thermal energy
Solar radiation converted into usable heat energy. Solar heat is absorbed by a solar collector and used for heating domestic hot water or to support space heating in buildings.

Solar thermal gains [kWh/m²a]
Short-wave solar radiation enters the building through transparent components such as windows. It strikes the floor and is converted into long-wave radiation. The long-wave radiation remains trapped in the building because glass is non-transparent for this part of the spectrum (similar to a greenhouse). The amount of gain depends on the size and orientation, the solar energy transmittance, and the shading and dirtiness of the window.

Sorption
Sorption allows a material to store moisture from the air in its surface layer. The absorption and release of moisture depends on the air moisture content.

Specific heat capacity [J/kgK]
This material-specific property gives the amount of energy required to raise the temperature of 1 kg of a material by 1 K. The specific heat capacity is an indication of a material's storage capacity. Because of their negligible weight, insulation materials usually have a very low specific heat capacity. Because of its high storage capacity, heavy insulation such as wood fibre insulation boards (bulk density > 100 kg/m³) can improve summer heat protection in areas of the building prone to overheating (e.g. rooms in the roof space).

Specific transmission heat loss
The specific transmission heat loss is the sum of the thermal transmittance losses of all the components of the building envelope. The calculation involves taking the U-value of a material and multiplying it by the area of the built-in surface and the temperature correction factor. If this sum is divided by the area of the overall building envelope, the result is the average U-value of the building. This value can be considered as the weighted U-value of the whole building envelope. The official description is "specific transmission heat loss related to the heat transmitting enveloping surface". For old buildings, this value is often greater than 1.00 W/m²K. New buildings must achieve values less than a specified minimum, which depends on the A/V ratio, and is normally between 0.50 and 0.60 W/m²K for detached, semi-detached and terraced houses.

Storage collectors
Flat-plate collectors with an integrated hot water tank.

Sufficiency
Sufficiency is an option for the energy- and resource-conscious consumer: individuals can replace energy-intensive services with ones that have a lower energy demand, and thus optimise their consumption behaviour, for example by video conferencing instead of flying, or by reducing their living area per person.

Supply air cooling
Just as a building can be heated by the supply air, it can also be cooled by the supply air in summer, e.g. using a compression refrigeration machine.

Swimming-pool collectors
Simple absorbers used to warm water, for example black hoses with no covers.

T

Tandem or triple cells
These solar cells consist of two or three thin coatings, which are placed successively onto the substrate. Each coating is optimised for a specific part of the light spectrum.

Thermal conductivity [W/mK]
Thermal conductivity, or the coefficient of thermal conductivity, is a material property. The thermal conductivity of a material indicates how much heat flows through an area A in time t at a temperature difference T.

Thermal energy [J]
Thermal energy is the energy stored in the random movement of the atoms or molecules of a material. It is a physical property and part of the material's internal energy. Thermal energy is measured in joules (abbreviation: J) in the SI system. In casual conversation, thermal energy is referred to rather inaccurately as heat, or heat energy, or even used interchangeably with temperature.

The introduction of heat increases the average kinetic energy of the molecules and therefore increases the thermal energy; taking heat away reduces it. If two systems with different temperatures come together, their temperatures equalise by the exchange of heat. Without external action, heat will never flow from a system of lower temperature to a system of higher temperature.

Thermal Insulation Ordinance 1977 | 1984 | 1995 (up to 2002) (WSchV)
Legislation on energy-saving heat insulation in buildings in Germany. Against a backdrop of increasing energy prices, the objective was to reduce energy consumption through improving construction, first in new buildings and then in the existing stock. The WSchV first applied in conjunction with the Heating Appliance Ordinance (HeizAnlV). This was superseded in 2002 by the German Energy Saving Ordinance (EnEV).

Thermal transmittance [W/m²K]
See U-value

Thermal resistance [m²K/W]
Thermal resistance is the reciprocal of the U-value. This is the resistance with which a component opposes the flow of heat at a temperature difference of 1 K on a surface of 1 m². The greater the thermal resistance, the better the heat insulation properties of the component.

Thermography
Thermography makes temperature distribution visible. It was originally based on a contact technique involving heat-sensitive paper that changed colour when in contact with warm surfaces. Today, the term is mainly used to refer to infrared thermography. See Infrared thermography.

Total solar energy transmittance (g-value)
The total solar energy transmittance (in %) of a pane of glass is the amount of solar energy it transmits (for use). For a pane of glass in a window with a g-value of 0.56, a maximum of 56 % of the solar radiation (energy) can be used.

Transmission
This is the transmission of heat through a building component by radiation and convection at its surfaces. It is calculated from the U-value and the surface area of the building.

Transmission heat losses Ht´ [kWh/a]
Transmission heat losses are also known as heat transfer losses. They comprise the amount of energy transmitted from inside to outside through the whole of the building envelope due to the temperature difference. Building components exhibit resistance against this passage of heat. This property is expressed as the thermal transmittance or U-value of the component.

U

U-value [W/m²K]
The U-value is the thermal transmittance (earlier: k-value). The U-value is a material- and component-specific property. It is a measure of the thermal transmittance of a component and indicates how much heat flows through 1 m² of wall surface when there is a 1 K difference in the air temperature on either side of the wall. U-values are measured in W/m²K. The smaller the U-value, the lower the thermal conductivity and the better the thermal insulation. Different forms of construction can be compared by their thermal insulation properties. The EnEV stipulates low maximum U-values for external components of buildings.

Useful energy [kWh]
Losses occur each time energy is converted or transported. Useful energy is the energy, excluding the plant and distribution losses, made available at the place of use, e.g. space heating.

Useful energy demand
Useful energy is the heating or cooling demand. It is the calculated heat or cold energy demand for maintaining the specified thermal room conditions within a building zone. There is also a useful energy demand for lighting. It is calculated as the energy demand to maintain the lighting quality required for a usage profile.

There is also a useful energy demand for domestic hot water. It is calculated as the energy demand for providing each building zone with the domestic hot water required for its usage profile.

Useful heat [kWh]
Heat that is available for use. The proportion of the final energy that is available to a room after all the losses of generation, storage, distribution and delivery.

V

Vacuum insulation panel (VIP)
A vacuum insulation panel provides highly efficient thermal insulation. The principle is similar to that of a vacuum flask; the vacuum inside the panel keeps the heat-conducting medium (air) away and thus drastically reduces heat transport by convection and conduction. VIPs have an open-pored material (e.g. fumed silica) at their core. The heat insulation properties are about 5–10 times better than those of conventional insulation materials. The disadvantage is the increased design input necessary to ensure that the panels can be prefabricated as accurately as possible, because modification on site is not possible.

Vacuum tube collectors
Vacuum tube collectors are components of solar thermal systems and are used to provide hot water. They consist of closely spaced glass tubes with diameters of between 65 and 100 mm, containing a selectively coated absorber.

Ventilation heat losses [kWh/m²a]
Ventilation heat losses are the losses that result from the ventilation of a building: warm indoor air is replaced by cooler outdoor air and must be warmed to room temperature. If the transmission heat losses are added, this gives the required heating energy demand.

Virtual power plant
This is the amalgamation of small, decentralised power plants, such as photovoltaic arrays, small hydropower plants and biogas plants, small wind turbines, and combined heat and power plants with low outputs, into one jointly controllable combined unit.

Volume flow V [m³/h]
Volume flow is the volume of a substance that flows in a unit of time, e.g. the flow of air in ventilation system. It should be designed optimally to provide the hygienic minimum.

W

Watts (peak) [kWp]
Peak refers to peak output. Watt (peak) is the peak output from photovoltaic modules. It is determined under standard conditions assuming radiation of 1,000 watts striking the panels at right angles. The electrical output from the panels is then taken as the standard output and expressed in watts (peak) or Wp. The sum of the outputs from all the panels in an array is then the standard output of that array, which would usually be a few kWp in a domestic installation.

About 8 m² of photovoltaic panels provide 1 kWp (at an efficiency of 12.5 %). The peak output gives no information about the yield from the array. A yield of 800 to 850 kWh can be expected for each kWp for arrays in the Rhine-Main area of northern Germany, over 1,000 kWh in Breisgau, southwest Germany, and less than 600 kWh in predominantly cloudy regions.

Z

Zoning
Zoning a building layout sensibly at an early stage in the design can lead to huge energy and cost savings. In arriving at a suitable zoning plan, the designer should consider not only each function, but also the fire, sound insulation, temperature and ventilation requirements.

Bibliography and Illustration credits

[001] Meadows, Donella; Meadows, Dennis; Randers, Jørgen; Behrens, William W.: The Limits To Growth. New York: Universe Books, 1972.

[002] Blaser, Werner; Heinlein, Frank: R 128 by Werner Sobek. Architecture in the 21st century. Basel: Birkhäuser, 2002.

[003] Arch + 157 (9/2001) (Special Issue about Haus R 128).

[004] Braungart, Michael; McDonough, William: Cradle to Cradle. Re-Making the Way We Make Things. London: Vintage, 2009.

[005] Michaely, Petra; Schroth, Jürgen; Schuster, Heide; Sobek, Werner; Thümmler, Thomas: F87: Mein Haus – mein Auto – meine Tankstelle. Greenbuilding 6/2012, 19–23.

[006] Sobek, Werner; Brenner, Valentin; Michaely, Petra: Das Gebäude als Ressourcenspeicher: Recyclinggerechtes Bauen in der Praxis. Munich: Detail Green Books, 1/2012, 48–52.

General References and Sources

Active House Alliance, "Active House – Ein Pflichtenheft. Gebäude, die mehr geben, als sie nehmen".

Daniels, Klaus: Technology in Ecological Building. The Fundamentals and Approaches, Examples and Ideas. Basel: Birkhäuser, 1997.

Daniels, Klaus: Low Tech – Light Tech – High Tech: Building in the Information Age. Basel: Birkhäuser, 2000.

Deloitte Touche Tohmatsu (DTT): Report for the World Economic Forum 2011, Risks Report: The Consumption Dilemma: Leverage Points for Accelerating Sustainable Growth. Davos, April 2011.

Directive 2010/31/EU of the European Parliament and of the council of 19 May 2010 "on the energy performance of buildings", Official Journal of the European Union.

Federal Ministry for the Environment, Nature, Conservation, Building and Nuclear Safety (BMUB), "What makes an Efficiency House Plus?", Berlin, 2011

Hausladen, Gerhard et al.: ClimaDesign. Solutions for Buildings that Can Do More with Less Technology. Basel: Birkhäuser, 2005.

Hausladen, Gerhard et al.: ClimaSkin. Building-skin Concepts that Can Do More with Less Energy. Basel: Birkhäuser, 2008.

Head, Peter/Arup: Entering the Ecological Age: The Engineer's Role, based on The Brunel Lecture Series. London, 2008.

Hegger, Manfred at al.: Energy Manual. Sustainable Architecture. Basel: Birkhäuser/ Edition Detail, 2012.

Herzog, Thomas et al.: Facade Construction Manual. Basel: Birkhäuser, 2012.

Hegger, Manfred et al.: Heat | Cool. Energy Concepts, Principles, Installations. Basel: Birkhäuser, 2012.

Jansen, Sylvia J. T., Coolen Henny C. C. H., Goetgeluk, R. (eds): The Measurement and Analysis of Housing Preferences and Choice. Heidelberg: Springer, 2011.

Lenz, Bernhard et al.: Sustainable Building Services: Principles – Systems – Concepts, Munich: Detail Green Books, 2011.

Lüling, Claudia: Energizing Architecture. Design and Photovoltaics. Berlin: Jovis, 2009.

Moewes, Guenther: Weder Hütten noch Paläste. Architektur und Ökologie in der Arbeitsgesellschaft. Basel: Birkhäuser, 1995.

www.klimahaus.it/en/climatehouse/1-0.html

www.minergie.ch (German, French, Italian)

http://passivehouse.com/

Photographs

BMVBS 113 1st row left

Bruno Klomfar 43, 65

Christel Derksen 105 bottom, 176, 178, 179

CLAYTEC 116

Constantin Meyer 25, 105 middle, 107 middle, 107 bottom, 113 1st row right and 2 row left, 120, 200, 202, 203, 220

dadarchitekten GmbH, bern 105 top, 107 top, 168, 170, 171

Department of Energy, SD, Kaye Evans Lutherodt 88

Eibe Soennecken 48, 65 left

Getty Images 127

Hannes Guddat 95 bottom right

IBA Hamburg GmbH/ Bernadette Grimmenstein 225

Jens Willebrand 126

Johannes Hegger 113 2nd row right, 3rd row left and right, 113 4 row left until 5th row

kämpfen für architektur ag, Zurich 46, 188, 190, 191, 196, 198, 199

Koch + Partner Architekten und Stadtplaner, Munich 223

Kontarframe 180, 182, 183

Matthias Koslick 13 top

o5 architekten bda 96 top

opus Architekten, Darmstadt 222

Roland Halbe 12

Sanjo Group, Altendorf 52, 192, 194, 195

Schüco International KG 209, 210, 211

Sebastian Schels 5, 102

SMA Solar Technology AG 148

Thomas Ott 83, 85, 115, 131

TU Darmstadt, FGee, Leon Schmidt 40, 84, 86, 87, 99 left, 90–92, 93 bottom, 95 bottom left

TU Darmstadt, FG ee, Simon Schetter 92, 93 top

Velux 50, 172, 177 bottom left

VELUX/Adam Mørk 65 middle, 119, 174, 175 top left, 177 right, 184, 186, 187, 214 left, 215

VELUX/Torben Eskerod 214 right

Verband Privater Bauherren (VPB)-Regionalbüro 114

Viessmann 216, 217

Walter Unterrainer 125 top

WOLF Anlagen-Technik GmbH & Co. KG 144

Zooey Braun 11

Sources

Active House Alliance 51

Federal Institute for Geosciences and Natural Resources; 2007; BMWi Arbeitsgruppe Energierohstoffe 2006 19

dadarchitekten GmbH, CH-Bern 171

Daniels, Klaus, Technology in Ecological Building. The Fundamentals and Approaches, Examples and Ideas. Birkhäuser, 1997 63 (right)

DGNB German Sustainable Building Council 59

Katharina Fey, Entwurfsverfasserin/VELUX Deutschland GmbH 175

grab architekten ag, CH-Altendorf 195

Hegger, Manfred et al.: Energy Manual. Sustainable, Architecture. Birkhäuser/ Edition Detail, 2012. 64, 74, 77

Hegger, Manfred et al.: Energy Manual. Sustainable, Architecture. Birkhäuser/ Edition Detail, 2012. Based on: Pistohl, Wolfram: Handbuch der Gebäudetechnik. Planungsgrundlagen und Beispiele. Düsseldorf, 2007 79 (bottom)

Hegger, Manfred et al.: Energy Manual. Sustainable, Architecture. Birkhäuser/ Edition Detail, 2012. Based on: Behling, Sophia et al.: Sol power. Die Evolution der solaren Architektur. Munich, 1996 62 (left)

Hegger, Manfred et al.: Energy Manual. Sustainable, Architecture. Birkhäuser/ Edition Detail, 2012. Based on: BINE Informationsdienst: Basis Energie 21. Kraft und Wärmekoppeln. Bonn, 2006 136

Hegger, Manfred et al.: Energy Manual. Sustainable, Architecture. Birkhäuser/ Edition Detail, 2012. Based on: Daniels, Klaus. Low Tech – Light Tech – High Tech. Bauen in der Informationsgesellschaft. Basel, 1998 66, 71 (top)

Hegger, Manfred et al.: Energy Manual. Sustainable, Architecture. Birkhäuser/ Edition Detail, 2012. Based on: Stark, Thomas: Untersuchungen zur aktiven Nutzung erneuerbarer Energie am Bespiel eines Wohn- und eines Bürogebäudes. Stuttgart, 2004 122 (bottom)

Hegger, Manfred et al.: Energy Manual. Sustainable, Architecture. Birkhäuser/ Edition Detail, 2012. Based on: Stark, Thomas. Wirtschaftsministerium Baden-Württemberg (ed.): Architektonische Integration von Photovoltaik-Anlagen. Stuttgart, 2005 79 (top), 122 (top)

Hegger, Manfred et al.: Heat I Cool. Energy Concepts, Principles, Installations. Birkhäuser, 2012 218

HHS Planer + Architekten AG, TU Darmstadt FG ee 121, 148

ina Planungsgesellschaft mbH 32, 34, 35, 36, 37, 38, 39 top, 41, 42, 43, 45, 47, 49, 58, 60, 68–69, 70, 75, 76, 78, 98, 99 (bottom and right)

ina Planungsgesellschaft mbH. TU Darmstadt, FG ee. Eigene Darstellung auf Basis von Daten des Statistischen Bundesamtes 57

ina Planungsgesellschaft mbH. TU Darmstadt, FG ee. Based on: Staubli/OeJ 53

ina Planungsgesellschaft mbH. TU Darmstadt, FG ee. Based on: Staubli/OeJ 53, 54

IPCC Expert Meeting Report: Towards New Scenarios (2007) 18 (bottom)

Leitfaden für das Monitoring der Demonstrationsbauten im Förderkonzept EnBau und EnSan 165

Lenz, Bernhard et al.: Sustainable Building Services: Principles – Systems – Concepts. Detail Green Books, 2011 156 (middle and bottom)

Lenz, Bernhard et al.: Sustainable Building Services: Principles – Systems – Concepts. Detail Green Books, 2011. Based on: Wolfgang Rönspiess, Berlin 156 (top)

Lüling, Claudia: Energizing Architecture. Design and Photovoltaics. Jovis, 2009 123, 124 (bottom)

o5 architekten bda – raab hafke lang 96 (middle and bottom), 97

Prof. Dr. Lutz Katzschner, University of Kassel, Urban Planning. Based on: Welsch 1985 73 (bottom)

Steinbeis Transfer Center Energy, Building and Solar Engineering Stuttgart; TU Darmstadt, FG ee 220 (bottom)

TU Darmstadt, FG ee. ina Planungsgesellschaft mbH 24, 61, 65, 71 (bottom)

TU Darmstadt, FG ee. Ina Planungsgesellschaft mbH. Based on: Adolf-W. Sommer, Passivhauser – Planung-Grundlagen-Details-Beispiele, Verlagsgesellschaft Rudolf Müller GmbH & Co. KG, Cologne 2008 44

TU Darmstadt, FG ee. Based on: Datenreport 2011 der Stiftung Weltbevölkerung 17

TU Darmstadt, FG ee. Based on: Hauser, Gerd. Zertifizierungssysteme für Gebäude. Der aktuelle Stand der internationalen Gebäudezertifizierung. Detail Green Books, 2010 22

TU Darmstadt, FG ee. Nach: Hegger, Manfred et al.: Energy Manual. Sustainable Architecture. Basel: Birkhäuser/Edition Detail, 2012. 101, 103, 117 (bottom)

TU Darmstadt, FG ee. Based on: Hegger, Manfred et al.: Energy Manual. Sustainable Architecture. Basel: Birkhäuser/Edition Detail, 2012. Based on: Stark, Thomas. Wirtschaftsministerium Baden-Württemberg (ed.): Architektonische Integration von Photovoltaik-Anlagen. Stuttgart, 2005 118

TU Darmstadt, FG ee. Based on: Institute of Energy Economics and Rational Energy Use (IER); Universität Stuttgart; German Solar Association; U.S. Solar Photovoltaic Manufacturing: Industry Trends, Global Competition, Federal Support 21 (top)

TU Darmstadt, FG ee. Based on: Intergovernmental Panel on Climate Change; Jean Robert Petit, Jean Jouzel, et al.: Climate and atmospheric history of the past 420,000 years from the Vostok ice core in Antarctica. 18 (top)

TU Darmstadt, FG ee. Based on: Statistisches Bundesamt 17

TU Darmstadt, FG ee. Based on: Statistisches Bundesamt, Statistisches Jahrbuch 2011; et al. 17

TU Darmstadt, FG ee. Based on: Stiebel Eltron aus "Energie für den täglichen Bedarf gewinnen" 134 (left)

TU Darmstadt, FG ee. Based on: Vaillant Deutschland GmbH & Co. KG 124 (left)

TU Darmstadt, FG ee. Based on: www.chemie-am-auto.de/brennstoffzelle, 20.01.2013 139

TU Darmstadt, FG ee. Based on: www.erdoelzeitalter.de 73 (top)

TU Darmstadt, FG ee. Based on: www.umweltbewusstheizen.de 81

TU Darmstasdt, FG ee. Based on: http://www.oocities.org/peterfette/histo.htm, 20.01.2013 137

US Department of Energy 72

United Nations, World Population Prospects: The 2010 Revision, 2011; Statista 2012; Murck, Environmental Science; Energy Watch Group 16

Zee Architekten, Utrecht (NL) 179

All plans in the "Projects" chapter were provided by the respective architectural offices. The site plans in that chapter were prepared on the basis of those documents by Jens Schiewe of Nuremberg.
All graphic illustrations not otherwise identified were created expressly for this publication by Patrick Pick / TU Darmstadt. The sources used are listed in the illustration credits.

Index

Authors

This publication was produced in the Energy-Efficient Building Design Unit at Darmstadt University of Technology and in the office of HHS Planer + Architekten AG. This collaboration has proven to be essential for the subject matter covered, which involves theory and practice in equal measure.

The Energy-Efficient Building Design Unit was founded in 2001 in order to address the issues of sustainable and energy-efficient construction and to integrate their basic principles in the educational training of architects.

The firm HHS Planer + Architekten AG has existed since 1980. Its projects are arise from a high level of environmental awareness and from the local, cultural and climatic features of the specific sites.

Both the design unit and the office are headed by Professor Manfred Hegger. They each pursue sustainable construction as their guiding principle. The holistic perspective that goes along with this principle reveals new approaches in construction technology and architecture.

We wish to thank …

… Schüco, VELUX, and Viessmann Group for their support
… the authors for their fascinating essays
… the interview partners for insightful and interesting conversations
… the architects and clients for providing extensive data that – especially concerning our very specific questions on energy performance – were not always easy to ascertain
… our colleagues from the Energy-Efficient Building Design Unit for their proficient assistance
… Patrick Pick for his tireless and excellent graphics work
… Raymond Peat and David Koralek for the professional translation and Monica Buckland for the equally attentive editing and proofreading
… Kathleen Bernsdorf for the adept layout work

Manfred Hegger is an architect and author. From 2001 to 2014 he was Professor of Energy-Efficient Building Design in the Department of Architecture at TU Darmstadt, where he led an interdisciplinary research and development team of architects, urban planners and energy consultants. Prof. Hegger is also chairman of HHS Planer + Architekten AG in Kassel.

Caroline Fafflok studied architecture at TU Darmstadt and architecture media management at Bochum University of Applied Sciences. After working in various museums and in the field of press and public relations, she has been working since 2008 at TU Darmstadt, where she holds, inter alia, a post as research associate in the Energy-Efficient Building Design Unit.

Johannes Hegger completed his architectural studies in 2009 at the University of Stuttgart after several stays abroad. Since then he has been working as an architect and expert on energy-efficient and climate-conscious design and construction at HHS Planer + Architekten AG in Kassel.

Isabell Passig studied architecture at TU Darmstadt, graduating in 2006. She then worked until 2011 as a research associate in the Energy-Efficient Building Design Unit. Since 2011 she has been managing partner of ina Planungsgesellschaft mbH in Darmstadt.